# Nonlinear Systems Stability Analysis

## Lyapunov-Based Approach

# Nonlinear Systems Stability Analysis

## Lyapunov-Based Approach

Seyed Kamaleddin Yadavar Nikravesh

Professor, Electrical Engineering Department
Amirkabir University of Technology
Tehran, Iran

CRC Press is an imprint of the
Taylor & Francis Group, an **informa** business

CRC Press
Taylor & Francis Group
6000 Broken Sound Parkway NW, Suite 300
Boca Raton, FL 33487-2742

First issued in paperback 2017



ISBN-13: 978-1-4665-6928-7 (hbk)
ISBN-13: 978-1-138-07277-0 (pbk)



**Library of Congress Cataloging-in-Publication Data**

Nikravesh, Seyed Kamaleddin Yadavar.
Nonlinear systems stability analysis : Lyapunov-based approach / Seyed Kamaleddin Yadavar Nikravesh.
pages cm
Summary: "The dynamic properties of a physical system can be described in terms of ordinary differential, partial differential, difference equations or any combinations of these subjects. In addition, the systems could be time varying, time invariant and/or time delayed, continues or discrete systems. These equations are often nonlinear in one way or the other, and it is rarely possible to find their solutions. Numerical solutions for such nonlinear dynamic systems with the analog or digital computer are impractical. This is due to the fact that a complete solution must be carried out for every possible initial condition in the solution space. Graphical techniques which can be employed for finding the solutions of the special cases of first and second order ordinary systems, are not useful tools for other type of systems as well as higher order ordinary systems"-- Provided by publisher.
Includes bibliographical references and index.
ISBN 978-1-4665-6928-7 (hardback)
1. Nonlinear control theory. 2. Lyapunov stability. I. Title.

QA402.35.N555 2013
515'.392--dc23 2012026897

**Visit the Taylor & Francis Web site at**
**http://www.taylorandfrancis.com**

**and the CRC Press Web site at**
**http://www.crcpress.com**

# Contents

*In the name of GOD*

*The most gracious the most merciful*

*To all who serve Almighty GOD's creation for HIS satisfaction*

# Preface

The dynamic properties of a physical system can be described in terms of ordinary differential, partial differential, and difference equations, or any combination of these subjects. In addition, the systems can be time-varying, time-invariant and/or time-delayed, and continuous or discrete systems. These equations are often nonlinear in one way or the other and it is rarely possible to find their solutions. Numerical solutions for such nonlinear dynamic systems with an analog or digital computer are impractical. This is due to the fact that a complete solution must be carried out for every possible initial condition in the solution space. Graphical techniques, which can be employed for finding the solutions for special cases of first- and second-order ordinary systems, are not useful tools for other types of systems as well as higher-order ordinary systems. However, there are different theorems and methods concerning existence, uniqueness, stability, and other properties of nonlinear systems and/or their solutions. Among these qualitative properties, the stability of a given system is the most crucial systems issue. Without the guaranteed stability, the system will be of no value.

Many researchers have worked on stability robustness analysis for different systems. For a good list of these studies, one may read chapter five of sensitivity analysis by Eslami (e1). The aim of this book is to introduce some advanced tools for stability analysis of nonlinear systems. Toward this end, first, standard stability techniques are discussed with the shortcomings highlighted; then some recent developments in stability analysis are introduced, which can improve the applicability of standard techniques. Finally, stability analysis of special classes of nonlinear systems, for example, time-delayed systems and fuzzy systems, are proposed.

This book is organized as follows: In the first chapter, the stability of ordinary time-invariant differential equations will be considered. In Chapter 2, Lyapunov stability analysis will be studied. The subject of the third chapter is time-invariant systems. Chapter 4 deals with time-delayed systems. The stability analysis of fuzzy linguistic systems models is considered in Chapter 5.

This book is intended for graduate students of all disciplines who are involved in stability analysis of dynamic systems.

**S.K.Y. Nikravesh**

*September 2010*

*(50th anniversary of the establishment of Amirkabir University of Technology [AUT])*

# Acknowledgments

I would like to express my gratitude to my colleagues, Dr. H.A. Talebi and Dr. A. Dostmohammadi, for their great assistance in editing the manuscript. Also, I would like to express my gratitude to some of my former PhD students for their contributions to dynamic system stability theorems that constitute the main part of this book; namely, Dr. Suratgar, Dr. Vali, Dr. Fathabadi, Dr. Dehghani, Dr. Meigoli, and Dr. Mahboobi, who are presently academic members of various universities in Iran.

I am also indebted to some of my former MSc and/or current PhD students whose work have enriched this book; namely, Aghili Ashtiani, Shamaghdari, Sangrody, and Alaviani.

I would also like to express my deepest thanks to Dr. V. Maghsoodi and M.M. Ganji for their editing of this book. Moreover, I need to thank Haghshenoo, S. Emyaiee, and M. Mashhadi for their patience in typing this book.

# 1 Basic Concepts

**Introduction**: In this chapter, the stability analysis of a system, the dynamics of which are represented in time domain by nonlinear time-invariant ordinary differential equations, is considered. This chapter consists of the following subsections:

1.1 Mathematical model for nonlinear systems.
1.2 Qualitative behavior of second-order linear time-invariant systems (LTI).

## 1.1 MATHEMATICAL MODEL FOR NONLINEAR SYSTEMS

A nonlinear system may mathematically be represented in the following form:

$$\begin{aligned}
\dot{x}_1 &= f_1(x_1,x_2,\dots,x_n,u_1,u_2,\dots,u_m,t),\\
\dot{x}_2 &= f_2(x_1,x_2,\dots,x_n,u_1,u_2,\dots,u_m,t),\\
&\vdots\\
\dot{x}_n &= f_n(x_1,x_2,\dots,x_n,u_1,u_2,\dots,u_m,t),
\end{aligned} \tag{1.1}$$

where $\dot{x}_i$, $i=1,2,\dots,n$ denotes the derivative of $x_i$ (the $i$th state variable) with respect to the time variable $t$ and $u_j$, $j = 1, 2, \dots,m$ denote the input variables. Equation (1.1) could be written in the following state-space form:

$$\dot{x} = f(x,u,t), \tag{1.2}$$

where,

$$x = \begin{pmatrix} x_1 \\ x_2 \\ \vdots \\ x_n \end{pmatrix}, \quad u = \begin{pmatrix} u_1 \\ u_2 \\ \vdots \\ u_m \end{pmatrix} \quad \text{and} \quad f(x,u,t) = \begin{pmatrix} f_1(x,u,t) \\ f_2(x,u,t) \\ \vdots \\ f_n(x,u,t) \end{pmatrix}.$$

The measurable outputs (a p-dimensional vector) are functions of the states, the inputs, and the time such that:

$$\begin{aligned}
y_1 &= h_1(x_1,x_2,\dots,x_n,u_1,u_2,\dots,u_m,t),\\
y_2 &= h_2(x_1,x_2,\dots,x_n,u_1,u_2,\dots,u_m,t),\\
&\vdots\\
y_p &= h_p(x_1,x_2,\dots,x_n,u_1,u_2,\dots,u_m,t).
\end{aligned} \tag{1.3}$$

or, in the following general form:

$$y = h(x,u,t). \tag{1.4}$$

Equations (1.2) and (1.4) together are called the *mathematical dynamic equations*, or:

$$\begin{aligned} \dot{x} &= f(x,u,t), \\ y &= h(x,u,t). \end{aligned} \tag{1.5}$$

These equations could be simulated using operational amplifiers (integrators) and function generators as shown in Figure 1.1 (a) and (b).

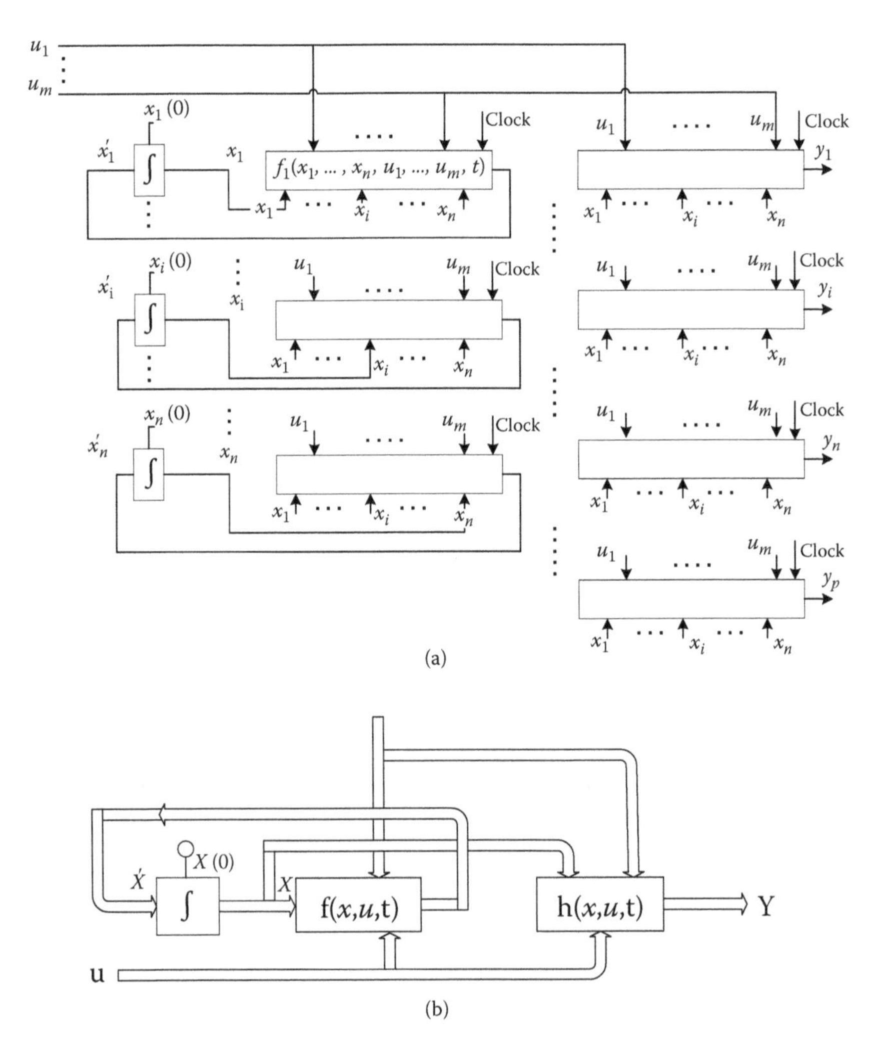


**FIGURE 1.1** System dynamic simulation.

It seems the dynamic systems could be simulated to obtain their responses, having signal generators $f_i$ and $h_i$. However, there is a drawback with this approach, since for each initial condition the simulation must be repeated. To have the actual response of dynamic systems, (1.2) and (1.4), the system must be at least locally Lipschitz in $x, \forall\, x \in D \quad R^n$ and continuous in $t$, for every $t$.

Throughout this book, wherever this type of dynamic equation occurs, the satisfaction of these conditions is assumed. The Lipschitz conditions are discussed shortly in this chapter.

Although in theory, the simulation could be proposed as a solution for the stability analysis, it is impractical or impossible, since in nonlinear system studies, every initial condition should be used.

**Special Cases:** If a system is a feedback system, then the system's inputs would be functions of the states, thus:

$$u \triangleq g(x,t). \tag{1.6}$$

Substituting (1.6) into (1.5) yields the following unforced dynamic equations:

$$\dot{x} = f(x,u,t) = F(x,t) \triangleq f(x,t), \;\; y = h(x,u,t) = H(x,t) \triangleq h(x,t), \tag{1.7a}$$

If the dynamic system (1.7a) is time invariant, then the system is called an *autonomous* (either *forced* or *unforced*) system.

$$\begin{aligned} \dot{x} &= f(x,u), \quad \text{or} \quad f(x) \\ y &= h(x,u), \quad \text{or} \quad h(x). \end{aligned} \tag{1.7b}$$

If the linearization technique is used in dynamic equations (1.5) or (1.7b), then linear time-varying (1.8) or linear time-invariant (forced or unforced) (1.9) equations yield:

$$\begin{aligned} \dot{x}_n &= \left( \frac{\partial f}{\partial x} \Big|_{\substack{x0 \\ u0}} \right) x_n + \left( \frac{\partial f}{\partial u} \Big|_{\substack{x0 \\ u0}} \right) u_n \triangleq A(t)\, x_n + B(t)\, u_n, \\ y_n &= \left( \frac{\partial h}{\partial x} \Big|_{\substack{x0 \\ u0}} \right) x_n + \left( \frac{\partial h}{\partial u} \Big|_{\substack{x0 \\ u0}} \right) u_n \triangleq C(t) x_n + D(t) u_n, \end{aligned} \tag{1.8}$$

or:

$$\begin{aligned} \dot{x}_n &\triangleq A x_n + B u_n, \\ y_n &\triangleq C x_n + D u_n. \end{aligned} \tag{1.9}$$

The index "n" stands for new variable. Note that (1.8) or (1.9) can only predict the local behavior of the nonlinear system of (1.5) or (1.7), respectively.

### 1.1.1 EXISTENCE AND UNIQUENESS OF SOLUTIONS [K1]

The existence and uniqueness of the solution of (1.6) are given by the following theorem.

---

**Theorem 1.1:**

Let *f(x,t)* be a single valued continuous function in a region defined by $|x_i - x_i(o)| < h_i,\ i = 1,2,...,n$ and $o \le t - t_1 < T$ in which $|f(x,t)| < M$ for some $o < M < \infty$, and $t_1$ is the domain of piecewise continuity of *f(x,t)*. If *f(x,t)* satisfies the following Lipschitz condition in *x*:

$$\|f(x_1,t) - f(x_2,t)\| \le L\|x_1 - x_2\| \quad ,o < L < \infty,$$

$$\forall x_1, x_2 \in B = \{x \in R^n \ \|x - x_o\| \le r\}, \forall t \in (t_o, t_1), \quad r > o,$$

then there exists some $\delta > o$ such that the state equation $\dot{x} = f(x,t)$ with $x(t_o) = x_o$ has a unique solution over $[t_o, t_o + \delta]$; $\delta = \min(T, \frac{h_i}{M})$. ■

When n = 1 and *f(x)* is autonomous, then the Lipschitz condition implies,

$$\frac{|f(x_1) - f(x_2)|}{|x_1 - x_2|} \le L,$$

that is, in a plane of *f(x)* versus *x*, a straight line joining any two points of *f(x)* cannot have a slope with absolute value greater than *L*. Therefore, a discontinuous function is not locally Lipschitz at the points of discontinuity.

More generally, if for $t \in I \subset R$ and $x \in D \subset R^n$, *f* (*x*,t) and its partial derivatives $\partial f_i / \partial x_j$ are continuous, then *f* (*x*,t) is locally Lipschitz in *x* on *D*. *f*(*x*,t) is globally Lipschitz in *x* if and only if (iff) $\frac{\partial f_i}{\partial x_j}$ are globally uniformly bounded in *t*.

**Example 1.1:**

Note that $\dot{x} = f(x) = x^{1/3}$ is not locally Lipschitz, at x = *o* since:

$$f'(x) = \frac{1}{3}x^{-2/3} \to \infty \ \ as\ x \to 0 \quad x(t) = \left(\frac{2t}{3}\right)^{\frac{3}{2}} \quad \text{and } x(t) = 0$$

are the two different solutions for this differential equation, when the initial state is

$$x(0) = 0.$$

■

Also, $\dot{x} = f(x) = -x^2$ is locally Lipschitz for all $x$ but not globally Lipschitz, because $f'(x) = -2x$ is not globally bounded.

Note that the linear time-varying system:

$$\dot{x} = A(t)x + b(t)u,$$

is globally Lipschitz if and only if (iff) the elements of *A(t)* are piecewise continuous and bounded. Therefore, the linear time-invariant systems are all globally Lipschitz.

In the following, it is assumed that the systems under consideration satisfy the Lipschitz conditions. If the equilibrium state is at $x_e \neq 0$, then let

$$y \triangleq x - x_e,$$

thus:

$$\dot{y} = \dot{x} = f(x) = f(y + x_e) \triangleq f_e(y),$$

where $f_e(0) = 0$. Therefore, without loss of the generality, the origin could be considered as an isolated equilibrium state.

**Equilibrium States:** These are the states "$x_e$" that if an unforced system (with neither control inputs nor disturbance) reaches every one of these states, it will stay there forever; therefore,

$$\dot{x}_e = f(x_e, u = 0, t) = 0, \qquad \forall t. \tag{1.10}$$

In a linear system with nonsingular *A(t)*, the sole equilibrium state is the origin. In the nonlinear case, the equilibrium state could be an isolated one, similar to a linear system, or infinitely many isolated equilibrium states, or there could be a continuum of equilibrium states.

## 1.2 QUALITATIVE BEHAVIOR OF SECOND-ORDER LINEAR TIME-INVARIANT SYSTEMS

Consider the unforced system:

$$\dot{x} = Ax. \tag{1.11}$$

The eigenvalues of $A$ may satisfy one of the following situations:

$a_1$. Both real negative with $\lambda_2 < \lambda_1 < 0$.
$a_2$. Both real positive with $\lambda_2 > \lambda_1 > 0$.

b. Real eigenvalues with opposite signs, that is, $\lambda_2 < 0 < \lambda_1$.
c. Complex eigenvalues $\lambda_{1,2} = \alpha \pm j\omega$.

The typical family of trajectories of these situations is shown accordingly in Figure 1.2. The proof is omitted here and interested readers are referred to the literature for classical nonlinear control systems.

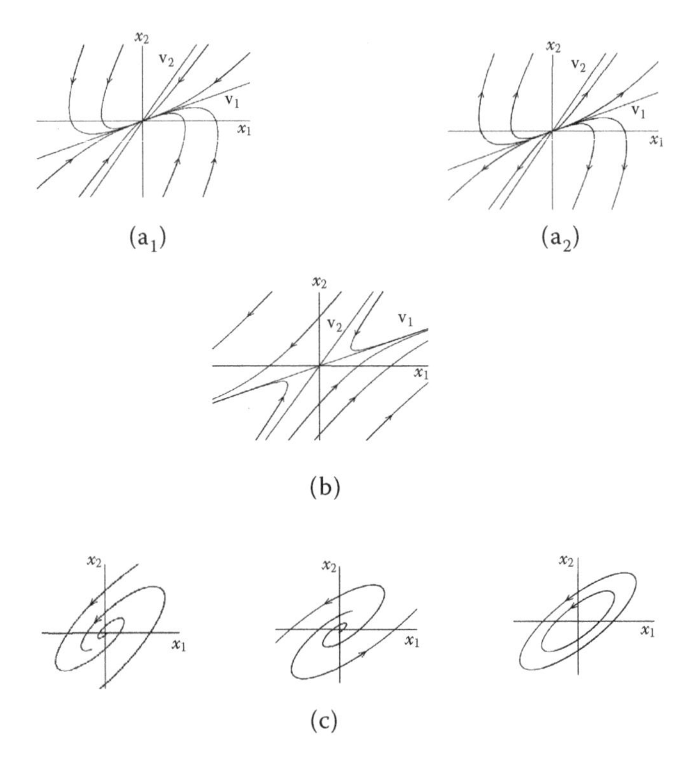


**FIGURE 1.2** Typical trajectories of second-order dynamic systems.

Note that when $\lambda_{1,2} = \pm j\omega$ (i.e., $Re\ [\lambda_i] = o$ for some i) the linearization technique does not work [k1].

### Example 1.2 [k1]:

Consider the following inverted pendulum equation with friction:

$$\dot{x}_1 = x_2,$$

$$\dot{x}_2 = -a\sin x_1 - bx_2,$$

The equilibrium states are as follows:

$$x_1 = k\pi, \qquad k = 0,1,\dots,$$

$$x_2 = 0.$$

The unforced linearized system would be as follows:

$$\begin{pmatrix} \dot{x}_1 \\ \dot{x}_2 \end{pmatrix} = \begin{pmatrix} 0 & 1 \\ -a\cos x_1 & -b \end{pmatrix}_{0,0} \begin{pmatrix} x_1 \\ x_2 \end{pmatrix} = \begin{pmatrix} 0 & 1 \\ -a & -b \end{pmatrix} \begin{pmatrix} x_1 \\ x_2 \end{pmatrix},$$

therefore:

$$\det(\lambda 1 - A) = \det\begin{pmatrix} \lambda & -1 \\ a & \lambda + b \end{pmatrix} = \lambda^2 + b\lambda + a = 0 \qquad \lambda = -\frac{b}{2} \pm \frac{1}{2}\sqrt{b^2 - 4a},$$

thus, for both a and b positive, the eigenvalues have negative real parts. Therefore, the origin is asymptotically stable (node).

To determine the stability of the equilibrium state at $(\pi,0)$, the Jacobian matrix would be evaluated at that state.

$$A = \begin{bmatrix} 0 & 1 \\ a & -b \end{bmatrix} \qquad \lambda^2 + b\lambda - a = 0,$$

$$\lambda_{1,2} = -\frac{1}{2}b \pm \frac{1}{2}\sqrt{b^2 + 4a},$$

For positive scalars $a$ and $b$, one of the eigenvalues is in the open right-half plane, which implies unstable equilibrium state. Figure 1.3 represents the phase plane, separatrices, and equilibrium state (stable nodes and saddle points).

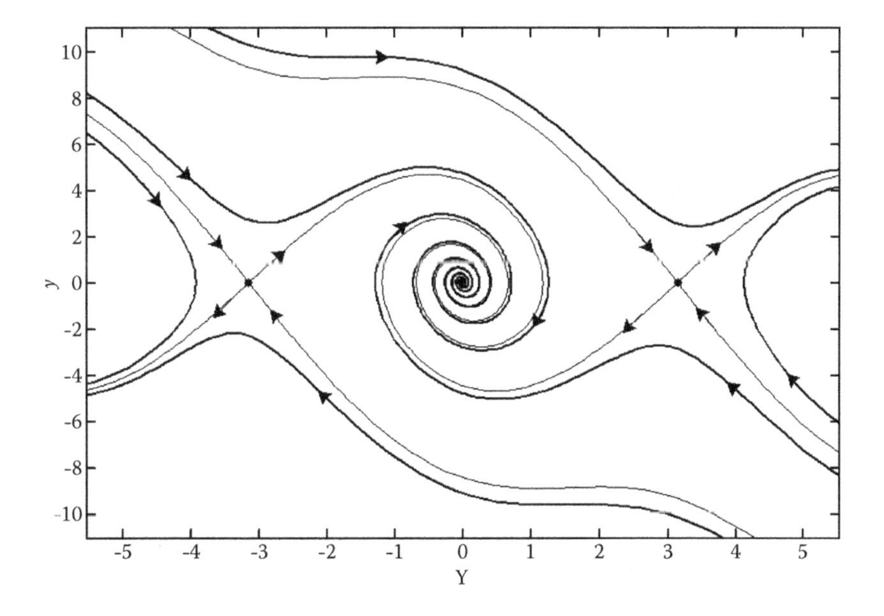


**FIGURE 1.3** Phase plane of Example 1.2.

**PROBLEMS:**

**1.1:** Consider the following scalar system:

$$\dot{x} = -(1 + x^2)\, x^3 \triangleq f(x)$$

a. Is $f(x)$ locally Lipschitz?
b. Is $f(x)$ globally Lipschitz? ■

**1.2:** Consider the following vector-valued system:

$$f(x) = \begin{bmatrix} x_2 \\ -sat(x_1 + x_2) \end{bmatrix}$$

where:

$$sat(x) = \begin{cases} -1 & x < -1 \\ x & |x| \le 1 \\ +1 & x > 1 \end{cases}$$

Does $f(x)$ satisfy the Lipschitz condition? ■

**1.3:** The nonlinear dynamic equation for a pendulum is given by:

$$ml\ddot{\theta} = -mg \sin\theta - kl\dot{\theta},$$

where $l$ is the length of the pendulum, $m$ is the mass of the ball, and $\theta$ is the angle suspended by the rod and the vertical axis through the pivot point.

a. Choose appropriate state variables and write down the state equation.
b. Find all equilibrium states of the system.
c. Linearize the system around the equilibrium states, and determine whether the system equilibrium states are stable or not.
d. Rewrite the pendulum model into the feedback connection form.
e. Make a simulation model of the system in Simulink®. Simulate the system from various initial states. Is the equilibrium state of the system stable? Are the equilibrium states unique? Explain the physical intuition behind your findings.
f. Use the function Linmod in MATLAB® to find the linearized models for the equilibrium states. Compare with the linearization that you derived. ■

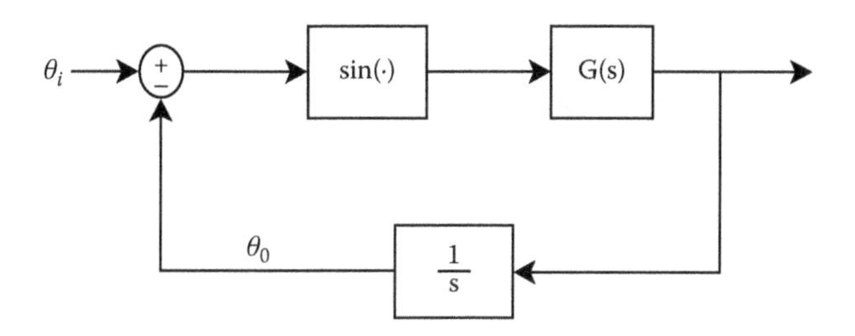


**FIGURE P1.4** The phase-locked loop.

**1.4:** A phase-locked loop system can be represented by the block diagram of Figure P1.4. Let $\{A,B,C\}$ be a state-space representation of the transfer function $G(s)$. Assume that all eigenvalues of $A$ have negative real parts, $G(0)\neq 0$ and $\theta_i$ is constant. Let $z$ be the state of the realization $\{A,B,C\}$.

a. Show that:

$$\dot{z} = Az + B\sin e$$

$$\dot{e} = -Cz$$

is a state equation for the closed-loop system.

b. Find all equilibrium states of the system.

c. Show that if $G(s) = 1/(\tau s+1)$, the closed-loop model coincides with the model of a pendulum equation. ■

**1.5:** A synchronous generator connected to an infinite bus can be modeled by:

$$M\ddot{\delta} = P - D\dot{\delta} - \eta E_q \sin\delta$$

$$\tau\ddot{E}_q = -\eta_2 E_q + \eta_3 \cos\delta + E_{fd},$$

where $\delta$ is the angle in radians, $E_q$ is voltage, $P$ is mechanical input power, $E_{fd}$ is field voltage (input), $D$ is damping coefficient, $M$ is inertial coefficient, $\tau$ is the time constant, and $\eta_1, \eta_2$, and $\eta_3$ are constant parameters.

a. Using $\delta$, $\dot{\delta}$, and $E_q$ as state variables, find the state equation.

b. Suppose that $\tau$ is relatively large so that $E_q \approx 0$. Show that assuming $E_q$ to be constant reduces the model to a pendulum equation.

c. For the simplified model, derived in Problem 1.5(b), find all equilibrium points. ■

**1.6:** A mass-spring system is shown in Figure P1.6. The displacement, $y$, from a reference point is given by :

$$m\ddot{y} + F_f + F_{sp} = F$$

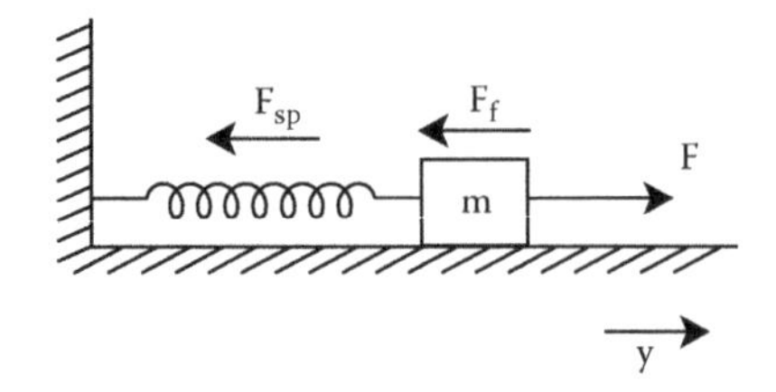


**FIGURE P1.6** The mass and spring system.

where $F_f$ is the friction force, $F_{sp}$ is the spring force and $F$ is the applied force. In reality both $F_f$ and $F_{sp}$ are nonlinear functions of $y$; however, we will use somewhat idealized relations here. First, we assume a linear spring such that $F_{sp} = ky$. Then, we assume that the friction force can be divided into two parts: friction due to viscosity $F_{fv}$ (the mass moves in air) and dry friction against the surface, $F_{fd}$.
We further assume that $F_{fv} = c\dot{y}$ and

$$F_{fd} = \begin{cases} -\mu_k mg & \dot{y} < 0 \quad \text{slipping friction} \\ F_s & \dot{y} = 0 \quad \text{rest friction} \\ \mu_k mg & \dot{y} > 0 \quad \text{slipping friction} \end{cases}$$

a. Assume $F = 0$, write the resulting state-space equation when:

$$F_{sp} = ky$$

and

$$F_f = F_{fv} + F_{fd} = c\dot{y} + \eta(y, \dot{y})$$

where

$$\eta(y, \dot{y}) = \begin{cases} \mu_k mg\,sign(\dot{y}) & |\dot{y}| > 0 \quad \text{slipping friction} \\ -ky & \dot{y} = 0 \quad \text{rest friction} \end{cases}$$

b. Characterize the equilibrium.
c. How can we benefit from the idealized computation above? ■

# 2 Stability Analysis of Autonomous Systems

## 2.1 SYSTEM PRELIMINARIES

It is obvious that stability plays an important role in system design. Obviously, an equilibrium state is said to be stable if all solutions of its dynamic equations starting at nearby states stay nearby, otherwise it is unstable. One of the most important properties of a system, that is, stability, can in principle be predicted or evaluated without having any knowledge about the solutions of the differential equations from given initial conditions with given inputs.

There are different stability analysis methods for linear systems that can be found in classical literature of linear systems. In this chapter, we consider a very important technique for nonlinear as well as linear system stability analysis in time domain; that is, stability in the sense of Lyapunov. This method is precise and involves no approximation. The idea involved is a generalization of the concept of energy for a conservative dynamic system. In such a system, the energy is a positive function, which decreases to zero as a stable equilibrium state is approached. Then, for such a general system, if a function with properties similar to those of an energy function could be found, the stability of the system would be guaranteed.

A.M. Lyapunov (in 1890) considered the stability of typical dynamic systems described by nonlinear ordinary differential equations. A general dynamic system among others may be described by ordinary and/or partial differential equations, time delay-continuous and/or discrete form. The original Lyapunov stability method is applicable to unforced autonomous ordinary differential equations (UAODE), that is,

$$\dot{x} = f(x). \tag{2.1}$$

It would be helpful here if we return for a while to a linear time invariant case. A linear system with constant coefficients is defined to be stable if and only if (iff) for every bounded input, the output response remains bounded. Note that a linear system is stable; it is so for any input regardless of its size, but this is not the case for nonlinear systems, where stability is a local concept and possibly a function of the input.

If $f(x) = Ax$, then the stability of the system would be obtained finding the eigenvalues of $A$.

---

**Theorem 2.1 [k1]:**

The equilibrium state $x_e = 0$ of $\dot{x} = Ax$ is stable iff all eigenvalues of $A$ satisfy $Re[\lambda_i] \leq 0$, and for every eigenvalue with $Re[\lambda_i] = 0$ and algebraic multiplicity $q_i \geq 2$;

rank $(A-\lambda_i I)=n-q_i$, where n is the dimension of $x$. The equilibrium state at $x_e=0$ is globally asymptotically stable iff all eigenvalues of $A$ satisfy $Re[\lambda_i]<0$. In this case, $A$ is called a *Hurwitzian matrix*. Also, the equilibrium state $x_e=0$ is said to be exponentially stable if:

$$\|x(t)\| \leq k\|x(0)\|e^{-\lambda t}, \ \forall\|x(0)\|<\infty, \quad \forall t\geq 0, \quad k\geq 1, \ \lambda>0. \tag{2.2}$$

■

It is obvious that exponential stability implies asymptotic stability, but the converse may not be true. As a counter example: $f(x)=-x^2$, $x(0)=a$.

If any $Re[\lambda_i]>0$, then, the system is unstable. Theorem 2.1 sometimes is called the *first method of Lyapunov*.

## 2.2 LYAPUNOV'S SECOND METHOD FOR AUTONOMOUS SYSTEMS

Let us start with a mathematical definition of stability (in the sense of Lyapunov).

---

**Definition 2.1 [12,s15]:**

The equilibrium state of (2.1) is stable if for every given $\varepsilon>0$, there is a $\delta=\delta(\varepsilon)>0$ such that:

$$\|x(0)-x_e\|<\delta \qquad \|x(t)-x_e\|<\varepsilon, \qquad \forall t\geq 0. \tag{2.3}$$

Otherwise it is unstable.

The equilibrium state of (2.1) would be asymptotically stable if it is stable and if:

$$t\to\infty \qquad x(t)\to x_e. \tag{2.4}$$

■

For any positive scalar $R$, if a spherical $R$ neighborhood of the equilibrium state is denoted by $S(R)$, then Definition 2.1 may be stated as follows:

> The equilibrium state is said to be stable in the sense of Lyapunov, or simply stable, if corresponding to each $S(R)$, there is an $S(r)$ such that every solution of (2.1) starting in $S(r)$ does not leave $S(R)$ as $t\to\infty$.

In addition, the system is called *asymptotically stable* if every solution starting in $S(r)$ not only stays in $S(R)$ but also approaches the equilibrium state as $t\to\infty$. If $S(r)$ is the whole state space then the stability or asymptotical stability would be global.

---

**Definition 2.2:**

The region of attraction (or in some sense, the region of asymptotic stability) is the set of all states such that the solution of (2.1) starting at those states converges towards $x_e$ as $t\to\infty$. ■

If the region of attraction is the whole state space, then $x_e$ is globally asymptotically stable.

Presenting different stability definitions, and finding ways to determine the stability using these definitions is a very tedious task. If the actual energy function of the system under stability consideration could be recognized, then the stability property of the system could be predicted by finding the overall slope of this function. If the energy function is a decreasing function as time increases, then the system is guaranteed to reach a relaxed position or an equilibrium state, thus the system would be stable. If the actual energy function is an increasing function as time increases, then the system is an unstable system.

The construction of an energy function for complex systems is a very complicated task if not impossible, perhaps even more difficult than finding the solutions of the dynamic equations themselves.

The beauty of the Lyapunov method is in replacing the energy-like function with the energy function in stability analysis of a dynamic system. The following theorem is due to Lyapunov. There exist also a large number of theorems, which are related to the second method of Lyapunov [yl], but again only the following are more important for engineering applications.

---

**Theorem 2.2 [k1]:**

Let $x_e = 0$ be an equilibrium state of (2.1) and $D \quad R^n$ be a domain containing $x_e = 0$. Let $V\colon D \to R$ be a continuously differentiable function such that:

$$V(0)=0 \qquad \text{and} \qquad V(x)>0, \quad \forall x \quad \text{in} \quad D-\{0\}. \tag{2.5}$$

The derivative of $V(x)$ along the trajectories of (2.1) is:

$$\dot{V}(x)=\frac{dV(x)}{dt}=\sum_{i=1}^{n}\frac{\partial V(x)}{\partial x_i} \qquad \dot{x}_i=\sum_{i=1}^{n}\frac{\partial V(x)}{\partial x_i}f_i(x)\leq 0 \text{ in } D, \tag{2.6}$$

then, $x_e = 0$ is stable. Moreover, if,

$$\dot{V}(x)<0 \quad \text{in} \quad D-\{0\}, \tag{2.7}$$

$x_e = 0$ is asymptotically stable. If the conditions for asymptotic stability hold globally and $V(x)$ is radially unbounded (RU) (*i.e.* $\|x\| \to \infty$, $V(x) \to \infty$) then, $x_e = 0$ is globally asymptotically stable.

For the proof, see Khalil [k1] and [12]. If for some neighborhoods of $x_e = 0$ where $V(0) = 0$, $V(x) > 0$ and also $\dot{V}(x) > 0$ $\forall x$ then, $x_e$ is an unstable equilibrium state of the system. ■

The function $V(x)$ in Theorem 2.2 is called a *Lyapunov function.* If $V(x) = c_i > 0$ then "$c_i$" is called Lyapunov surface or "level." The radial unboundedness condition is needed to ensure the boundedness of $V(x) \le c_i$ for every $c_i > 0$. For example, $V(x) = \dfrac{x_1^2}{1+x_1^2} + x_2^2$, as shown in Figure 2.1, is not bounded when $x_1 \to \infty$ [h1].

Note that these theorems give sufficient but not necessary conditions for stability, asymptotic stability, and global asymptotic stability and/or instability. Thus, failure of a particular $V(x)$ function to prove stability does not imply that the system in question is unstable or vice versa.

**Example 2.1:**

Consider Example 1.2 again,

$$\begin{aligned} \dot{x}_1 &= x_2, \\ \dot{x}_2 &= -a\sin x_1 - bx_2, \qquad a > 0,\ b > 0. \end{aligned} \tag{2.8}$$

Choose:

$$V(x) = a(1 - \cos x_1) + \frac{1}{2}x_2^2, \tag{2.9}$$

This function is positive definite because $V(x \ne x_e) > 0$ and $V(x = x_e) = 0$, then:

$$\dot{V}(x) = a\dot{x}_1 \sin x_1 + x_2\ \dot{x}_2 = -bx_2^2 \le 0. \tag{2.10}$$

Since $\dot{V}(x)$ in (2.10) is negative semidefinite while $V(x)$ in (2.9) is positive definite, then $x_e = 0$ is stable. If (2.9) is the actual energy function of (2.8), then (2.10) implies asymptotic stability of $x_e = 0$. ■

Note that choosing such a $V(x)$ function is not an easy job. Lyapunov function (we may use LF as well) selection is a drawback in Lyapunov direct method. In the following section, this shortcoming is tackled a bit.

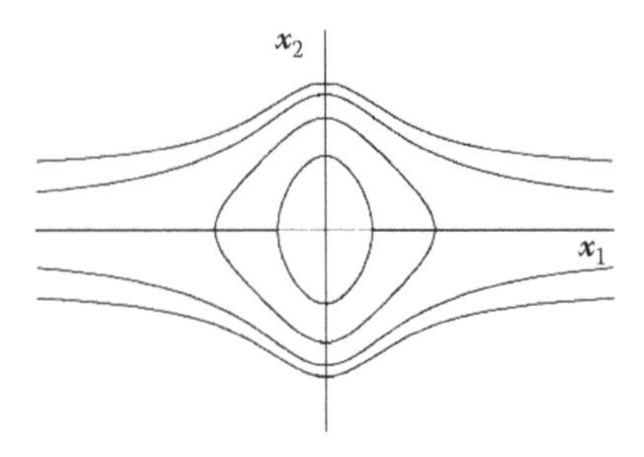


**FIGURE 2.1** The unbounded Lyapunov level.

### 2.2.1 Lyapunov Function Generation for Linear Systems

The linear autonomous case is simple and the following theorem will take care of stability of the systems that are represented by:

$$\dot{x} = Ax. \tag{2.11}$$

Note that for this case and for nonsingular A matrix, there is only an equilibrium state, and $\dot{V}(x)$ only needs to be semidefinite, since the solutions to the $\dot{V}(x) = 0$ will satisfy the original system equation only at $x_e = 0$.

---

**Theorem 2.3 [n1,b1][1]:**

In Equation (2.11) the equilibrium state is asymptotically stable iff there exists a symmetric positive definite matrix *P*, which is the unique solution of the following matrix equation,

$$A^T P + P A = -Q, \tag{2.12}$$

for any given symmetric[2] positive definite matrix *Q*, not vice versa.

The Lyapunov Function (LF) for the linear system of (2.11) may then be given by:

$$V(x) = x^T P x, \tag{2.13}$$

therefore,

$$\dot{V}(x) = x^T (A^T P + PA)\, x = -x^T Q x. \tag{2.14}$$

Thus, satisfaction of (2.14) implies satisfaction of conditions of Theorem 2.3, which in turn implies asymptotic stability of (2.11). ■

Note that if *Q* is positive semidefinite rather than positive definite and *P* is positive definite, this yields only the stability of the origin but, the system may or may not be asymptotically stable. For example while,

$$A = \begin{pmatrix} -1 & 0 \\ 0 & -2 \end{pmatrix}, \quad Q = \begin{pmatrix} 3 & 6 \\ 6 & 12 \end{pmatrix}, \quad P = \begin{pmatrix} 3/2 & 2 \\ 2 & 3 \end{pmatrix},$$

(for this *A*, the system is asymptotically stable), since *Q* is positive semidefinite and *P* is positive definite, exhibits only stability and also for the case:

$$A = \begin{pmatrix} 0 & 1 \\ 0 & -1 \end{pmatrix}, \quad Q = \begin{pmatrix} 0 & 0 \\ 0 & +2 \end{pmatrix} \quad \text{and} \quad P = \begin{pmatrix} 1 & 1 \\ 1 & 2 \end{pmatrix},$$

with *P* is positive definite and *Q* is positive semidefinite, exhibits stability too.

There are several different numerical methods of solving (2.12) in the literature as well as in MATLAB toolbox. It is important to note that $P$ must be positive definite. The example,

$$A=\begin{pmatrix} 1 & -3 \\ 2 & -4 \end{pmatrix}, \quad Q=\begin{pmatrix} 2 & -2 \\ -2 & 2 \end{pmatrix} \text{ and } P=\begin{pmatrix} 1 & -1 \\ -1 & 1 \end{pmatrix}.$$

shows that even though $A$ is stable and $Q$ is positive semidefinite, $P$ is only positive semidefinite [n1,b1]. In this case, of course, (2.13) is not a Lyapunov function (LF) for (2.11).

**Example 2.2:**

Consider the following linear system.

$$\begin{aligned} \dot{x}_1 &= -x_1 - x_2, \\ \dot{x}_2 &= -x_2. \end{aligned} \tag{2.15}$$

For positive definite $Q = I$, the Equation (2.12) implies:

$$P=\begin{pmatrix} 1/2 & -1/4 \\ -1/4 & 3/4 \end{pmatrix},$$

which is positive definite. Thus, the equilibrium state is asymptotically stable by Theorem 2.3 and the Lyapunov function (LF) $V$ is then given by (2.13) as follows:

$$V=\frac{1}{4}\left(2x_1^2 - 2x_1x_2 + 3x_2^2\right). \tag{2.16}$$ ■

The main difficulty with the Lyapunov method is how to choose these functions. In fact, there are no given procedures to select a suitable Lyapunov function for a given dynamic system. What follows is only a small collection of methods for generating Lyapunov functions for some systems. Having this background of definitions and theorems, the different methods of generating Lyapunov functions could be understood. Note that if two different methods give two different regions of stability for the same nonlinear system, then the region of stability of the system may be considered as the union of the two regions.

## 2.3 LYAPUNOV FUNCTION GENERATION FOR NONLINEAR AUTONOMOUS SYSTEMS [n1][3]

This section is a survey of methods for generating Lyapunov functions. More specifically, an attempt has been made to bring together in one place, and relate to one another, some of the various techniques for generating Lyapunov functions. A review of the definitions and the theorems of stability in Section 2.2 is recommended for a better understanding of this section.

Before getting involved in generating Lyapunov functions for the following nonlinear systems:

$$\dot{x} = f(x), \tag{2.17}$$

it seems reasonable if we first try to linearize the nonlinear system of (2.17) about the equilibrium state. By expanding and evaluating the nonlinear vector function $f(x)$ in the Taylor series expansion about the equilibrium state, $x_e$, (2.17) becomes:

$$\dot{x} = J(x - x_e) + H(x)(x - x_e), \tag{2.18}$$

Letting:

$$x - x_e = y,$$

yields:

$$\dot{y} = Jy + G(y)\,y, \tag{2.19}$$

where:

$$J = \left.\begin{bmatrix} \dfrac{\partial f_1(x)}{\partial x_1} & \dfrac{\partial f_1(x)}{\partial x_2} & \cdots & \dfrac{\partial f_1(x)}{\partial x_n} \\ \vdots & \vdots & & \vdots \\ \dfrac{\partial f_n(x)}{\partial x_1} & \dfrac{\partial f_n(x)}{\partial x_2} & \cdots & \dfrac{\partial f_n(x)}{\partial x_n} \end{bmatrix}\right|_{x=x_e} \tag{2.20}$$

and $f_1, f_2, \ldots, f_n$ are the $n$ components of $f(x)$. Since $G(y)$ contains the higher-order term of $y$, its elements vanish at the equilibrium state and then (2.19) in a sufficiently small neighborhood of the equilibrium state becomes:

$$\dot{y} = Jy, \tag{2.21}$$

which is the same as (2.11) and can be treated by Theorem 2.3.

The problem under consideration was nonlinear; therefore, after obtaining the Lyapunov function, it must be tested with the original nonlinear system in order to obtain the region of stability.

### Example 2.3:

Consider the following second-order nonlinear system:

$$\begin{aligned} \dot{x}_1 &= -2x_1 + x_2^2, \\ \dot{x}_2 &= x_1^2 - 2x_2. \end{aligned} \tag{2.22}$$

The equilibrium states can be obtained by letting:

$$\dot{x}_1 = \dot{x}_2 = 0,$$

which implies $(x_1, x_2) = (0,0)$ and $(x_1, x_2) = (2,2)$.

First, consider $(x_1, x_2) = (0,0)$, then,

$$J = \begin{pmatrix} -2 & 2x_2 \\ 2x_1 & -2 \end{pmatrix}\Bigg|_{(0,0)} = \begin{pmatrix} -2 & 0 \\ 0 & -2 \end{pmatrix},^{4} \tag{2.23}$$

or a first approximation to the nonlinear system (2.22) is:

$$\dot{x} = Jx, \tag{2.24}$$

where $J$ is given by (2.23). For linear system stability analysis of (2.24), one may use the Theorem 2.3 where a symmetric positive definite matrix $P$ must be found, such that (2.12) holds for some given symmetric positive definite matrix $Q$. Let $Q = I$ and $P = \begin{pmatrix} P_{11} & P_{12} \\ P_{21} & P_{22} \end{pmatrix}$ with $p_{12} = p_{21.}$ Then Equation (2.12) becomes:

$$\begin{pmatrix} -2 & 0 \\ 0 & -2 \end{pmatrix}\begin{pmatrix} P_{11} & P_{12} \\ P_{21} & P_{22} \end{pmatrix} + \begin{pmatrix} P_{11} & P_{12} \\ P_{21} & P_{22} \end{pmatrix}\begin{pmatrix} -2 & 0 \\ 0 & -2 \end{pmatrix} = \begin{pmatrix} -1 & 0 \\ 0 & -1 \end{pmatrix},$$

from which:

$$P = \begin{bmatrix} 1/4 & 0 \\ 0 & 1/4 \end{bmatrix}.$$

It is possible to choose:

$$P_1 = 4P = \begin{bmatrix} 1 & 0 \\ 0 & 1 \end{bmatrix}. \tag{2.25}$$

which is positive definite and thus, asymptotic stability of the zero equilibrium state of the linear system is guaranteed. Also the Lyapunov function represented by (2.13) is given below:

$$V(x) = x^T P_1 x = x_1^2 + x_2^2, \tag{2.26}$$

according to (2.14) with $Q_1 = 4Q$, for the linear system, then,

$$\dot{V}(x) = -x^T Q_1 x = -4(x_1^2 + x_2^2),$$

which implies global[5] asymptotic stability of $x_e = (0,0)$ of the system. Note that, as mentioned before, the result was obvious since the eigenvalues of $J$ are negative.

However, this equilibrium state of the nonlinear system may not be globally asymptotically stable, but one might be able to obtain the region of asymptotic stability (or region of attraction) by evaluating the total derivative of the Lyapunov function with respect to the nonlinear system (2.22). Note that $\dot{V}(x)$ is given by:

$$\dot{V}(x) = 2(x_1\dot{x}_2 + x_2\dot{x}_2) = -4(x_1^2 + x_2^2) + 2x_1x_2(x_1 + x_2). \tag{2.27}$$

The closed region[6] in which $\dot{V}(x)$ given by (2.27) is negative definite, if exists, is a region of asymptotic stability of the zero equilibrium state of (2.22). For the region in which $\dot{V}(x)$ is negative semidefinite, the other conditions on Theorem 2.2 must be satisfied for asymptotic stability of $x_e$.

Now, consider the second equilibrium state $(x_1, x_2) = (2, 2)$. Let,

$$y_1 = x_1 - 2,$$

$$y_2 = x_2 - 2.$$

Then,

$$\dot{x}_1 = \dot{y}_1 = -2(y_1 + 2) + (y_2 + 2)^2 = -2y_1 + y_2^2 + 4y_2,$$

$$\dot{x}_2 = \dot{y}_2 = (y_1 + 2)^2 - 2(y_2 + 2) = y_1^2 + 4y_1 - 2y_2,$$

or,

$$\begin{aligned} \dot{y}_1 &= -2y_1 + 4y_2 + y_2^2, \\ \dot{y}_2 &= 4y_1 - 2y_2 + y_1^2. \end{aligned} \tag{2.28}$$

Note that (2.28) is a nonlinear system with an equilibrium state at the origin, and by the similar procedure, the Lyapunov function and the region of asymptotic stability could be obtained if it exists at all. Another approach to the linearization is given by LaSalle and Lefschetz [l2] and Leondes [l1]. ■

### 2.3.1 Aizerman's Method [n1,l1]

It is obvious from the previous discussion that the linearization technique does not provide a great deal of information about the region of asymptotic stability. Also, in the linearization technique, an analytical expression is necessary for the nonlinearity, otherwise the method is not applicable. Aizerman's technique attempts to cover both of these shortcomings.

In Aizerman's method, the nonlinearity is approximated by a suitable gain $k$. The gain k may be changed until the desired results are obtained.[7] The Lyapunov function for this method is, in general, a quadratic form as in (2.13) with an unspecified matrix $P$. The $P$ matrix will be specified in the process of forcing $\dot{V}(x)$ to be negative definite for the system with constant gain $k$. As mentioned before, the sign definiteness of $V(x)$ and $\dot{V}(x)$ imply global asymptotic stability for this system. The region of asymptotic stability will be defined as the region of negative definiteness of $\dot{V}(x)$ with respect to the nonlinear system.

The following example will illustrate the use of Aizerman's method.

**Example 2.4 [n1,l1]:**

Consider the following nonlinear system:

$$\begin{aligned} \dot{x}_1 &= -3x_2 - g(x_1)x_1, \\ \dot{x}_2 &= -2x_2 + g(x_1)x_1. \end{aligned} \tag{2.29}$$

Also assume $k = 1$ is the most suitable gain to replace the nonlinearity. Let,

$$V(x) = x^T P x = x^T \begin{bmatrix} p_{11} & p_{12} \\ p_{12} & p_{22} \end{bmatrix} x = p_{11}x_1^2 + 2p_{12}x_1x_2 + p_{22}x_2^2. \tag{2.30}$$

The total derivative of $V(x)$ with respect to a linearized system with $g(x_1) = k = 1$ is given by:

$$\dot{V}(x) = 2[x_1^2(p_{12} - p_{11}) + x_1x_2(p_{22} - 3p_{11} - 3p_{12}) + x_2^2(-3p_{12} - p_{22})]. \tag{2.31}$$

By choosing:

$$p_{11} = 2, \quad p_{12} = -1 \text{ and } \quad p_{22} = 3,$$

then,

$$V(x) = 2x_1^2 - 2x_1x_2 + 3x_2^2, \tag{2.32}$$

which is positive definite everywhere, and,

$$\dot{V}(x) = -6\left(x_1^2 + x_2^2\right), \tag{2.33}$$

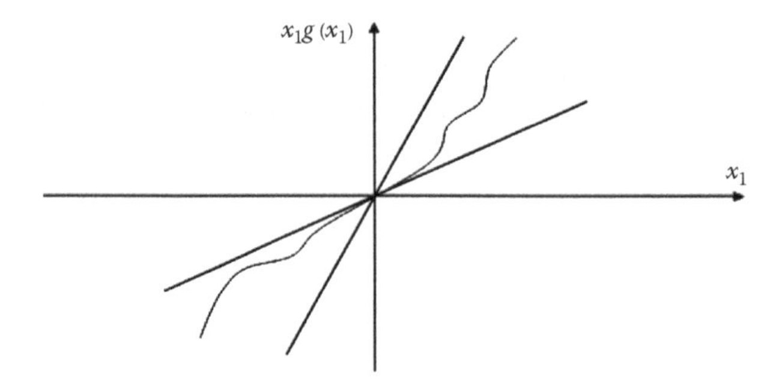


**FIGURE 2.2** Possible nonlinearities' limits for the system (2.29).

which is negative definite for the whole space. Now, for the $V(x)$ given by (2.32) and the nonlinear system given by (2.29), the total derivative of $V(x)$ is:

$$\frac{dV(x)}{dt} = \dot{V}(x) = -6\{g(x_1)x_1^2 + (4/3)[1 - g(x_1)]x_1 x_2 + x_2^2\}. \tag{2.34}$$

A region of asymptotic stability of the nonlinear system (2.29) with respect to the particular Lyapunov function of (2.32) is defined wherever $\dot{V}(x)$ given by (2.34) is negative definite. Use of Sylvester's theorem gives a required region. Thus:

$$\begin{vmatrix} g(x_1) & 2[1 - g(x_1)]/3 \\ 2[1 - g(x_1)]/3 & 1 \end{vmatrix} > 0, \tag{2.35}$$

which implies:

$$g(x_1) > 4[1 - g(x_1)]^2 / 9,$$

or,

$$1/4 < g(x_1) < 4. \tag{2.36}$$

Since $x_1 g(x_1) = 0$; when $x_1 = 0$, the equilibrium state, $x_e = 0$ of the system is globally asymptotically stable wherever the nonlinearity lies in the region between the following lines:

$$\begin{aligned} 0.25x_1 < g(x_1)x_1 < 4x_1; \quad x_1 \geq 0, \\ 0.25x_1 > g(x_1)x_1 > 4x_1; \quad x_1 < 0, \end{aligned} \tag{2.37}$$

This case is shown in Figure 2.2. ■

Note that in this method, the control system can at most have one block of mem oryless nonlinearity.

### 2.3.2 Lure's Method

This method is also applicable only to a certain class of systems with single memoryless nonlinearities. In addition, the nonlinearity should satisfy the following conditions [n1,h1]:

$$\begin{aligned} f(e) = 0, \qquad & e = 0, \\ \int_0^e f(u)\, du > 0, \quad & e \neq 0, \end{aligned} \tag{2.38}$$

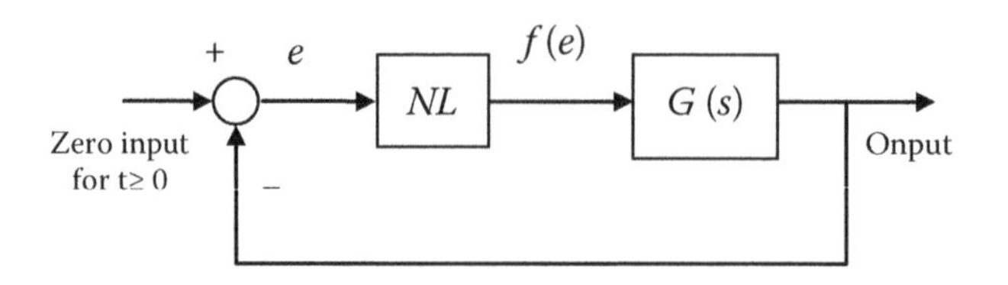


**FIGURE 2.3** Nonlinear control system.

where $f(e)$ is continuous except possibly at $e = 0$. Note that because of the conditions of (2.38), Lure's method includes the class of nonlinearities considered in Aizerman's method.

An advantage of this approach is that $V(x)$ functions have been developed for nonlinear systems in certain canonic forms. The differential equations may be represented in a canonic form as [o1,r1]:

$$\dot{z}_i = \lambda_i z_i + f(e), \tag{2.39}$$

in which

$$e = \sum_{i=1}^{n} \alpha_i z_i ,$$

where,

$z_i$: canonic variable,
e: input to the nonlinear element,
$f(e)$: output of the nonlinear element [the input to G(s)] (see Figure 2.3),
$\lambda_i$: $i$th pole of $G(s)$ ($\lambda_i$ can be either real or complex),
$\alpha_i$: negative of the residue at pole $\lambda_i$.

Note that:

$$\dot{e} = \sum_{i=1}^{n} \alpha_i \lambda_i z_i - rf(e), \tag{2.40}$$

where,

$$r = -\sum_{i=1}^{n} \alpha_i .$$

The necessary and sufficient conditions for the existence of this canonic form in closed-loop systems with a single nonlinear element are as follows [n1,r1]:

1. The poles of $G(s)$ must be simple.
2. The number of zeros of $G(s)$ must be less than the number of its poles.

The canonic variables become real or complex conjugate pairs, if the $\lambda$ are real or complex, respectively.

Consider the general case given in the vector form [n1,o1]:

$$\dot{z} = Az + bf(e), \qquad e = c^T z, \tag{2.41}$$

where, $z$: *n vector*, $A$ is an $n \times n$ diagonal matrix whose diagonal elements are $\lambda_1, \lambda_2, \ldots, \lambda_n$, where the $\lambda_i$'s are negative real quantities and/or complex conjugate pairs having negative real parts. $b = (1,1,\ldots,1)^T$ is an $n$ vector, $e$, the input to the nonlinear element is a scalar, $f(e)$, the output of the nonlinear element is also a scalar, and $c = (\bar{\alpha}_1, \bar{\alpha}_2, \ldots, \bar{\alpha}_n)^T$ is an $n$ vector.

Note that $\bar{\alpha}_i$ is a complex conjugate of $\alpha_i$, also $\dot{e}$ can be obtained from:

$$\dot{e} = c^T Az - rf(e) = z^T A^T c - rf(e). \tag{2.42}$$

Around 1950, using the following Lyapunov function,

$$V(z,e) = z^T S z + \int_0^e f(u)\,du, \tag{2.43}$$

Lure's sufficient conditions for asymptotic stability of the equilibrium state of the system given by (2.41) were obtained. In the above equation, $S$ is given by:

$$S = \begin{bmatrix} -\dfrac{\bar{a}_1 a_1}{\bar{\lambda}_1 + \lambda_1} & -\dfrac{\bar{a}_1 a_2}{\bar{\lambda}_1 + \lambda_2} & \cdots & -\dfrac{\bar{a}_1 a_n}{\bar{\lambda}_1 + \lambda_n} \\ \vdots & \vdots & & \\ -\dfrac{\bar{a}_n a_1}{\bar{\lambda}_n + \lambda_1} & -\dfrac{\bar{a}_n a_2}{\bar{\lambda}_n + \lambda_2} & \cdots & -\dfrac{\bar{a}_n a_n}{\bar{\lambda}_n + \lambda_n} \end{bmatrix}, \tag{2.44}$$

($a_i$ will be specified later). The following theorem summarizes his result:

---

**Theorem 2.4:**

(Lure's theorem): The origin of the system given by (2.41) is globally asymptotically stable if:

1. $r = -\sum_{i=1}^{n} \alpha_i > 0.$
2. $f(e)$ is continuous for all values of $e$, $ef(e) > 0$ for all values $e \neq 0_x$,

$$\int_0^e f(u)\,du \to \infty \text{ as } |e| \to \infty \quad \text{and} \quad f(0) = 0.$$

3. There exists at least one set of $a_1, a_2, ..., a_n$ which satisfies the following equation:

$$\begin{bmatrix} 2\sqrt{r}\,a_1 - 2 & \sum_{i=1}^{n} \frac{a_1 a_i}{\lambda_1 + \lambda_i} & +a_1\lambda_1 \\ \vdots & & \vdots \\ 2\sqrt{r}\,a_n - 2 & \sum_{i=1}^{n} \frac{a_n a_i}{\lambda_n + \lambda_i} & +a_n\lambda_n \end{bmatrix} = \begin{bmatrix} 0 \\ \vdots \\ 0 \end{bmatrix}, \tag{2.45}$$

where $a_j$ is real if $\lambda_j$ is real, and $a_k$ and $a_{k+1}$ are complex conjugates if $\lambda_k$ and $\lambda_{k+1}$ are complex conjugates.

If the first condition $r > 0$ is modified to $r \geq 0$, then the equilibrium state is globally stable (or stable in the large) if $\dot{V}$ does not vanish identically in $t \geq 0$ for any $x_0 \neq 0$. The proof is given by Ogata [o1] and a similar theorem is proved in [h1]. ■

**Example 2.5 [n1,o1]:**

Consider the nonlinear system of Figure 2.3 where:

$$G(s) = \frac{(s+4)(s+5)}{(s+1)(s+2)(s+3)},$$

and,

$$f(e) = e^3. \tag{2.46}$$

According to (2.39) the canonic form is:

$$\begin{aligned} \dot{z}_1 &= -z_1 + f(e), \\ \dot{z}_2 &= -2z_2 + f(e), \\ \dot{z}_3 &= -3z_3 + f(e), \end{aligned} \tag{2.47}$$

where,

$$e = -6z_1 + 6z_2 - z_3.$$

In this system the equilibrium state is the origin and,

$$\begin{aligned} &\lambda_1 = -1, \quad \lambda_2 = -2, \quad \lambda_3 = -3, \\ &\alpha_1 = -6, \quad \alpha_2 = 6, \quad \alpha_3 = -1, \\ &r = -(\alpha_1 + \alpha_2 + \alpha_3) = 1, \\ &ef(e) = e^4 \geq 0, \\ &\int_0^e f(e)de \to \infty \text{ as } |e| \to \infty. \end{aligned} \tag{2.48}$$

Equation (2.45) becomes:

$$\begin{bmatrix} 2a_1 - 2\left(-\dfrac{a_1^2}{2} - \dfrac{a_1a_2}{3} - \dfrac{a_1a_3}{4}\right) + 6 \\ 2a_2 - 2\left(-\dfrac{a_2a_1}{3} - \dfrac{a_2^2}{4} - \dfrac{a_2a_3}{5}\right) - 12 \\ 2a_3 - 2\left(-\dfrac{a_3a_1}{4} - \dfrac{a_3a_2}{5} - \dfrac{a_3^2}{6}\right) + 3 \end{bmatrix} = \begin{bmatrix} 0 \\ 0 \\ 0 \end{bmatrix}, \tag{2.49}$$

where the following set of $a_i$ will satisfy the above equations and satisfy the condition that $a_i$ must be real if $\lambda_i$ is real.

$$a_1 = -3.09, \quad a_2 = 5.55, \quad a_3 = -1.35.$$

Hence, by Theorem 2.4 the equilibrium state of the system is globally asymptotically stable. ■

Note that since Lure's theorem gives only a sufficient condition, therefore nonexistence of a set of $a_i$ satisfying (2.49) does not imply any conclusions on stability of the system under consideration.

The major disadvantages of this approach are that, this method can be applied to only a rather limited class of nonlinear systems, and that it involves the rather tedious task of finding a solution to algebraic equations, Gibson [g1, pp. 324–326] and Rekasius and Gibson [r1] developed a table which would facilitate the stability analysis of nonlinear systems in the canonic form of Lure. Another disadvantage is that Theorem 2.4 is for global asymptotic stability and thus local regions of stability, which might be the case in some systems, would be ignored.

### 2.3.3 Krasovskii's Method [a11,n1,o1]

Similar to Lure's method, Krasovskii's method also makes use of a theorem for showing the stability of the nonlinear system, in the proof of which, the second method of Lyapunov plays a significant role.

Consider the nonlinear system (2.17) and define the Jacobian matrix for this system as:

$$J(x) = \left[\frac{\partial(f_1,\ldots\ldots,f_n)}{\partial(x_1,\ldots\ldots,x_n)}\right] = \begin{bmatrix} \dfrac{\partial f_1}{\partial x_1} & \dfrac{\partial f_1}{\partial x_2} & \cdots & \dfrac{\partial f_1}{\partial x_n} \\ \dfrac{\partial f_2}{\partial x_1} & \dfrac{\partial f_2}{\partial x_2} & \cdots & \dfrac{\partial f_2}{\partial x_n} \\ \vdots & & & \\ \dfrac{\partial f_n}{\partial x_1} & \dfrac{\partial f_n}{\partial x_2} & \cdots & \dfrac{\partial f_n}{\partial x_n} \end{bmatrix}. \tag{2.50}$$

The following theorem, which is due to Krasovskii, gives a sufficient condition for the stability of the equilibrium state of a nonlinear system (2.17).

---

**Theorem 2.5:**

Consider nonlinear system (2.17) and assume $f(x)$ is differentiable with respect to $x_i$, $\forall i = 1,2,\ldots,n$. Define

$$\hat{J}(x) = J(x) + J^*(x), \tag{2.51}$$

where $J(x)$ is the Jacobian matrix and $J^*(x)$ is the conjugate transpose of $J(x)$. If $\hat{J}(x)$ is negative definite, then the equilibrium state $x_e = 0$ is asymptotically stable. ■

Note that, if $f(x)$ is real, then $J(x)$ is real and $J^*(x)$, which is real, can be written as $J^T(x)$. In this case $\hat{J}(x)$ is symmetric. $\hat{J}(x)$ is clearly Hermitian. Also notice that Krasovskii's theorem differs from usual linearization approaches. It is not limited to a small deviation from the equilibrium state.

The proof of this theorem is straightforward using the following Lyapunov function [o1], and is omitted here.

$$V(x) = f^*(x) f(x) = \|f(x)\|^2. \tag{2.52}$$

If, in addition $f^*(x)\, f(x) \to \infty$ as $\|x\| \to \infty$, then the equilibrium state is globally asymptotically stable. In this case, the theorem gives sufficient conditions for nonlinear systems and necessary and sufficient conditions for linear systems.

This method is relatively simple to apply, which is illustrated by the following example.

**Example 2.6 [n1, g2]:**

Consider the nonlinear system (2.53) with a > 1.

$$\begin{aligned} \dot{x}_1 &= -ax_1 + x_2. \\ \dot{x}_2 &= x_1 - x_2 - x_2^3, \end{aligned} \tag{2.53}$$

$x_e = 0$ is the equilibrium state of the system. Here:

$$J(x) = \begin{bmatrix} -a & 1 \\ 1 & -1-3x_2^2 \end{bmatrix},$$

and,

$$\hat{J}(x) = \begin{bmatrix} -2a & 2 \\ 2 & -2-6x_2^2 \end{bmatrix}.$$

Since $\hat{J}(x)$ is negative definite, then by Theorem 2.5, the zero equilibrium state of the system is asymptotically stable. ■

Another form of Krasovskii's theorem is given in Slotine and Li [s1].

### 2.3.4 Szego's Method [n1,l1,r1]

In contrast to Aizerman's method (Section 2.3.1), where the nonlinearity was assumed to be represented by a suitable constant $k$, it is approximated by a polynomial in Szego's method. Therefore, instead of having the same $V(x)$ for a variety of nonlinearities lying in the limit obtainable for the given $k$, Szego's approach is seeking to find a best $V(x)$ function for a given specific nonlinearity, and hence, the resulting region of stability could be the largest possible. As in Aizerman's method, although $V(x)$ is assumed to be of quadratic form, that is, $V(x) = x^T Px$, here the elements of $P$ are not necessarily assumed to be constants. Thus,

$$V(x) = \sum_{i,j=1}^{n} p_{ij}(x_i, x_j) x_i \, x_j, \qquad (P_{ij} = P_{ji}). \tag{2.54}$$

However, the variable coefficients $P_{ij}(x_i, x_j)$ are not allowed to be functions of $x_n$, but they can be general polynomials in $x_i$ and $x_j$. The latter assumption is justified by the fact that the limit cycles of the most general nonlinear system in the phase space have at most two real intersections with the hyper plane $x_i$ = const. ($i$ = 1, 2, …, $n$ – 1) [s2]. Since the coefficients $P_{ij}(x_i, x_j)$ do not contain $x_n$, then $V(x)$ is always an algebraic equation of second degree in $x_n$, and can be solved for $x_n$ by the quadratic formula. The solutions to the equation $\dot{V}(x)$ changes as these surfaces are crossed. For $\dot{V}(x)$ have the same sign throughout the whole phase space, these surfaces are forced to coincide. Figure 2.4 demonstrates the idea in the two-dimensional case. In Figure 2.4, the signs are arbitrarily assigned, and note that when the curves $\dot{V}(x) = 0$ coincide, $\dot{V}(x)$ is negative in the whole space. The last major difference in Aizerman's and Szego's methods can be considered as follows: In the latter case, since $\dot{V}(x)$ is negative or zero in the whole space, the region of stability is determined by the largest $V(x)$ surface that remains closed, while in the former case, the region of

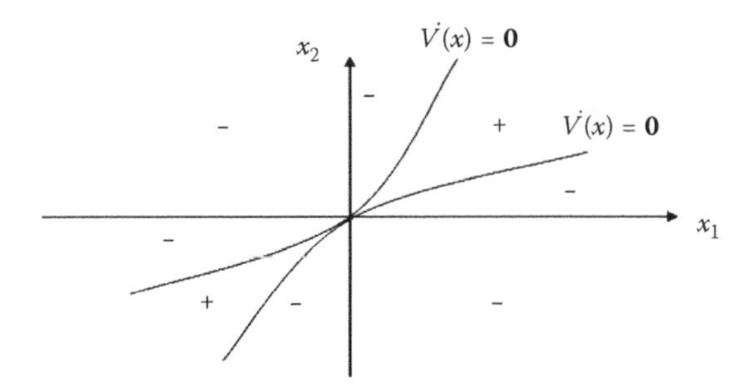


**FIGURE 2.4** Negative semi-definite $\dot{V}(x)$ function.

stability, if it is less than global, is the largest positive definite $V(x)$ region that always remains in the region in which $\dot{V}(x)$ is negative.

Szego's method is illustrated most easily in the second-order case. Consider the following second-order system [n1,s2]:

$$\begin{aligned} \dot{x}_1 &= x_2, \\ \dot{x}_2 &= f(x_1,x_2)x_2 + g(x_1)x_1, \end{aligned} \tag{2.55}$$

where $f(x_1,x_2)$ is a polynomial of at most first degree in $x_2$. Let,

$$V(x) = p_{11}(x_1)x_1^2 + 2p_{12}(x_1)x_1x_2 + p_{22}x_2^2, \tag{2.56}$$

where $p_{ij}(x_1)$ are unknown polynomial functions of $x_1$. By introducing the new quantities given below:

$$p'_{ij}(x_1) = p_{ij}(x_1) + \varepsilon_{ij}\frac{\partial P_{ij}(x_1)}{\partial x_1}x_1, \tag{2.57}$$

where,

$$\varepsilon_{ji} = \begin{cases} 1/2 & \text{for} \quad i = j, \\ 1 & \text{for} \quad i \neq j, \end{cases} \tag{2.58}$$

and $P'_{ji}(x_1) \neq P'_{ij}(x_1)$, then $\dot{V}(x)$ is as follows:

$$\begin{aligned} \dot{V}(x) = 2\,p'_{11}(x_1)x_1x_2 + 2p'_{12}(x_1)x_2^2 \\ + 2[p_{12}(x_1)x_1 + p_{22}x_2]\,[f(x_1,x_2)x_2 + g(x_1)x_1]. \end{aligned} \tag{2.59}$$

Note that $\dot{V}(x) = 0$ is second degree in $x_2$, and by choosing the coefficients $p_{ij}(x_1)$ in a proper way, it is possible to force the solutions to coincide. Then, the sign of $\dot{V}(x)$ will not change in the whole phase space as is apparent from Figure 2.4.

Now consider the following scalar function:

$$\begin{aligned} \psi(x) = 2p'_{11}(x_1)x_1x_2 + 2p'_{12}(x_1)x_2^2 \\ + 2[p_{12}(x_1)x_1 + p_{22}x_2]\,[f(x_1,x_2)x_2 + g(x_1)x_1]. \end{aligned} \tag{2.60}$$

Since the coefficients $p_{ij}(x_1)$ are polynomial functions, one might denote them as:

$$p_{ij}(x_1) = \sum_{k=1}^{n}(h_{ij})_k\ x_1^k. \tag{2.61}$$

The coefficients $p'_{ij}(x_1)$ will consequently have the following form:

$$p'_{ij}(x_1) = (h_{ij})_0 + \sum_{k=0}^{n} (h_{ij})_k \ (1 + \varepsilon_{ij} (\beta_{ij})_k) x_1^k,$$

where $(\beta_{ij})_k$ are positive constants. $p'_{ij}(x_1)$ could also be defined as:

$$p'_{ij}(x_1) = \sum_{k=0}^{n} (g_{ij})_k \ x_1^k. \tag{2.62}$$

The polynomial function $p'_{ij}(x_1)$ has consequently the same form as $p_{ij}(x_1)$ in the sense that $(g_{ij})_k = 0$, iff $(h_{ij})_k = 0$ and the $sign(g_{ij})_k = sign(h_{ij})_k$, since $\beta_{ij}$ and $\varepsilon_{ij}$ are positive constants. Expression (2.60) will consequently have the same form as (2.59).

Instead of forcing the solutions of $\dot{V}(x) = 0$ to coincide, as a first step, the solutions of $\psi(x) = 0$ are forced to coincide. Then, the $p'_{ij}(x_1)$ coefficients will have the following form:

$$p'_{ij}(x_1) = \sum_{k=0}^{n} (f_{ij})_k \, (p_{22}) x_1^k$$

In order to compute the coefficients $p_{ij}(x_1)$, which makes $\dot{V}(x)$ to have the desired form, according to the previous discussion on the relation between terms of $p_{ij}(x_1)$, it is necessary to replace the coefficients of the terms of $p'_{ij}(x_1)$ in $x_1^i$ $(i > 0)$ by some other constants as follows:

$$p''_{ij}(x_1) = (f_{ij})_o \, (p_{22}) + \sum_{k=1}^{n} (b_{ij})_k \ x_1^k, \tag{2.63}$$

where $(b_{ij})_k = 0$ for $(f_{ij})_o (p_{22}) = 0$ and $\text{sign}(b_{ij})_k = \text{sign}(f_{ij})_k (p_{22})$. By substituting $p_{ij}(x_1)$ into (2.56), it is possible to construct a new $V(x)$-function, denoted by $V'(x)$, where,

$$V'(x) = p''_{11}(x_1) x_1^2 + 2 p''_{12}(x_1) x_1 x_2 + p_{22} x_2^2. \tag{2.64}$$

The new $V'(x)$ function, in general, is no longer a quadratic form, but it is still at most of second degree in $x_2$. The coefficients of (2.64) can be determined such that $\dot{V}(x)$ is at least negative semidefinite. The following example will clarify Szego's method for a second-order case. For extended discussions as well as more examples of general cases, refer to Szego [s2].

**Example 2.7 [n1,s2]:**

Consider the following second-order system:

$$\begin{aligned} \dot{x}_1 &= x_2, \\ \dot{x}_2 &= \varepsilon(1 - x_1^2 + x_1^4)x_2 - x_1^3. \end{aligned} \tag{2.65}$$

Let the $V(x)$ function be as follows:

$$V(x) = p_1(x_1)x_1^2 + x_2^2, \tag{2.66}$$

then (2.60) takes the form,

$$\psi(x) = 2p_1'(x_1)x_1x_2 + 2x_2[\varepsilon(1 - x_1^2 + x_1^4)x_2 - x_1^3]. \tag{2.67}$$

Thus, $\psi(x) = 0$ implies:

$$x_2 = 0, \tag{2.68}$$

and,

$$x_2 = \frac{x_1^3 - p_1'(x_1)x_1}{\varepsilon(1 - x_1^2 + x_1^4)}. \tag{2.69}$$

Now these two curves, (2.68) and (2.69), should coincide. Thus, from (2.69):

$$x_1^3 - p_1'(x_1)x_1 = 0,$$

or,

$$p_1'(x_1) = x_1^2, \tag{2.70}$$

which is a polynomial in $x_1$. By introducing a new constant in (2.70) one may obtain:

$$p_1'(x_1) = b_1x_1^2. \tag{2.71}$$

Substituting (2.71) for $p_1(x_1)$ into (2.66) leads to:

$$V'(x) = b_1x_1^4 + x_2^2. \tag{2.72}$$

This is a positive definite function if $b_1 > 0$. Then,

$$\dot{V}'(x) = 4b_1x_1^3x_2 + 2\varepsilon x_2^2(1 - x_1^2 + x_1^4) - 2x_1^3x_2. \tag{2.73}$$

Only $b_1 > 0$ is left to be chosen. Let $b_1 = 1/2$, then (2.73) takes on the following form:

$$\dot{V}''(x) = 2\varepsilon x_2^2(1 - x_1^2 + x_1^4), \tag{2.74}$$

which is a semidefinite function and is not identically equal to zero on any non-trivial solution of the system (2.65) thus, the system is globally stable if ε < 0, and completely unstable if ε > 0, (see the last part of Theorem 2.2). ■

Since the coefficients of $V(x)$ or $\psi(x)$ are representable as a power series (see [2.61] and [2.63]), it seems reasonable if one tries to computerize this method. Numerical application of Szego's method for a particular class of systems is given in Hermit and Story [h2], and will be considered in some detail here [n1].

Consider the autonomous restricted class of a nonlinear system given by (2.75):

$$\begin{aligned} \dot{x} &= y, \\ \dot{y} &= f(x, y), \end{aligned} \tag{2.75}$$

also assume that $f(x, y)$ is of at most first degree in $y$ or:

$$f(x, y) = \sum_{i=1}^{n} \alpha_i x^i + y \sum_{i=1}^{n} \beta_i x^{i-1}. \tag{2.76}$$

Equations (2.56) and (2.57) for the system (2.76) are as follows, respectively:

$$V(x, y) = p_{11}(x)x^2 + 2p_{12}(x)xy + y^2, \tag{2.77}$$

$$\begin{aligned} \dot{V}(x, y) = &\left[2p_{11}(x)x + \frac{dp_{11}(x)}{dx}x^2\right]y + 2\left[p_{12}(x) + \frac{dp_{12}(x)}{dx}x\right]y^2 \\ &+ [2p_{12}(x)x + 2y]f(x, y). \end{aligned} \tag{2.78}$$

From (2.61), one can get:

$$p_{ij}(x) = \sum_{k=0}^{n} (h_{ij})_k x^k. \tag{2.79}$$

Substituting (2.79) and (2.76) into (2.78) yields:

$$\dot{V}(x, y) = A(x)y^2 + B(x)y + C(x), \tag{2.80}$$

where,

$$A(x)=\sum_{k=0}^{n}2(h_{12})_k x^k+\sum_{k=1}^{n}2k(h_{12})_k x^k+\sum_{k=1}^{n}2\beta_k x^{k-1}, \tag{2.81}$$

$$B(x)=\sum_{k=0}^{n}2(h_{11})_k x^{k+1}+\sum_{k=1}^{n}k(h_{11})_k x^{k+1}$$

$$+\sum_{k=0}^{n}\sum_{j=1}^{n}2(h_{12})_k\beta_i x^{k+i}+\sum_{k=1}^{n}2\alpha_k x^k, \tag{2.82}$$

$$C(x)=2\sum_{k=0}^{n}\sum_{i=1}^{n}(h_{12})_k\alpha_i x^{k+i+1}. \tag{2.83}$$

To coincide the solution curves of (2.80) it is necessary that:

$$B^2(x)=4A(x)\,C(x) \tag{2.84}$$

This condition may be achieved by setting $A(x)$ and $B(x)$ identically to zero. For $A(x)\equiv 0$, each coefficient, in the expansion (2.81), should be zero, or:

Terms: Coefficients:

constant $(h_{12})_0+\beta_1=0,$

$x$: $2(h_{12})_1+\beta_2=0,$

$x^2$: $3(h_{12})_2+\beta_3=0,$

$x^3$: $4(h_{12})_3+\beta_4=0,$

or, in general:

$$(h_{12})_k=\begin{cases}-\dfrac{\beta_{k+1}}{k+1}, & \text{for } k=0,1,2,\ldots,N,\\ 0, & \text{for } k>N.\end{cases} \tag{2.85}$$

Similarly, for $B(x)\equiv 0$,

Terms: Coefficients:

$x$: $2(h_{11})_0+2(h_{12})_0\beta_1+2\alpha_1=0,$

$x^2$: $3(h_{11})_1+2(h_{12})_0\beta_2+2(h_{12})_1\beta_1+2\alpha_2=0,$

$x^3$: $4(h_{11})_2+2(h_{12})_0\beta_3+2(h_{12})_1\beta_2+2(h_{12})_2\beta_1+2\alpha_3=0,$

or, in general:

$$(h_{11})_k=\begin{cases}\dfrac{-1}{k+2}\left(2\alpha_{k+1}+\displaystyle\sum_{j=0}^{k}2(h_{12})_j\beta_{k-j+1}\right), & \\ \quad j=0,1,\ldots.,N, & \\ \quad k-j+1\le N, & \\ \quad k=0,1,2,\ldots, & \\ 0, \quad k>N. & \end{cases} \tag{2.86}$$

Equations (2.85) and (2.86) provide an algorithm for calculating the coefficients of the Lyapunov function (2.77). The time derivative, $\dot{V}(x)$, is then given by:

$$\dot{V}(x) = C(x), \tag{2.87}$$

which is a semidefinite function. Note that $V(x)$ is a Lyapunov function in a $\Omega$ neighborhood of the origin. If it satisfies the conditions of Theorem 2.3 then, the equilibrium state of the system is asymptotically stable and $\Omega$ is a region of asymptotic stability. Also, note that for the nonlinear system (2.75), a direct calculation of $\dot{V}(x)$ is possible; however, the above procedure is useful when manipulation is not possible.

**Example 2.8 [n1,h2]:**

Consider the following nonlinear system:

$$\begin{aligned} \dot{x} &= y, \\ \dot{y} &= -x - y + x^3, \end{aligned} \tag{2.88}$$

thus:

$$f(x,y) = \sum_{i=1}^{3} \alpha_i x^i + y \sum_{1}^{3} \beta_i x^{i-1} = (-1)x + (0)x^2 + (1)x^3 + (-1)y,$$

or:

$$\begin{array}{lll} \alpha_1 = -1, & \alpha_2 = 0, & \alpha_3 = 1, \\ \beta_1 = -1, & \beta_2 = 0, & \beta_3 = 0. \end{array}$$

Therefore, from (2.85) and (2.86):

$$(h_{12})_0 = -\frac{\beta_1}{1} = -\beta_1 = 1,$$

$$(h_{11})_0 = -\frac{1}{2}[2\alpha_1 + 2(h_{12})_0\beta_1] = -\frac{1}{2}[-2 - 2] = 2,$$

$$(h_{11})_1 = -\frac{1}{3}[2(0) + 2(0)\ (0)] = 0,$$

$$(h_{11})_2 = -\frac{1}{4}[2\ (1)] = -1/2,$$

$$(h_{11})_3 = 0,$$

then:

$$p_{11}(x)=\sum_{k=0}^{3}(h_{11})_k\, x^k = 2-\frac{1}{2}x^2,$$

$$p_{12}(x)=\sum_{k=0}^{3}(h_{12})_k\, x^k = 1.$$

A Lyapunov function, which is given by (2.77) could be suggested as follows:

$$V(x,y)=\left(2-\frac{x^2}{2}\right)x^2+2xy+y^2=2x^2+2xy+y^2-\frac{x^4}{2}$$

$$=\begin{pmatrix} x & y \end{pmatrix}\begin{pmatrix} 2-\frac{x^2}{2} & 1 \\ 1 & 1 \end{pmatrix}\begin{pmatrix} x \\ y \end{pmatrix} \tag{2.89}$$

which is positive definite for $-\sqrt{2}<x<\sqrt{2}$. Also, using (2.83) and (2.87) or directly differentiating (2.89) with respect to "$t$" and using (2.88), yields:

$$\dot{V}(x,y)=2x^2(x^2-1), \tag{2.90}$$

which is negative definite for $-1 < x < 1$. Thus, the reign of stability would be $-1 < x < 1$ for all $y$. ■

### 2.3.5 Ingwerson's Method [n1,i1,b1,g2,l3,r2]

One of the general Lyapunov functions is a quadratic form, in all or in some of the state variables. In Aizerman's method, a quadratic form with constant coefficients was used and in Szego's method a quadratic form with variable coefficients was used, see (2.30) and (2.54).

The quadratic form of the function $V(x)$ could be modified when an odd nonlinearity is involved within the system. This modification could be an additional integral of the nonlinearity, such as in Lure's method, see (2.43).

In Ingwerson's method, once again a quadratic form is the choice for the $V(x)$ function. Consider the nonlinear system (2.17) and rewrite it as the following form:

$$\dot{x}=f(x)=J(x)x. \tag{2.91}$$

For the above system, the matrix relation (2.12) becomes:

$$J(x)^T P(x)+P(x)\,J(x)=-Q(x). \tag{2.92}$$

For linear systems with a scalar $V(x)$ function given by (2.13), holding the following relation can be shown in a straightforward manner:

$$p_{ij}(x) = \frac{\partial^2 V(x)}{\partial x_i \, \partial x_j},$$

so that the element of $P(x)$ are related to the second derivative of $V(x)$. Thus, once $P(x)$ is available, the first integration yields the gradient of $V(x)$, $\nabla V(x)$, and the second integration yields the $V(x)$ itself.

The same idea might be used for the nonlinear system. However, in this case, certain conditions as follows, are necessary for the elements of a matrix to be the second partial derivative of a scalar function:

(i) $P(x)$ must be symmetric,

(ii) $\dfrac{\partial p_{ij}}{\partial x_k} = \dfrac{\partial p_{ik}}{\partial x_j}. \qquad j,k \neq i.$

The first condition is always met by the solution of (2.92) since the matrix $Q(x)$ is symmetric. The second condition, which is the necessary and sufficient condition for a vector function to be the gradient of a scalar function, is not generally met. This relation can be satisfied if the $p_{ij}(x)$ are allowed to contain only the variables $x_i$ and $x_j$. Thus, a matrix $p(x_i, x_j)$ can be formed simply by letting all variables in each element $p_{ij}$ of $P(x)$ vanish except $x_i$ and $x_j$, where $i$ and $j$ are the respective indices of the row and column containing the element. Once $Q(x)$ is chosen, $P(x)$ and consequently $p(x_i, x_j)$ can be obtained and then, the integration leading to $V(x)$ and $\nabla V(x)$ can be carried out. Thus, first let,

$$\begin{aligned} V(x) &= \int_0^x P(x_i, x_j)\, dx \\ &= \int_0^{x_1} p_{i1}\, dx_1 + \int_0^{x_2} p_{i2} dx_2 + \cdots + \int_0^{x_n} P_{in}\, dx_n, \end{aligned} \tag{2.93}$$

and then,

$$V(x) = \int_0^x V(x)^T \, dx. \tag{2.94}$$

In this last integral, the unique scalar function $V(x)$ is obtained by a line integration of $\nabla V(x)$ along any path. To have this integral independent of the path, it is sufficient that the matrix of partial derivatives of the vector $\nabla V(x)$ have a vanishing skew symmetric part. In three-dimensional space, this is equivalent to the vanishing of the curl of $\nabla V(x)$, or satisfying condition (ii) [13], which already is satisfied because of the way $P(x_i, x_j)$ has been formed. Thus, the simplest path for the integration might be used, or:

$$V(x) = \int_0^{x_1} V_1(u_1, 0, \ldots, 0)\, du_1 + \int_0^{x_2} V_2(x_1, u_2, 0, \ldots, 0)\, du_2$$
$$+ \cdots + \int_0^{x_n} V_n(x_1, x_2, \ldots, x_{n-1}, u_n)\, du_n. \tag{2.95}$$

An example will serve to illustrate the method.

**Example 2.9 [n1,i1]:**

Consider the following nonlinear system:

$$\begin{aligned} \dot{x}_1 &= x_2, \\ \dot{x}_2 &= x_3, \\ \dot{x}_3 &= -(x_1 + cx_2)^3 - bx_3. \end{aligned} \tag{2.96}$$

Then, let $J(x)$ be $\dfrac{\partial f(x)}{\partial x}$ or:

$$J(x) = \begin{bmatrix} 0 & 1 & 0 \\ 0 & 0 & 1 \\ -3(x_1 + cx_2)^2 & -3c(x_1 + cx_2)^2 & -b \end{bmatrix},$$

By choosing $Q(x)$ as:

$$Q(x) = \begin{bmatrix} 0 & 0 & 0 \\ 0 & 2(q_1 q_2 - q_3) & 0 \\ 0 & 0 & 0 \end{bmatrix},^8$$

where

$$q_1 = b, \quad q_2 = 3c(x_1 + cx_2)^2, \quad q_3 = 3(x_1 + cx_2)^2.$$

The matrix $P(x)$ could be calculated from (2.92), that is:

$$P(x) = \begin{bmatrix} 3b(x_1 + cx_2)^2 & 3(x_1 + cx_2)^2 & 0 \\ 3(x_1 + cx_2)^2 & b^2 + 3c(x_1 + cx_2)^2 & b \\ 0 & b & 1 \end{bmatrix},$$

Then only by deleting the prespecified terms, one has:

$$P(x_i, x_j) = \begin{bmatrix} 3bx_1^2 & 3(x_1 + cx_2)^2 & 0 \\ 3(x_1 + cx_2)^2 & b^2 + 3c^3 x_2^2 & b \\ 0 & b & 1 \end{bmatrix}, \tag{2.97}$$

Performing the integration (2.93) yields the following gradient:

$$V(x)=\begin{bmatrix} bx_1^3+\frac{1}{c}(x_1+cx_2)^3 \\ (x_1+cx_2)^3+b^2x_2-c^3x_2^3+bx_3 \\ bx_2+x_3 \end{bmatrix}. \tag{2.98}$$

As a result, *V*(*x*), given by the line integral (2.94), is:

$$V(x)=\int_0^x V(x)^T\,dx=\frac{bx_1^4}{4}+\frac{(x_1+cx_2)^4}{4c}-\frac{x_1^4}{4c}+\frac{b^2x_2^2}{2}+bx_2x_3+\frac{x_3^2}{2}. \tag{2.99}$$

It follows that if b, c, and bc-1 are positive, *V*(*x*) is positive definite. Under the same conditions:

$$\dot{V}(x)= V(x).\dot{x}=-(bc-1)(3x_1^2+3cx_1x_2+cx_2^2)x_2^2, \tag{2.100}$$

is negative semidefinite; therefore, the equilibrium state of the system, that is, the origin, is stable. ■

Before leaving this method, it is worthwhile summarizing the formal procedure of the above construction in the following steps:

(a) Determine the matrix:

$$J(x)=\left(J_{ij}(x)=\frac{\partial f_i(x)}{\partial x_j}\right).$$

(b) Choose a symmetric matrix *Q*(*x*) which is either positive definite or positive semidefinite.
(c) Obtain *P*(*x*) from (2.92).
(d) Construct $P(x_i,x_j)$ matrix by setting, in each element of $P_{ij}$, all variables to zero except $x_i$ and $x_j$.
(e) Carry out the integration $V(x)=\int_0^x P(x_i,x_j)dx$ over each component of x treating the other components as constants.
(f) Evaluate the line integral $V(x)=\int_0^x V(x)^T\,dx$ along a path parallel to the coordinate axes, which is the simplest path.
(g) Determine if *V*(*x*) is positive definite. If not, step (b) through step (g) should be repeated.

Note that the most important part of this procedure is step (b), since the choice of *Q*(*x*) completely determines *V*(*x*). As Rodden [r2] pointed out, the step-by-step procedure suggests the computer generation of *V*(*x*) functions using Ingwerson's method.

This has been done for third-order systems whose differential equations are general polynomials and of the following form:

$$
\begin{aligned}
\dot{x}_1 &= x_2\,, \\
\dot{x}_2 &= x_3\,, \\
\dot{x}_3 &= \sum_{i=1}^{hd} \sum_{j=1}^{i+1} \sum_{k=1}^{i+m-j} R_{i,j,k}\, x_1^{\;(i+m-j-k)}\, x_2^{\;(j-1)}\, x_3^{\;(k-1)},
\end{aligned}
\tag{2.101}
$$

where $hd$ is the highest degree of the polynomial. The coefficients $R_{i,j,k}$ are inputs to the machine as data. By following the above steps, the first formal operation should be the construction of the $J(x)$ matrix. Since the equations of motion are of the form (2.101), then the $J(x)$ matrix has the following simple form:

$$
J(x) = \begin{bmatrix} 0 & 1 & 0 \\ 0 & 0 & 1 \\ j_1 = \dfrac{\partial \dot{x}_3}{\partial x_1} & j_2 = \dfrac{\partial \dot{x}_3}{\partial x_2} & j_3 = \dfrac{\partial \dot{x}_3}{\partial x_3} \end{bmatrix}. \tag{2.102}
$$

The coefficients of the three polynomials in the third row are computed from the closed-form formula for the derivative of a power and then stored. For step (b) one might choose the following $Q(x)$ matrix. Note that the coefficients $c_i$, $(i = 1,\ldots,6)$ are arbitrary, but not all zero.

$$
Q(x) = \begin{bmatrix} c_3[2j_3(j_1 j_2 - j_3) & c_4 & c_5 \\ c_4 & c_2[2(j_1 j_2 - j_3)] + 2c_5 & c_6 \\ c_5 & c_6 & c_1[2(j_1 j_2 - j_3)] \end{bmatrix}. \tag{2.103}
$$

Then, the result of step (c) is as given by (2.104). As shown by (2.104) each of the elements of the resulting $P(x)$ is made up of algebraic relations of the polynomial elements in $J(x)$. Step (d) is straightforward. Steps (e) and (f) are just routine integration techniques. With the operation of step (f), the Lyapunov function construction is completed and must be tested for definiteness:

$$
P(x) = \begin{bmatrix} \begin{aligned} & c_1 j_3^2 + c_2 j_1 j_3 \\ & \quad + c_3(j_1 j_2^2 - j_2 j_3 + j_1^2 j_3) - c_4 \end{aligned} & c_1 j_2 j_3 + c_2 j_3 + c_3 j_1^2 j_2 - c_5 & c_3(j_1 j_2 - j_3) \\ c_1 j_2 j_3 + c_2 j_3 + c_3 j_1^2 j_2 - j_5 & \begin{aligned} & c_1(j_1 j_3 + j_2^2) + c_2(j_1^2 + j_2) \\ & \quad + c_3(j_1^3 + j_3) - c_6 \end{aligned} & c_1 j_3 + c_2 j_1 + c_3 j_1^2 \\ c_3(j_1 j_2 - j_3) & c_1 j_3 + c_2 j_1 + c_3 j_1^2 & c_1 j_2 + c_2 + c_3 j_1 \end{bmatrix}. \tag{2.104}
$$

As seen from the above procedure, Ingwerson's method is straightforward for lower-order systems. In higher-order systems however, solving the matrix equation (2.92) is a tedious task. This method, which is well motivated, can often be carried out analytically for lower-order systems, and often gives excellent results. Ingwerson's LF generating method is applicable to systems with continuous and single-valued functions in the system differential equations. However, in principle, it is also applicable to the piecewise linear systems. In Ingwerson's method, the equations of motion need not have a polynomial expansion in terms of the state variables as is required with many other LF generating methods.

So far, only the quadratic form, in one way or the other, was used as a $V(x)$ function. There are additional methods with quadratic forms as a choice for $V(x)$ functions; among those are Rosenbrock's method [r3], Ku–Puri's method [k2], Puri–Weygandt's method [p1], the Kinnen–Chen method [k3], and many others.

### 2.3.6 Variable Gradient Method of Schultz and Gibson [b5, n1,l1,o1,g1,g2,l3,s3]

From now on, methods will be considered in which the choice for a $V(x)$ function is not necessarily a quadratic form. The first of such methods is the variable gradient method, which is in many ways similar to that of Ingwerson's (see Section 2.3.5). In Ingwerson's method, by constructing $P(x_i, x_j)$, then the vector gradient of the Lyapunov function, $\nabla V(x)$, was obtainable from the following integration:

$$V(x) = \int_0^x P(x_i, x_j)\,dx.$$

Instead, in the variable gradient method, $\nabla V(x)$ is assumed to be known and the rest of the procedure is very similar to Ingwerson's method.

By assuming an arbitrary form for $\nabla V(x)$, $\dot{V}(x)$ will be given by:

$$\dot{V}(x) = \quad V(x)^T f(x). \tag{2.105}$$

By forcing $\dot{V}(x)$ to be at least negative semidefinite, some of the parameters in $V(x)$ will be specified. The symmetry conditions of the Jacobin matrix of $V(x)$, which are necessary and sufficient conditions for a vector function to be the gradient of a scalar function, specify the rest of $V(x)$'s parameters. Then $V(x)$ would be given by the line integral (2.106) which is independent of the integration path:

$$V(x) = \int_0^x \quad V(x)^T . dx.^9 \tag{2.106}$$

See (2.95).

For the sake of comparison of the two methods, the problem that was solved by Ingwerson's method will be solved again using the variable gradient method.

**Example 2.10 [n1,o1]:**

Consider the following nonlinear system:

$$\begin{aligned}\dot{x}_1 &= x_2,\\ \dot{x}_2 &= x_3,\\ \dot{x}_3 &= -(x_1 + bx_2)^3 - cx_3.\end{aligned} \tag{2.107}$$

Let the vector gradient of $V(x)$ be defined by:

$$\nabla V(x) = \begin{bmatrix} a_{11}(x)x_1 + a_{12}(x)x_2 + a_{13}(x)x_3 \\ a_{21}(x)x_1 + a_{22}(x)x_2 + a_{23}(x)x_3 \\ a_{31}(x)x_1 + a_{32}(x)x_2 + x_3 \end{bmatrix}, \tag{2.108}$$

then, (2.105) becomes:

$$\begin{aligned}\dot{V}(x) = \ &\nabla V(x)^T f(x) = (a_{11}(x)x_1 + a_{12}(x)x_2 + a_{13}(x)x_3)x_2 \\ &+ (a_{21}(x)x_1 + a_{22}(x)x_2 + a_{23}(x)x_3)x_3 - (a_{31}(x)x_1 + \\ &a_{32}(x)x_2 + x_3)cx_3 - (a_{31}(x)x_1 + a_{32}(x)x_2 + x_3)(x_1 + bx_2)^3,\end{aligned} \tag{2.109}$$

$\dot{V}(x)$ might be constrained in a number of ways in order to prove asymptotic stability. For instance, the following choices of parameters guarantee the asymptotic stability of the equilibrium state of the system:

$$a_{13}(x) = 0, \quad a_{22}(x) = c^2, \quad a_{23}(x) = c, \quad a_{31}(x) = 0, \quad a_{32}(x) = c.$$

Then (2.108) becomes

$$\nabla V(x) = \begin{bmatrix} a_{11}(x)x_1 + a_{12}(x)x_2 \\ a_{21}(x)x_1 + c^2x_2 + cx_3 \\ cx_2 + x_3 \end{bmatrix}. \tag{2.110}$$

The symmetry conditions of the Jacobin matrix of $\nabla V(x)$ in this case are:

$$\begin{aligned}\frac{\partial\ \nabla V_1(x)}{\partial x_2} &= \frac{\partial\ \nabla V_2(x)}{\partial x_1},\\ \frac{\partial\ \nabla V_1(x)}{\partial x_3} &= \frac{\partial\ \nabla V_3(x)}{\partial x_1},\\ \frac{\partial\ \nabla V_2(x)}{\partial x_3} &= \frac{\partial\ \nabla V_3(x)}{\partial x_2},\end{aligned} \tag{2.111}$$

thus,

$$\frac{\partial(a_{11}(x)x_1)}{\partial x_2} + \frac{\partial(a_{12}(x)x_2)}{\partial x_2} = \frac{\partial(a_{21}(x)x_1)}{\partial x_1},$$
$$\frac{\partial(a_{11}(x)x_1)}{\partial x_3} + \frac{\partial(a_{12}(x)x_2)}{\partial x_3} = 0, \tag{2.112}$$
$$\frac{\partial(a_{21}(x)x_1)}{\partial x_3} + c = c.$$

From the last relation of (2.112) it is obvious that $a_{21}(x)$ does not contain $x_3$. Thus, a choice for $a_{21}(x)$ might be as follows (see [2.121]):

$$a_{21}(x) = \frac{1}{x_1}(x_1 + bx_2)^3.$$

Then:

$$\frac{\partial(a_{21}(x)x_1)}{\partial x_1} = 3(x_1 + bx_2)^2.$$

By choosing:

$$a_{12}(x) = \frac{1}{bx_2}(x_1 + bx_2)^3,$$

then:

$$\frac{\partial(a_{12}(x)x_2)}{\partial x_2} = 3(x_1 + bx_2)^2.$$

These choices will satisfy (2.112) if $a_{11}(x)$ is a function of $x_1$ only, then (2.110) becomes:

$$V(x) = \begin{bmatrix} a_{11}(x_1)x_1 + \frac{1}{b}(x_1 + bx_2)^3 \\ (x_1 + bx_2)^3 + c^2x_2 + cx_3 \\ cx_2 + x_3 \end{bmatrix} \tag{2.113}$$

Then (2.109) becomes:

$$\dot{V}(x) = \nabla V(x)^T f(x) = \left[ a_{11}(x_1)x_1 + \frac{1}{b}(x_1 + bx_2)^3 \right] x_2$$
$$+ [(x_1 + bx_2)^3 + c^2 x_2 + cx_3]x_3$$
$$+ (cx_2 + x_3)[-(x_1 + bx_2)^3 - cx_3]$$
$$= (b^2 - cb^3)x_2^4 + 3(1 - bc)x_1^2 x_2^2 + 3(b - b^2 c)x_1 x_2^3$$
$$+ \left( \frac{1}{b} - c \right) x_2 x_1^3 + a_{11}(x_1)x_1 x_2.$$

By choosing:

$$a_{11}(x_1) = \left( c - \frac{1}{b} \right) x_1^2, \tag{2.114}$$

then:

$$\dot{V}(x) = x_2^2(1 - bc)(b^2 x_2^4 + 3x_1^2 + 3bx_1 x_2), \tag{2.115}$$

which is negative semidefinite if:

$$bc > 1. \tag{2.116}$$

$V(x)$ is given by (v2.106) where $\nabla V(x)$ is given by (2.113) and $a_{11}(x_1)$ given by (2.114), thus:

$$V(x) = \int_0^x \nabla V(x)^T . dx = \int_0^{x_1(x_2 = x_3 = 0)} c x_1^3 \, dx_3$$
$$+ \int_0^{x_2(x_1 = x_1, x_3 = 0)} [(x_1 + bx_2)^3 + c^2 x_2] dx_2 + \int_0^{x_3(x_1 = x_1, x_2 = x_2)} (cx_2 + x_3) dx_3$$
$$= \frac{c}{4} x_1^4 + \frac{1}{4b}(x_1 + bx_2)^4 + \frac{1}{2}(cx_2 + x_3)^2, \tag{2.117}$$

which is positive definite if,

$$b > 0, c > 0. \tag{2.118}$$

Note that $V(x) \to \infty$ as $\|x\| \to \infty$, therefore the system is globally asymptotically stable if (2.116) and (2.118) are satisfied simultaneously. ■

Obviously, the Ingwerson and variable gradient methods are quite similar, thus both methods have the same advantages and disadvantages, as compared with other methods of generating Lyapunov functions.

Computer generation of Lyapunov functions for a particular class of second-order system based on the variable gradient method system is given in Hang and Chang [h3], and is considered here in some detail [n1].

Consider the particular class of nonlinear system given by (2.119) which is at most first degree in $y$:

$$\begin{aligned} \dot{x} &= y, \\ \dot{y} &= -f(x) - y\, g(x). \end{aligned} \tag{2.119}$$

For the origin to be an equilibrium state, it is necessary to have $f(0) = 0$. Let:

$$V(x, y) = \begin{cases} a_{11}(x)x_1 + a_{12}(x)x_2, \\ a_{21}(x)x_1 + 2x_2, \end{cases} \tag{2.120}$$

where $a_{ij}$ are only functions of $x$ except $a_{22} = 2$. The symmetry condition of the Jacobin matrix $\nabla V(x, y)$ implies

$$a_{21}(x) = \frac{1}{x}\int_0^x a_{12}(\lambda)\, d\lambda. \tag{2.121}$$

Choosing the following for the rest of $a_{ij}$:

$$\begin{aligned} a_{11}(x) &= a_{21}(x)\, g(x) + \frac{2f(x)}{x}, \\ a_{12}(x) &= 2g(x), \end{aligned} \tag{2.122}$$

and using (2.105) and (2.106), the Lyapunov function and its total derivative would be as follows:

$$V(x, y) = (y + E(x))^2 + D(x), \tag{2.123}$$

$$\dot{V}(x, y) = -2E(x)\, f(x), \tag{2.124}$$

where,

$$\begin{aligned} E(x) &= \int_0^x g(\lambda)\, d\lambda, \\ D(x) &= 2\int_0^x f(\lambda)\, d\lambda. \end{aligned} \tag{2.125}$$

Note that $\dot{V}(x,y)$ is only a semidefinite function. Assuming $f(x)$ lies only in the first and third quadrants of the $x$–$f(x)$ plane, and $g(x)$ lies only in the first two quadrants of the $x$–$g(x)$ plane, then,

$$V(x,y) > 0, \quad \dot{V}(x,y) \leq 0, \tag{2.126}$$

thus, according to Theorem 2.2, the equilibrium state of (2.119) is globally asymptotically stable.

Equation (2.123) through Equation (2.125) provide the algorithm for constructing the Lyapunov function. For systems, which are only asymptotically stable in some regions, for example, $\Omega$, the boundary of the region is given by:

$$V = V_{max} = D(\varepsilon), \tag{2.127}$$

where $\varepsilon$ is the root of $E(x)f(x) = 0$ which has the smallest amplitude (excluding $x = 0$).

To improve the stability region, the following could be chosen instead of what is given by (2.122):

$$a_{12}(x) = \alpha g(x), \quad 0 \leq \alpha \leq 2, \tag{2.128}$$

and then the rest of the calculations could be modified accordingly.

**Example 2.11 [n1,h3]:**

Consider the system of (2.129):

$$\begin{aligned} \dot{x} &= y, \\ \dot{y} &= -x - y + x^3. \end{aligned} \tag{2.129}$$

Thus,

$$f(x) = x - x^3 \quad \text{and} \quad g(x) = 1.$$

Then,

$$D(x) = x^2 - x^4 / 2 \quad \text{and} \quad E(x) = x,$$

and finally from (2.123) and (2.124) one has:

$$\begin{aligned} V(x,y) &= (y + x)^2 + x^2 - \frac{x^4}{2}, \\ \dot{V}(x,y) &= -2x^2(1 - x^2), \end{aligned} \tag{2.130}$$

but, since:

$$E(x) \neq 0, \qquad \text{for all } x \neq 0,$$

$$f(x) = 0, \qquad \text{for } x = \pm 1,$$

then:

$$\varepsilon = \pm 1,$$

$$V_{max} = D(\varepsilon) = 0.5.$$

Thus, $V_{max} = 0.5$ is the stability boundary. ■

### 2.3.7 Reiss–Geiss's Method [n1,r4]

This method is applicable to $n$th-order nonlinear systems. In Ingwerson's method, $Q(x)$ was chosen as a positive definite or semidefinite matrix, then $\nabla V(x)$ was calculated. Finally, $V(x)$ and $\dot{V}(x)$ were obtained. In the variable gradient method, $\nabla V(x)$ was chosen; from that $\dot{V}(x)$ and $V(x)$ were obtained. In the present method, $\dot{V}(x)$ is chosen and from that $V(x)$ is derived as follows:

$$V(x) = \int \dot{V}(x)\, dt. \tag{2.131}$$

This integration can be carried out by parts with respect to the nonlinear system and the constraint $V(0) = 0$.

According to Theorem 2.2, $\dot{V}(x)$ needs to be at least negative semidefinite. Thus, the first choice for $\dot{V}(x)$ could be $\dot{V}_1(x) = \dot{V}_{11}(x) = -x_n^2$. If this choice is insufficient, then an additional term might be chosen, $\dot{V}_{12}(x) = -x_{n-1}^2$, and thus:

$$\dot{V}_2(x) = \dot{V}_1(x) + \alpha \dot{V}_{12}(x) = \dot{V}_{11}(x) + \alpha \dot{V}_{12}(x). \tag{2.132}$$

This procedure is continued until a suitable Lyapunov function is found.

**Example 2.12 [n1,r4]:**

Again the same example as in Ingwerson's method and the variable gradient methods is used. Consider the following system:

$$\begin{aligned} \dot{x}_1 &= x_2, \\ \dot{x}_2 &= x_3, \\ \dot{x}_3 &= -(x_1 + cx_2)^n - bx_3. \end{aligned} \tag{2.133}$$

By choosing $\dot{V}_{11}(x) = -x_3^2$ then,

$$-V_1(x) = -V_{11}(x) = \int x_3^2 dt = x_2x_3 - \int x_2\dot{x}_3 dt$$

$$= x_2x_3 + \int x_2[(x_1 + cx_2)^n + bx_3]dt$$

$$= x_2x_3 + \frac{bx_2^2}{2}\int (x_2 + cx_3)(x_1 + cx_2)^n dt$$

$$-\int cx_3(x_1 + cx_2)^n dt$$

$$= x_2x_3 + \frac{bx_2^2}{2} + \frac{(x_1 + cx_2)^{n+1}}{n+1} - \int cx_3(-\dot{x}_3 - bx_3)dt$$

$$= x_2x_3 + \frac{bx_2^2}{2} + \frac{(x_1 + cx_2)^{n+1}}{n+1} + \frac{cx_3^2}{2} + bc\int x_3^2 dt. \quad (2.134)$$

Let:

$$V_2(x) = -V_1(x) - bc\int x_3^2 dt$$

$$= \frac{bc-1}{2c}x_2^2 + \frac{1}{2c}(x_2 + cx_3)^2 + \frac{(x_1 + cx_2)^{n+1}}{n+1}, \quad (2.135)$$

then:

$$\dot{V}_2(x) = -(bc-1)x_3^2. \quad (2.136)$$

By Theorem 2.2, the zero equilibrium state of the system is globally asymptotically stable as long as:

$$bc-1 > 0,\ c > 0,\ n = 2k+1,\ k = 0, 1, 2, \ldots . \quad (2.137)$$

■

### 2.3.8 Infante–Clark's Method [n1,i2]

This method considers the second-order systems. The essence of the method consists of trying to determine a nontrivial time-independent integral of the differential equations under study. If such an integral existed, then the stability problem could be solved; if not, then the differential equations are modified such that the time independent integral could easily be found for the new system. The stability properties of this new system are determined, and by a vector operation, which compares the two systems' trajectories, the stability properties of the original system is obtained.

For the general second-order autonomous system:

$$\begin{aligned} \dot{x}_1 &= f_1(x_1,x_2), \\ \dot{x}_2 &= f_2(x_1,x_2), \end{aligned} \tag{2.138}$$

a sufficient condition for the existence of a time independent integral is:

$$\frac{\partial f_1(x_1,x_2)}{\partial x_1} + \frac{\partial f_2(x_1,x_2)}{\partial x_2} = 0. \tag{2.139}$$

When the physical system does not satisfy (2.139), then, a modification is necessary.

For a less general case, consider the system (2.140):

$$\dot{x}_0 = \begin{bmatrix} \dot{x}_{10} \\ \dot{x}_{20} \end{bmatrix} = \begin{bmatrix} \dot{x}_1 \\ \dot{x}_2 \end{bmatrix} = \begin{bmatrix} x_2 \\ f_2(x_1,x_2) \end{bmatrix}. \tag{2.140}$$

If,

$$\frac{\partial f_2(x_1,x_2)}{\partial x_2} = 0, \tag{2.141}$$

then, a time-independent integral of (2.140) could be an appropriate Lyapunov function.

However, if:

$$\frac{\partial f_2(x_1,x_2)}{\partial x_2} = f_3(x_1,x_2) \neq 0, \tag{2.142}$$

then a modification is required. The system (2.143) would be a reasonable first step modification for (2.140).

$$\begin{aligned} \dot{x}_1' &= x_2' - \int_0^{x_1'} f_3(u,x_2')\,du, \\ \dot{x}_2 &= f_2(x_1',x_2'). \end{aligned} \tag{2.143}$$

Since (2.143) and (2.140) are not identical, the following modification is made to assure that the nature of their solutions is the same.

$$\dot{x}_m = \begin{bmatrix} \dot{x}_{1m} \\ \dot{x}_{2m} \end{bmatrix} = \begin{bmatrix} \dot{x}_1 \\ \dot{x}_2 \end{bmatrix} = \begin{bmatrix} x_2 - \int_0^{x_1} f_3(u,x_2)\,du + f_4(x_1,x_2) \\ f_2(x_1,x_2) + f_5(x_1,x_2) \end{bmatrix}. \tag{2.144}$$

The functions $f_4(x_{1m}, x_{2m})$ and $f_5(x_{1m}, x_{2m})$ satisfy the following condition:

$$\frac{\partial f_4(x_{1m}, x_{2m})}{\partial x_{1m}} + \frac{\partial f_5(x_{1m}, x_{2m})}{\partial x_{2m}} = 0. \tag{2.145}$$

The time independent integral would be as follows:

$$h(x_{1m}, x_{2m}) = \oint_C [\dot{x}_{1m}\, dx_{2m} - \dot{x}_{2m}\, dx_{1m}] = c,$$

$c$ is a constant and the curve $h$ must be closed and the vectors $\dot{x}_o$ and $\dot{x}_m$, which are the velocity vectors of the original system (2.140) and the modified system (2.144), are represented in Figure 2.5. The sign of the cross product of these two velocity vectors determines the stability of the system. Define $x$ as:

$$x = (\dot{x}_0) \times (\dot{x}_m). \tag{2.146}$$

The vector $x = \ h(x_1, \dot{x}_2) \cdot \dot{x}_2$ is negative if the direction of the velocity vectors are as indicated in Figure 2.5, zero if the vectors coincide, and positive if $\dot{x}_0$ were an outward pointing vector relative to vector $\dot{x}_m$. Obviously, the desired sign for this vector product is the negative sign, which guarantees asymptotic stability.

The following example illustrates the method just described:

### Example 2.13 [n1,r3]:

Consider the van der Pol's equation:

$$\ddot{x} + \varepsilon(1 - x^2)\dot{x} + x = 0,$$

which is equivalent to:

$$\dot{x}_0 = \begin{bmatrix} \dot{x}_{10} \\ \dot{x}_{20} \end{bmatrix} = \begin{bmatrix} x_2 \\ -x_1 - \varepsilon(1 - x_1^2)x_2 \end{bmatrix}. \tag{2.147}$$

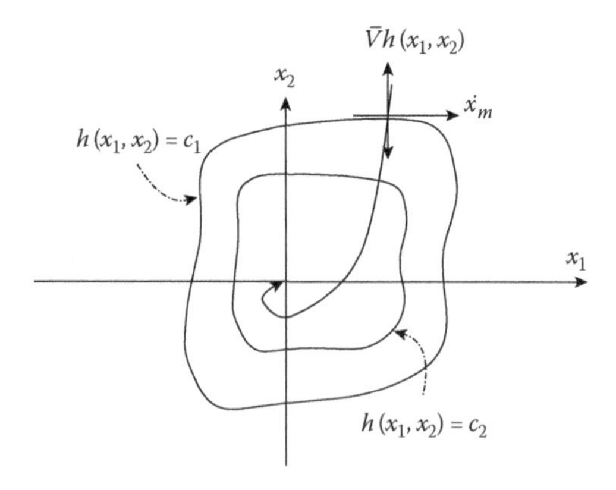


**FIGURE 2.5** Geometric representation of Infante–Clark's methods.

Since the system (2.147) does not satisfy the condition (2.139) or in this case (2.141), a time-independent integral does not exist, and the modified system of (2.144) should be used.

$$\dot{x}_m = \begin{bmatrix} \dot{x}_{1m} \\ \dot{x}_{2m} \end{bmatrix} = \begin{bmatrix} \dot{x}_1 \\ \dot{x}_2 \end{bmatrix} = \begin{bmatrix} x_2 + \varepsilon\left(x_1 - \dfrac{x_1^3}{3}\right) + f_4(x_1, x_2) \\ -x_1 - \varepsilon(1 - x_1^2)x_2 + f_5(x_1, x_2) \end{bmatrix}, \tag{2.148}$$

where,

$$\frac{\partial f_4(x_{1m}, x_{2m})}{\partial x_{1m}} + \frac{\partial f_5(x_{1m}, x_{2m})}{\partial x_{2m}} = 0. \tag{2.149}$$

The vector product of (2.146) in this case is as follows:

$$\begin{aligned} x = \dot{x}_0 \times \dot{x}_m &= \begin{bmatrix} \dot{x}_{10} \\ \dot{x}_{20} \end{bmatrix} \times \begin{bmatrix} \dot{x}_{1m} \\ \dot{x}_{2m} \end{bmatrix} = \begin{vmatrix} \dot{x}_{10} & \dot{x}_{1m} \\ \dot{x}_{20} & \dot{x}_{2m} \end{vmatrix} \\ &= \dot{x}_{10}\dot{x}_{2m} - \dot{x}_{20}\dot{x}_{1m} \\ &= \varepsilon x_1^2\left(1 - \frac{x_1^2}{3}\right) + \varepsilon^2 x_1 x_2\left(1 - \frac{x_1^3}{3}\right)(1 - x_1^2) \\ &\quad + x_2 f_5(x_1, x_2) + x_1 f_4(x_1, x_2) \\ &\quad + \varepsilon(1 - x_1^2)x_2 f_4(x_1, x_2). \end{aligned} \tag{2.150}$$

The choice of function $f_4(x_1, x_2)$ is subject to the following constraints:

(i) Equation (2.149) must be satisfied.
(ii) The $x$ given by (2.150) should be at least negative semidefinite in some neighborhood of the equilibrium state.

The following choices satisfy both of these conditions:

$$\begin{aligned} f_4(x_{1m}, x_{2m}) &= 0, \\ f_5(x_{1m}, x_{2m}) &= -\varepsilon^2 x_1\left(1 - \frac{x_1^3}{3}\right)(1 - x_1^2). \end{aligned} \tag{2.151}$$

With these choices, (2.148) and (2.150) become:

$$\begin{aligned} \dot{x}_{1m} &= \dot{x}_1 = x_2 + \left(x_1 - \frac{x_1^3}{3}\right), \\ \dot{x}_{2m} &= \dot{x}_2 = -x_1 - (1 - x_1^2)x_2 - \varepsilon^2\left(x_1 - \frac{x_1^3}{3}\right)(1 - x_1^2), \end{aligned} \tag{2.152}$$

and:

$$x = \varepsilon x_1^2\left(1-\frac{x_1^2}{3}\right). \tag{2.153}$$

Then the integral of Equation (2.152), which is the Lyapunov function for (2.147), would be as follows:

$$\begin{aligned} h(x_{1m}, x_{2m}) &= \oint_C [\dot{x}_{1m}\, dx_{2m} - \dot{x}_{2m}\, dx_{1m}] \\ &= x_1^2 + \left[x_2 + \varepsilon\left(x_1 - \frac{x_1^3}{3}\right)\right]^2. \end{aligned} \tag{2.154}$$

The region of asymptotic stability of the system is the region of negative semi-definiteness of (2.147) which for $\varepsilon > 0$ would be:

$$x_1^2 + \left[x_2 + \varepsilon\left(x_1 - \frac{x_1^3}{3}\right)\right]^2 < 3, \tag{2.155}$$

and is indicated in Figure 2.6. ■

The advantages of this method are given below:

(i) The use of the geometric properties of the function $V(x)$, which makes it very understandable; meanwhile, its application is very straightforward.
(ii) The estimates of the domain of asymptotic stability are, in general, very reasonable and they require little time and no deep insight to the problem.
(iii) Finally, the method is very flexible.

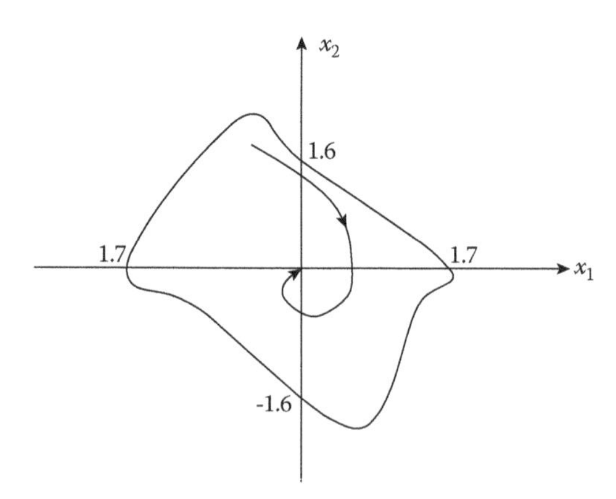


**FIGURE 2.6** Region of asymptotic stability for the system (2.147).

The disadvantages of this method are:

(i) It is only applicable to second-order autonomous systems.
(ii) Their nonlinearity should be continuous except possibly at the origin.
(iii) The technique is far from being an algorithm and this is because of the requirement for proper selection of the unknown functions $f_4(x_1,x_2)$ and $f_5(x_1,x_2)$.
(iv) The $V(x)$ functions usually depend on the nonlinearities.

The idea of using the modified system for obtaining the Lyapunov function in higher-order systems is given by Walker and Clark [w1], and Kinnen and Chen [k4].

### 2.3.9 Energy Metric of Wall and Moe [n1,w2,p2]

This method, which is somewhat restrictive, uses the functional relation to the original system to generate the $V(x)$ function. The procedure consists of the following steps:

(a) Describe the $n$th-order system as a first-order vector system:

$$\dot{x}_i = f_i(x),\ i = 1,2,\dots,n. \tag{2.156}$$

(b) Form a set of differential equations of integral curves, or,

$$\frac{dx_i}{dx_j} = \frac{f_i(x)}{f_j(x)},\quad j > i. \tag{2.157}$$

There are $\frac{n(n-1)}{2}$ such equations.

(c) Change the form (2.157) to:

$$f_j(x)dx_i - f_i(x)dx_j = 0. \tag{2.158}$$

(d) By addition and substitution operations in (2.158), obtain the single equation as:

$$\omega = \omega_1(x)dx_1 + \omega_2(x)dx_2 + \dots + \omega_n(x)dx_n. \tag{2.159}$$

(e) Obtain the $V(x)$ function by line integration of (2.159) as follows:

$$V(x) = \int \omega(x)dx = \int_0^{x_1} \omega_1(\tau_1,0,\dots,0)d\tau_1$$
$$+ \int_0^{x_2} \omega_2(x_1,\tau_2,0,\dots,0)d\tau_2$$
$$+ \cdots + \int_0^{x_n} \omega_n(x_1,x_2,\dots,x_{n-1},\tau_n)d\tau_n. \tag{2.160}$$

Note that $V(x)$ in the first trial might be indefinite, then the choice for $V(x)$ should be changed for other trials.

(f) Take the total derivative of $V(x)$, that is, $\dot{V}(x)$, to determine the region of stability.

**Example 2.14 [n1,w2]:**

Consider the following equation:

$$\dddot{x} + (1 + \dot{x}^2)\ddot{x} + \dot{x} + x = 0, \tag{2.161}$$

(a)

$$\dot{x}_1 = x_2,$$

$$\dot{x}_2 = x_3,$$

$$\dot{x}_3 = -x_2^2 x_3 - x_3 - x_2 - x_1. \tag{2.162}$$

(b)

$$\frac{dx_1}{dx_2} = \frac{x_2}{x_3},$$

$$\begin{aligned} \frac{dx_2}{dx_3} &= \frac{x_3}{-x_2^2 x_3 - x_3 - x_2 - x_1}, \\ \frac{dx_1}{dx_3} &= \frac{x_2}{-x_2^2 x_3 - x_3 - x_2 - x_1}. \end{aligned} \tag{2.163}$$

(c)

$$x_3 dx_1 - x_2 dx_2 = 0, \tag{2.164}$$

$$x_3 dx_3 + (x_2^2 x_3 + x_3 + x_2 + x_1) dx_2 = 0, \tag{2.165}$$

$$x_2 dx_3 + (x_2^2 x_3 + x_3 + x_2 + x_1) dx_1 = 0. \tag{2.166}$$

(d) Substitute (2.164) into (2.166) and add (2.165) and (2.166). Then:

$$\begin{aligned} \omega(x) = {} & (x_2 + x_3) dx_3 + (x_2^2 x_3 + 2x_2 + x_1 + x_2^3 + x_3) dx_2 \\ & + (x_2 + x_1) dx_1. \end{aligned} \tag{2.167}$$

Thus,

$$\omega_1(x) = x_1 + x_2,$$

$$\omega_2(x) = x_2^2 x_3 + 2x_2 + x_1 + x_2^3 + x_3, \tag{2.168}$$

$$\omega_3(x) = x_3 + x_2.$$

(e)

$$\begin{aligned} V(x) &= \int_0^x \omega(x)dx \\ &= \int_0^{x_1} \tau_1 d\tau_1 + \int_0^{x_2} (x_1 + 2\tau_2 + \tau_2^3)d\tau_2 + \int_0^{x_3} (x_2 + \tau_3)d\tau_3 \\ &= \frac{1}{2}(x_1 + x_2)^2 + \frac{1}{2}(x_2 + x_3)^2 + \frac{x_2^4}{4}. \end{aligned} \tag{2.169}$$

(f)

$$\dot{V}(x) = -x_2^2 x_3^2. \tag{2.170}$$

Therefore, due to Theorem 2.2, the origin which is the equilibrium state of the system is globally asymptotically stable. ■

The present form of the method has a straightforward application to nonlinear systems with lower-order nonlinear terms. Additional work is required to extend this method for studying the system of the form:

$$\dddot{x} + a_1 f(x, \dot{x}, \ddot{x})\ddot{x} + a_2\dot{x} + a_3 f(x, \dot{x}, \ddot{x})x = 0.$$

## 2.3.10 Zubov's Method [n1,b1,o1,g2,l3,r2,m1,z2]

In Zubov's method, the Lyapunov function is a solution of a partial differential equation by a power series expansion. Obviously, the nonlinearity should have an analytic form, otherwise, this method is not applicable. This method is applicable to systems that are asymptotically stable in some region.

For the nonlinear system (2.17), if there exists a Lyapunov function, its total derivative would be given by:

$$\dot{V}(x) = \quad V(x)^T f(x). \tag{2.171}$$

Let $W(x)$ be a positive definite scalar function, then it is desired to have:

$$\dot{V}(x) = \quad V(x)^T f(x) = -W(x). \tag{2.172}$$

Let $W(x)$ be of the following form:

$$W(x) = U(x)\,[1 - V(x)], \tag{2.173}$$

where $U(x)$ is assumed to be a positive definite quadratic form. Then (2.172) becomes:

$$\dot{V}(x) = V(x)^T f(x) = -U(x)[1 - V(x)]. \tag{2.174}$$

Assume $V(x)$ and $f(x)$ are given by:

$$V(x) = \sum_{k=2}^{\infty} V_k(x), \qquad f(x) = \sum_{k=1}^{\infty} f_k(x), \tag{2.175}$$

where $V_k(x)$ and $f_k(x)$ are homogeneous of degree $k$; that is,

$$V_k(\lambda x) = \lambda^k V_k(x), \quad \text{and} \quad f_k(\lambda x) = \lambda^k f_k(x). \tag{2.176}$$

Then, (2.174) becomes:

$$\left[\sum_{k=2}^{\infty} V_k(x)\right]^T \left[\sum_{i=1}^{\infty} f_i(x)\right] = -U(x)\left[1 - \sum_{k=2}^{\infty} V_k(x)\right]. \tag{2.177}$$

Equating the terms with the same degrees of order would yield:

$$\begin{aligned}
&V_2(x)^T f_1(x) = -U(x),\\
&V_3(x)^T f_1(x) + V_2(x)^T f_2(x) = 0,\\
&V_4(x)^T f_1(x) + V_3(x)^T f_2(x) + V_2(x)^T f_3(x) = U(x)V_2(x),\\
&\vdots\\
&V_k(x)^T f_1(x) + V_{k-1}(x)^T f_2(x) + \ldots + V_2(x)^T f_{k-1}(x) = U(x)V_{k-2}(x).
\end{aligned} \tag{2.178}$$

The system of Equations (2.178) is a recursive equation, and every term in $V(x)$ is obtained from the previous terms.

**Example 2.15 [n1,l3]:**

Consider the following system:

$$\dot{x} = -x + y + x(x^2 + y^2), \qquad \dot{y} = -x - y + y(x^2 + y^2). \tag{2.179}$$

The linearized system is asymptotically stable; thus Zubov's method is applicable here. The right-hand side of system (2.179) is of the following form:

$$f(x,y) = f_1(x,y) + f_3(x,y),$$

where,

$$f_1(x,y) = \begin{bmatrix} -x+y \\ -x-y \end{bmatrix} \quad \text{and} \quad f_3(x,y) = \begin{bmatrix} x^3 + xy^2 \\ x^2y + y^3 \end{bmatrix}. \tag{2.180}$$

Let,

$$U(x,y) = 2(x^2 + y^2). \tag{2.181}$$

From (2.175) and (2.176) the following general form for $V_k(x,y)$ should be assumed:

$$\begin{aligned} V_k(x,y) = a_k x^k + a_{k-1}x^{k-1}y + a_{k-2}x^{k-2}y^2 + \dots \\ + a_2 x^2 y^{k-2} + a_1 x y^{k-1} + a_0 y^k, \end{aligned} \tag{2.182}$$

thus,

$$V_2(x,y) = a_2 x^2 + a_1 xy + a_0 y^2. \tag{2.183}$$

Substituting from (2.183), (2.181), and (2.179) into the first equation of (2.178) and choosing $a_1 = 0,\ a_2 = a_0 = 1$ yields:

$$V_2(x,y) = x^2 + y^2. \tag{2.184}$$

Continuing in this way, and using the $i$th equation of (2.178), the results are:

$$V_i(x,y) = 0, \qquad i = 3,4,\dots,n, \tag{2.185}$$

thus, from (2.184) and (2.185), one can get the following:

$$V(x,y) = x^2 + y^2. \tag{2.186}$$

Thus,

$$\dot{V}(x,y) = -2(x^2+y^2)[1-(x^2+y^2)].$$

Since $\dot{V}(x,y)$ must be negative semidefinite, from (2.174) the region of asymptotic stability is given by:

$$V(x,\ y) < 1. \tag{2.187}$$

Equation (2.187) for the system (2.179) yields:

$$x^2 + y^2 \leq 1.$$

The equality gives the boundary of the region, outside of which, the equilibrium state of the system is unstable. (Why?) ■

In the previous example, a closed-form solution of $V(x)$ is achievable, which allows one to obtain the region of asymptotic stability. If, however, this was not the case, and an approximation was used to obtain $V(x)$, then the region of asymptotic stability is an approximate one. Thus, the main advantage of Zubov's method is that the whole region of asymptotic stability is given. For higher-order systems, solutions to (2.178) are not easily obtainable.

### 2.3.11 Leighton's Method [n1,l4,a1]

Consider the following second-order system:

$$\ddot{x} = r(x, \dot{x}), \tag{2.188}$$

which is equivalent to the following system:

$$\begin{aligned} \dot{x} &= y, \\ \dot{y} &= r(x, y). \end{aligned} \tag{2.189}$$

System (2.189) will be called regular if:

$$r_y(0,0) \neq 0. \tag{2.190}$$

The idea of regularity can be generalized for the following general second-order system:

$$\begin{aligned} \dot{x} &= f(x, y), \\ \dot{y} &= g(x, y), \end{aligned} \tag{2.191}$$

as

$$f_x(0,0) + g_y(0,0) \neq 0. \tag{2.192}$$

The following theorem is due to Leighton, the proof of which is omitted here.

---

**Theorem 2.6 [n1,l4]:**

Consider system (2.189) and assume $r(x, y)$ is of class $C^1$ (continuous up to the first derivative) in some neighborhood $N$ of the equilibrium state of the system. Let $N' = N - \{0\}$, and assume the origin is an isolated equilibrium state (ZES) of (2.189).

If $r_y > 0$ in $N'$, the equilibrium state is unstable. If $r_y \leq 0$ in $N'$ and $xr(x,0) < 0$ for all $x$ in $N'$, the equilibrium state is stable. This stability is asymptotic provided that the strict inequality $r_y < 0$ holds. If $r_y$ is indefinite in every neighborhood of the equilibrium state, this state may be either stable or unstable. ■

The last statement of the theorem will be clarified as follows. The equilibrium state of the following system is unstable:

$$\begin{aligned} \dot{x} &= y, \\ \dot{y} &= x + xy. \end{aligned} \tag{2.193}$$

This is a consequence of the fact that one of the characteristic roots of the linearized system corresponding to (2.193) is positive. On the other hand, the equilibrium state of the following system:

$$\begin{aligned} \dot{x} &= y, \\ \dot{y} &= -x + xy, \end{aligned} \tag{2.194}$$

is stable. The stability of the equilibrium state of the nonlinear system (2.194) can be investigated by the following Lyapunov function:

$$V(x,y) = x^2 - 2[y + \ln(1-y)], \tag{2.195}$$

which has $\dot{V}(x,y) = 0$ along trajectories. Thus, from Theorem 2.2 the equilibrium state of (2.194) is stable. ■

In the proof of Theorem 2.6, Leighton constructed the following Lyapunov function:

$$V(x,y) = y^2 + 2\int_0^x r(u,0)\,du. \tag{2.196}$$

For general second- and third-order systems, the reader is referred to Leighton [14] and Anderson and Leighton [a1].

Application of Leighton's method is straightforward, and his theorem gives stability or instability. The following example will clarify these points.

**Example 2.16 [n1,l4]:**

In the following van der Pol equation:

$$\ddot{x} = \varepsilon(1 - x^2)\dot{x} - x. \tag{2.197}$$

$r(x,\dot{x})$ is given by:

or,
$$r(x,\dot{x}) = \varepsilon(1-x^2)\dot{x} - x,$$
$$r(x,y) = \varepsilon(1-x^2)y - x,$$

and,

$$r(x,0) = -x, \quad xr(x,0) = -x^2 \le 0$$
$$r_y(x,y) = \varepsilon(1-x^2).$$

Thus,

$$r_y(0,0) = \varepsilon, \tag{2.198}$$

which implies instability of the equilibrium state of (2.197) for $\varepsilon > 0$, and asymptotic stability of that state for $\varepsilon < 0$ in the region: $x^2 + y^2 \le 1$. ■

Extensions of Lyapunov's method were first considered by LaSalle and Lefschetz [12]. They also presented some theorems for the Zero Equilibrium State (ZES) stability of the nonlinear system in finite time interval, ultimate bounded and practical system's stability analysis. In order to prove their theorems, they used a positive definite $V(x(t),t)$ function and its derivative, which was given by $\dot{V} \le g(v,t)$ for all $t$. As a matter of fact, if $\dot{V} \ge g(v,t)$ for all $t$, then the ZES of the system would be unstable. These inequality relations could be considered as nonlinear inequality relations between the Lyapunov function and its derivative. This might be considered as relaxed Lyapunov stability conditions. Some more extensions of the Lyapunov method are considered in the following.

## 2.4 RELAXED LYAPUNOV STABILITY CONDITIONS

This section considers the stability analysis of the following nonlinear time-invariant system:

$$\dot{x} = f(x), \quad x \in R^n. \tag{2.199}$$

The Lyapunov theorem explains the two following properties for a Lyapunov Function (LF) $V(x)$:

1. $V(x)$ is a Positive Definite Function (PDF).
2. $\dot{V}(x)$ is a Negative Definite Function (NDF).

These two properties are based on the concept of decrement of energy in a dynamic system. This concept assumes energy in a dynamical system is a continuous nonnegative function $V(x)$, while the zero energy is only achieved at zero equilibrium state ($V(0) = 0$), thus the decrement of energy means approaching the stable zero equilibrium state.

The positive definiteness (PD-ness) of $V(x)$ is a basic property, because using this property and the property that $V(x) \rightarrow 0$ implies $x(t) \rightarrow 0$; however, the negative definiteness (ND-ness) of $\dot{V}(x)$ is only a simplifying property for the consequence of the Lyapunov theorem; this property implies $V(x) \rightarrow 0$.

Consider a nonmonotone signal $f(t)$ such that $\dot{f}(t) > 0$, for every $t$, for example, $f(t) = (1.1 + \sin t)e^{-t}$. Although $\dot{f}(t)$ is sometimes positive, but negativeness of $\ddot{f}(t)$ makes $\dot{f}(t)$ to be a decreasing function, and $f(t)$ decreases in average.

It seems that the PD-ness of $V(x)$ could not be removed, but there is a question about the second property of LF: Is the ND-ness of $\dot{V}(x)$ the only way to show that $V(x) \rightarrow 0$? In other words, could $V(x) \rightarrow 0$ without strictly decreasing of $V(x)$? To answer this question some research removed the ND-ness of $\dot{V}(x)$ according to the following classifications:

1. Invariance Principle.
2. Average decrement of $V(x)$ function.
3. Vector Lyapunov function.
4. Higher-order derivatives of LF candidate.

Some of these methods are used only for autonomous systems and are introduced in the following subsections, but others are primarily introduced for time-varying systems; the time-invariant version of those is presented here.

### 2.4.1 LaSalle Invariance Principle

The LaSalle Invariance Principle uses $\dot{V}(x) \le 0$ along the solutions of $\dot{x} = f(x)$ to conclude $x(t)$ approaches to some bounded region M [15]. In the following, this principle is reviewed, then its application for globally asymptotical stability of zero equilibrium state is considered. For proof the interested reader is refer to Khalil [k1] and LaSalle [15].

---

**Theorem 2.7 [15]: (Invariance Principle)**

Consider a given $C^1$ indefinite function $V(x)$, the $\Omega_c = \{x : V(x) \le c\}$ is a bounded region and $\dot{V}(x) \le 0$ along the solutions of $\dot{x} = f(x)$. Then, any solution starting in $\Omega_c$ converges to the largest invariant set M in $S = \{x : \dot{V}(x) = 0\} \cap \Omega_c$.

This kind of stability combines two properties:

1. Marginal stability of the zero equilibrium state.
2. Region of attraction of the equilibrium state.

In the above principle, $V(x)$ could be a sign indefinite function and M could be any set. But when considering asymptotic stability (AS) of the zero equilibrium state,

M must coincide with {0} to prove attraction of the equilibrium state. On the other hand, the invariance principle may not conclude marginal stability of the equilibrium state. Thus, we need the Lyapunov theorem for marginal stability. The following theorem combines the invariance principle of LaSalle and the Lyapunov theorem. ■

---

**Theorem 2.8 [15]:**

(a) Consider a given nonlinear system $\dot{x} = f(x)$ and a given class $C^1$ $V(x)$ function, which is Locally Positive Definite Function (LPDF). Moreover, consider a bounded region $\Omega_c = \{x : V(x) \le c\}$ for some $c > 0$ and let M be the largest invariant set in $S = \{x : \dot{V}(x) = 0\} \cap \Omega_c$. If $\dot{V}(x) \le 0$ in $\Omega_c$ and $M = \{0\}$, then the zero equilibrium state of $\dot{x} = f(x)$ is asymptotically stable.

(b) If all the previous conditions of this theorem hold for a Radially Unbounded (RU) and PD $V(x)$ function, for any $c > 0$, then the zero equilibrium state of $\dot{x} = f(x)$ is globally asymptotically stable. ■

**Example 2.17 [k1]:**

Consider the following system:

$$\dot{x}_1 = x_2,$$

$$\dot{x}_2 = -h_1(x_1) - h_2(x_2),$$

where $h_1(\cdot)$ and $h_2(\cdot)$ are locally Lipschitz and satisfying the following:

$$h_i(\cdot) = 0, \qquad yh_i(y) > 0, \quad \forall y \neq 0 \text{ and } y \in (-a, a)$$

The system has an isolated equilibrium state at the origin. Depending upon the functions $h_i(\cdot)$, it might have other equilibrium states. The system can be viewed as a generalized pendulum with $h_2(x_2)$ as the friction term. Therefore, a Lyapunov function candidate may be taken as the following energy-like function:

$$V(x) = \int_0^{x_1} h_1(y)dy + \frac{1}{2}x_2^2.$$

Let $D \triangleq \{x \in R^2 \mid -a < x_i < a\}$, $V(x)$ be positive definite in D, and

$$\dot{V}(x) = h_1(x_1)x_2 + x_2[-h_1(x_1) - h_2(x_2)] = -x_2h_2(x_2) \le 0$$

is negative a semidefinite function. To find $S = \{x \in D \mid \dot{V}(x) = 0\}$, note that:

$$\dot{V}(x) = 0 \;\Rightarrow x_2h_2(x_2) = 0 \;\Rightarrow x_2 = 0, \text{ since } -a < x_2 < a.$$

Hence,

$$S = \{x \in D \mid x_2 = 0\}.$$

Let x(t) be a solution that belongs identically to S:

$$x_2(t) \equiv 0 \Rightarrow \dot{x}_2(t) \equiv 0 \Rightarrow h_1(x_1(t)) = 0 \quad x_1(t) = 0.$$

Therefore, the only solution that can stay identically in $S$, is the trivial solution $x(t) \equiv 0$. Thus, the origin is asymptotically stable. ■

### 2.4.2 Average Decrement of the V(x) Function

Let $V(x)$ be LPDF[10] and $\dot{V}(x)$ along the solutions of $\dot{x} = f(x)$ be not NDF, but $V(x)$ decreases in average, then asymptotic stability of the zero equilibrium state may be obtained as follows:

---

**Theorem 2.9 [m2]:**

Let $\dot{x} = f(x)$ be a nonlinear system with the zero equilibrium state. If there exists an LPDF, a given $T > 0$ and a $\gamma \in \mathcal{K}$[11] such that:

$$V(x(t+T)) - V(x(t)) \le -\gamma(\|x(t)\|) < 0 \quad ,\forall t, \tag{2.200}$$

then, the zero equilibrium state of $\dot{x} = f(x)$ is asymptotically stable. ■

This theorem requires an average decrement of the $V(x)$ function on every time interval $[t, t+T]$. However, the following theorem reduces the set of such intervals to an infinitely countable set.

---

**Theorem 2.10 [m3,p3]:**

Let $\dot{x} = f(x)$ be a nonlinear system with the zero equilibrium state. If there exists a LPDF $V(x)$, a given $T > 0$, $\gamma \in \mathcal{K}$, and strictly increasing sequence of times $\{t_k^*\}_{-\infty}^{+\infty}$ with the properties $0 < t_{k+1}^* - t_k^* < T$, $t_k^* \to \pm\infty$ as $k \to \pm\infty$, such that:

$$V(x(t_{k+1}^*)) - V(x(t_k^*)) \le -\gamma(\|x(t_k^*)\|) < 0, \tag{2.201}$$

then the zero equilibrium state of $\dot{x} = f(x)$ is asymptotically stable. ■

The following theorem is for exponential stability of the zero equilibrium state using the average decrement method.

**Theorem 2.11 [m3,a2]:**

If in addition to conditions of Theorem 2.10, there exist positive numbers $r, \lambda_{min}$, and $\lambda_{max}$ such that:

$$\begin{cases} V(x(t_{k+1}^*)) - V(x(t_k^*)) \le -r\|x(t_k^*)\|^2 < 0, \\ \lambda_{min}\|x\|^2 \le V(x) \le \lambda_{max}\|x\|^2, \end{cases} \tag{2.202}$$

then, the zero equilibrium state of $\dot{x} = f(x)$, $x(t_0) = x_0$, $x \in R^m$ is exponentially stable. ■

Note that all three theorems need some upper bound of the system's solution, which is the drawback for application of these theorems. It could be shown that this is not so for the nonlinear homogenous systems.

### 2.4.3 Vector Lyapunov Function

According to the inverse Lyapunov theorem, all practical dynamic systems with the stable zero equilibrium state have a valid LF. However, finding a suitable LF for a given dynamic system is not a simple task, especially in the case of large-scale systems.

Assume a given system $\dot{x} = f(x)$ could be decomposed into $m \in N$ subsystems, as follows:

$$\dot{x}_i = f_i(x_i) + h_i(x_1, x_2, \ldots x_{i-1}, x_{i+1}, \ldots, x_m), \quad i = 1, 2, \ldots, m, \tag{2.203}$$

where $x = (x_1^T, \ldots, x_m^T)^T$ is the state vector of all subsystems of a given large-scale system. The $h_i(\cdot)$ terms denote the interaction effects on the $i$th subsystem from other subsystems. Let an LF candidate for each individual subsystem $\dot{x}_i = f_i(x_i)$ be given by $v_i(x_i)$, that is, $v_i(0) = 0$ and:

$$v_i(x_i) \ge \phi_i(\| x_i \|), \quad \phi_i \in \mathcal{K} \tag{2.204}$$

then a Vector Lyapunov Function (VLF) is defined as follows.

$$V(x) \triangleq [v_1(x_1), v_2(x_2), \ldots, v_m(x_m)]^T \tag{2.205}$$

where every $v_i(x_i)$ should be at least positive semidefinite, and an LF candidate for $\dot{x} = f(x)$ could be introduced as:

$$V(x) = \sum_{i=1}^{m} k_i v_i(x_i), \quad k_i > 0 \tag{2.206}$$

To use the Lyapunov theorem, the values of $k_i$ must be determined such that $V(x)$ be PDF and $\dot{V}(x)$ be NDF or at least NSDF [h7,16]. The stability analysis of a (large-scale) nonlinear system $\dot{x} = f(x)$ using the vector Lyapunov function (VLF) approach is presented in Dehghani and Nikravesh [d1] and Dehghani and Nikravesh [d2] and is given in Appendix A1.

However, some papers use VLFs of the form (2.205) in different ways. Some of them use the following inequality component-wise, that is, each $\dot{v}_i(x)$ could be sign indefinite, in other words:

$$\dot{V}(x) \le g[V(x)]. \tag{2.207}$$

To demonstrate their theorems, one needs the following definition and the lemma.

---

**Definition 2.3 [m2]:**

Let $\underline{a} = (a_1,\ldots,a_m)^T \in R^m$ and $b = (b_1,\ldots,b_m)^T \in R^m$. The mapping $g: R^m \to R^m$ is of class $W$ (quasi-monotone nondecreasing) if:

$$\forall i = 1,2,\ldots,m \quad , \quad \begin{pmatrix} a_i = b_i \\ a_j \le b_j \ , \forall_j, j \ne i \end{pmatrix} \quad g_i(a) \le g_i(b). \tag{2.208}$$

■

This definition could be used in the time-varying case ($\forall t > 0$) as well.

**Example 2.18:**

Consider the following LTI system:

$$\dot{u} = \begin{pmatrix} 1 & 1 \\ 0 & 2 \end{pmatrix} u.$$

Let $a = (1 \quad 1)^T$ and $b = (1 \quad 2)^T$ then, $g(a) = (2 \quad 2)^T$ and $g(b) = (3 \quad 4)^T$. Thus, for $a_1 = b_1$ one has $g_1(a) \le g_1(b)$ therefore, this system is of the class $W$. ■

**Exercise 2.1:**

Show that $g(u) = \begin{pmatrix} 1 & -1 \\ 0 & 2 \end{pmatrix} u$ is not of the class $W$. ■

**Exercise 2.2:**

Verify if the following $g(t, x)$ is of the class $W$.

$$g(t,x) = \begin{pmatrix} x_1^3 \sin t + 4x_2 \cos t \\ x_2^2 + \dfrac{1}{t+1} x_1^3 \end{pmatrix}$$ ■

---

**Lemma 2.1 [m2][12]:**

Consider the following vector differential equation:

$$\dot{u} = g(u), \quad u(t_0) = u_0, \quad u \in R^m \tag{2.209}$$

where $g(u)$ is locally Lipschitz. Let $V(x)$ be a continuous vector function whose upper right-hand derivative $D^+V(x)$[13] satisfies the following differential inequality component-wise:

$$D^+V(x) \le g(V(x)), \quad V(x_0) \le u_0, \quad V \in R^m \tag{2.210}$$

If $g(u)$ is of the class, then $V(x(t)) \le u(t)$ for all $t \ge t_0$. ■

**Proof:**

The proof of this lemma is given in Meigoli and Nikravesh [m2]. ■

---

**Theorem 2.12 [m2]:**

Consider a nonlinear system $\dot{x} = f(x)$ and the VLF given in (2.205), that is, $v_i(0) = 0$ and (2.204) is satisfied. Moreover, let the inequality $\dot{V}(x) \le g[V(x)]$[14] be satisfied component-wise and the mapping $g: R^m \to R^m$ be of the class, then the following are equivalent:

If the zero equilibrium state of $\dot{u}(t) = g[u(t)]$ is:

(i) Stable in the sense of Lyapunov, then the zero equilibrium state of $\dot{x} = f(x)$ is also stable in the sense of Lyapunov.
(ii) Asymptotically stable, then the zero equilibrium state of $\dot{x} = f(x)$ is also asymptotically stable.

**Proof:**

The proof of this theorem is omitted here and the interested reader is referred to Meigoli and Nikravesh [m2]. Also, a proof for a more completed case, that is, proof of Theorem 3.8, which is the time-varying version of this theorem, is given in Appendix A2. ■

Despite the direct Lyapunov method, the VLF could use positive Lyapunov functions with sign indefinite time-derivative functions.

**Remark 2.1:**

It is clear that an LTI, which maps $g(u) = Au$, is of class iff: each off-diagonal element of $A_{m\times m}$ will be nonnegative scalar. ■

**Example 2.19:**

Let $g(u) = Au = \begin{pmatrix} a & b \\ c & d \end{pmatrix} u$. Assume $u = [u_1, u_2]^T, u_1 = u_1^*$ and $u_2 \le u_2^*$. Then, $g_1(u) = au_1 + bu_2$.and $g_2(u) = au_1 + bu_2$. If $b \ge 0$ then, $g_1(u) \le g_1(u^*)$. On the other hand, if $u_2 = u_2^*$ and $u_1 \le u_1^*$, then $g_2(u) \le g_2(u^*)$ whenever $c \ge 0$, since $g_2(u) = cu_1 + du_2$. Thus, $g(u)$ is of the class. ■

Also see Exercise 2.1.

**Example 2.20:**

Consider the following nonlinear system:

$$\dot{x} = f(x) \triangleq \begin{cases} \dot{x}_1 = -x_1 + x_2^2 \\ \dot{x}_2 = -x_2 \end{cases}$$

From (2.203) one has

$$\dot{x}_i = f(x_i) \triangleq \begin{cases} \dot{x}_1 = -x_1 \\ \dot{x}_2 = -x_2 \end{cases}.$$

Let $V(x_i) = [x_1^2, x_2^2]^T$,

where

$$V(0) = [0, 0]^T$$

then,

$$\dot{V}(x_i) = \left[-2x_1^2, \ -2x_2^2\right].$$

Note that (2.207) yields:

$$_i = \left[\ _1, \ _2\right]^T = \left[x_1^2, x_2^2\right]^T \in \mathcal{K}.$$

Obviously, the linear system $\dot{u} = g(u) = \begin{pmatrix} -1 & 0 \\ 0 & -2/3 \end{pmatrix}\begin{pmatrix} u_1 \\ u_2 \end{pmatrix}$ is asymptotically stable, then if g is of the class, therefore $\dot{x} = f(x)$ is also asymptotically stable. From Remark (2.1) it is obvious that g is of the class. ■

Note that using either the linearizing technique or, for example, the Krasovskii method, one would be able to prove the asymptotic stability of this nonlinear system.

### Example 2.21 [m7]:

Consider the longitudinal motion of an aircraft, which can be represented by the following system of differential equations:

$$\frac{dx_i}{dt} = -\rho_i x_i + \sigma, \quad i = 1, 2, \ldots, n,$$

$$\frac{d\sigma}{dt} = \sum_{i=1}^{n} a_i x_i - p\sigma - f(\sigma), \tag{2.211}$$

where $\rho_i > 0, p > 0, \sigma f(\sigma) > 0$ for $\sigma \neq 0$ and $f(0) = 0$.

Let us illustrate the estimations of the domain of the parameters values for which the state ($x = 0$, $\sigma = 0$) of the system is asymptotically stable. By means of a VLF; $V(x) \triangleq [v_1, v_2, \ldots, v_n, v_{n+1}]^T$ with the following components:

$$v_i \triangleq \tfrac{1}{2} x_i^2, \quad i = 1, 2, \ldots, n, \quad v_{n+1} \triangleq \tfrac{1}{2}\sigma^2 \tag{2.212}$$

and by some manipulations (e.g., $\dot{x}^2 \geq 0 \quad \frac{1}{2\rho_i}(-\rho_i x_i + \sigma)^2 \geq 0$), the system (2.211) is reduced to the following form:

$$\frac{dv_i}{dt} = -\rho_i x_i^2 + x_i \sigma \leq -\frac{1}{2}\rho_i x_i^2 + \frac{1}{2}\frac{\sigma^2}{\rho_i}, \quad i = 1, 2, \ldots, n,$$

$$\frac{dv_{n+1}}{dt} = \sum_{i=1}^{n} a_i x_i \sigma - p\sigma^2 - f(\sigma)\sigma \leq \sum_{i=1}^{n} |a_i| \rho_i \frac{x_i^2}{2}$$

$$-\left(2p - \sum_{i=1}^{n} \frac{|a_i|}{\rho_i}\right)\frac{\sigma^2}{2} - f(\sigma)\sigma. \tag{2.213}$$

For the second set of equations, that is, $\frac{dv_{n+1}}{dt}$, one should use the inequality

$$x_i\sigma \le \frac{\rho_i x_i^2}{2} + \frac{\sigma^2}{2\rho_i},$$

which is obtained from the first set of inequality. Now:

$$\begin{aligned} \frac{dv_i}{dt} &\le -\rho_i v_i + \frac{v_{n+1}}{\rho_i} \quad i = 1,2,\ldots,n, \\ \frac{dv_{n+1}}{dt} &\le \sum_{i=1}^{n} |a_i|\rho_i v_i - \left(2p - \sum_{i=1}^{n} \frac{|a_i|}{\rho_i}\right) v_{n+1} - g(v_{n+1}). \end{aligned} \tag{2.214}$$

This inequality is of the form (2.207). The following nonlinear system:

$$\begin{aligned} \frac{du_i}{dt} &= -\rho_i u_i + \frac{u_{n+1}}{\rho_i}, \quad i = 1,2,\ldots,n, \\ \frac{du_{n+1}}{dt} &= \sum_{i=1}^{n} |a_i|\rho_i u_i - \left(2p - \sum_{i=1}^{n} \frac{|a_i|}{\rho_i}\right) u_{n+1} - g(u_{n+1}), \end{aligned} \tag{2.215}$$

is of the class *W*. Moreover, it was shown in Martynyuk [m7] that the zero equilibrium state of (2.215) is asymptotically stable if:

$$\frac{1}{2}\sum_{i=1}^{n} \frac{|a_i|}{\rho_i} \le p. \tag{2.216}$$

Therefore Theorem 2.12 implies that the zero equilibrium state of (2.211) is also asymptotically stable if (2.216) holds true. ■

### 2.4.4 Higher-Order Derivatives of a Lyapunov Function Candidate [m13]

Consider a nonlinear system $\dot{x} = f(x)$ with an asymptotically stable zero equilibrium state and a LF candidate $V(x)$. Some recent works assumed $\dot{V}(x)$ is not NDF and focused on the sign of higher-order time derivatives of $V(x)$ to prove asymptotic stability of the zero equilibrium state. These derivatives are computed iteratively using the following relation:

$$V^{(i)}(x) \triangleq [\partial V^{(i-1)}/\partial x]^T f(x) \quad , \quad i = 1,2,\ldots,m \tag{2.217}$$

It is clear that $V(x)$ and $f(x)$ must be smooth enough such that the above relation could be iterated m times, otherwise $\partial V^{(i-1)}/\partial x$ could be discontinuous and (2.217) would not hold.[15]

It was first proved that if for some $h > 0$ the following inequality holds:

$$\min\{\dot{V}(x), h\ddot{V}(x)\} < -\phi_2(\| x \|), \quad \phi_2 \in \mathcal{K} \tag{2.218}$$

that is, min $\{\dot{V}(x), h\ddot{V}(x)\}$ is NDF, then the zero equilibrium state of $\dot{x} = f(x)$ is asymptotically stable. However, it was shown later that this is an empty criterion, because if (2.218) holds, then for a LF $V(x)$ in a given region around the zero equilibrium state, $\dot{V}(x)$ must be NDF in that region [m2]. Therefore, the role of $\dddot{V}(x)$ in stability analysis was added and the following theorem was introduced.

---

**Theorem 2.13 [a3]:**

Let $\dot{x} = f(x)$ be a nonlinear system with a zero equilibrium state, and $V(x)$ be LPDF. If,

$$a_3\dddot{V}(x) + a_2\ddot{V}(x) + \dot{V}(x) < 0, \forall x \neq 0, \qquad a_3 \geq 0 \quad \text{and} \quad a_2 \geq 0 \tag{2.219}$$

hold, then the zero equilibrium state of $\dot{x} = f(x)$ is asymptotically stable. ■

**Proof:**

The proof of this theorem is omitted here and the interested reader is referred to Aeyels [a3]. ■

The following example shows the application of this theorem. Even though the stability analysis of this linear system could be done more easily using linear methods.

**Example 2.22 [b2]:**

Consider the following LTI system with a given quadratic LF candidate:

$$\dot{x} = Ax,$$
$$V(x) = \tfrac{1}{2}x^T Px, \quad P > 0. \tag{2.220}$$

Then,

$$\dot{V}(x) = \tfrac{1}{2}x^T Qx\,; \quad Q = PA + A^T P,$$

$$\ddot{V}(x) = \tfrac{1}{2}x^T Rx\,; \quad R = QA + A^T Q,$$

$$\dddot{V}(x) = \tfrac{1}{2}x^T Sx\,; \quad S = RA + A^T R. \tag{2.221}$$

If the stable system and the LF candidate of (2.220) is chosen with:

$$A = \begin{bmatrix} -4 & -5 \\ 1 & 0 \end{bmatrix}, \quad P = \begin{bmatrix} 1 & \frac{1}{2} \\ \frac{1}{2} & 1 \end{bmatrix}, \tag{2.222}$$

then, one has:

$$Q = \begin{bmatrix} -7 & -6 \\ -6 & -5 \end{bmatrix}, \quad R = \begin{bmatrix} 44 & 54 \\ 54 & 60 \end{bmatrix}, \quad S = \begin{bmatrix} -244 & -376 \\ -376 & -540 \end{bmatrix}. \tag{2.223}$$

Therefore, $Q$ is not negative semidefinite. Thus, one could not use a Lyapunov direct method with this $P$ matrix for the given LTI system stability analysis. Further calculation gives:

$$a_3 S + Q < 0, \quad \text{for} \quad 0.00217 < a_3 < 0.0485, \tag{2.224}$$

that is, $a_3 \dddot{V}(x) + \dot{V}(x)$ is NDF, therefore, using Theorem 2.13, the given LTI system is asymptotically stable.[16] ■

The following theorem uses Lemma 2.1 (generalized comparison principle) for higher-order derivatives of an LF candidate Equation (2.220).

---

**Theorem 2.14 [m2]:**

Let $\dot{x} = f(x)$ be a nonlinear system with the zero equilibrium state, and $V(x)$ be a function satisfying:

$$V^{(m)}(x) \leq g_m(V, \dot{V}, \ldots, V^{(m-1)}). \tag{2.225}$$

If the following conditions hold:

(i) $\alpha_1(\| x \|) \leq V(x) \leq \alpha_2(\| x \|), \quad \| x \| < r, \quad \alpha_1, \alpha_2 \in \mathcal{K}$.
(ii) The following nonlinear system, that is, Comparison Equation (CE) is of the class.

$$u^{(m)}(t) = g_m(u, \dot{u}, \ldots, u^{(m-1)}). \tag{2.226}$$

(iii) The solutions of (2.226) for identically equal initial conditions, that is,

$$u^{(i)}(0) = V^{(i)}(x_0) \quad , \quad i = 0, 1, \ldots, m-1 \quad \forall \| x_0 \| < r, \tag{2.227}$$

satisfy:

$$u(t) < \alpha_3(u(0)) \quad t \geq 0, \quad \alpha_3 \in \mathcal{K} \tag{2.228}$$

then the zero equilibrium state of $\dot{x} = f(x)$ is stable in the sense of Lyapunov. ■

**Proof:**

Since (2.226) is of the class and (2.227) is satisfied, using Lemma 2.1, then:

$$V^{(i)}(x(t)) \leq u^{(i)}(t), \quad i = 0,1,\ldots,m-1. \tag{2.229}$$

Combining (2.229), (2.228), and (2.227) for $i = 0$ yields:

$$V(x(t)) \leq u(t) < \alpha_3(u(0)) = \alpha_3(V(x_0)). \tag{2.230}$$

Then, implementing part (i) of this theorem twice yields:

$$\begin{aligned} &\alpha_1(\| x(t) \|) \leq V(x(t)), \\ &V(x_0) \leq \alpha_2(\| x_0 \|). \end{aligned} \tag{2.231}$$

Thus, substituting (2.231) in (2.230) implies:

$$\begin{aligned} &\alpha_1(\| x(t) \|) \leq V(x(t)) < \alpha_3(V(x_0)) \leq (\alpha_3 \circ \alpha_2)(\| x_0 \|) \\ &\| x(t) \| < (\alpha_1^{-1} \circ \alpha_3 \circ \alpha_2)(\| x_0 \|), \end{aligned} \tag{2.232}$$

Therefore, the zero equilibrium state of $\dot{x} = f(x)$ is stable in the sense of Lyapunov if $\alpha = (\alpha_1^{-1} \circ \alpha_3 \circ \alpha_2) \in \mathcal{K}$. ■

---

**Lemma 2.2 [m2]:**

Let $\dot{x} = f(x)$, $f(0) = 0$ and $V(x)$ be smooth enough such that: $V(x) \triangleq [v(x), \dot{v}(x), \ldots, v^{(m-1)}(x)]^T$ for some $m \in N$ is of the class $C^1$, then:

(i) $V(0) = 0$.
(ii) Since $V(0) = 0$, there exists some $\gamma \in \mathcal{K}$ such that $\|V(x)\| \leq \psi(\|x\|)$.

**Proof:**

Since $f(0) = 0$, using (2.217) iteratively yields $V^{(i)}(0) = 0$ for $i = 0,1, \ldots, m-1$, that is, $V(0) = 0$. Since $V(x)$ is continuous everywhere, the function $\nu(r) \triangleq r + \sup_{\|x\| \leq r} \|V(x)\|$ is well-defined, continuous, and strictly increasing on $[0,+\infty)$. Moreover, $\psi(0) = 0$ since $V(0) = 0$, thus $\psi \in \mathcal{K}$ and $\|V(x)\| \leq \psi(\|x\|)$. ■

---

**Theorem 2.15 [m2]:**

Let $\dot{x} = f(x)$ with the zero equilibrium state and $V(x)$ be a function satisfying the conditions of parts (i) and (ii) of Theorem 2.14. Then, the zero equilibrium state of $\dot{x} = f(x)$ is asymptotically stable if any one of the following conditions hold:

(a) The solutions of (2.226) for identically equal initial conditions (2.227) satisfy:

$$u(t) < \alpha_3(u(0),t), \quad t \geq 0, \quad \alpha_3 \in \mathcal{KL} \tag{2.233}$$

(b) The zero equilibrium state of (2.226) is asymptotically stable.

**Proof:**

Part (a) Using this class condition and (2.227), implies (2.229). Then combining (2.229), (2.233) and (2.227) for $i = 0$ yields:

$$V(x(t)) \leq u(t) < \alpha_3(u(0),t) = \alpha_3(V(x_0),t), \alpha_3 \in \mathcal{KL} \tag{2.234}$$

then, substituting (2.231) into (2.234) implies:

$$\alpha_1(\| x(t) \|) \leq V(x(t)) < \alpha_3(V(x_0),t) \leq \alpha_3(\alpha_2(\| x_0 \|),t)$$

$$\| x(t) \| < \alpha_1^{-1}\left[\alpha_3(\alpha_2(\| x_0 \|),t)\right]. \tag{2.235}$$

Defining $\beta(\| x_0 \|,t) \triangleq \alpha_1^{-1}\left[\alpha_3(\alpha_2(\| x_0 \|),t)\right] \in \mathcal{KL}$, implies that the zero equilibrium state of $\dot{x} = f(x)$ is asymptotically stable.

Part (b) Let $U(t) \triangleq [u(t),\dot{u}(t),\ldots,u^{(m-1)}(t)]^T$. Since the zero equilibrium state of (2.226) is asymptotically stable, it implies:

$$\| \underline{U}(t) \| \leq \nu(\| U(0) \|,t), \forall t \geq 0, \forall \| U(0) \| < c, \nu \in \mathcal{KL}. \tag{2.236}$$

Substituting (2.227) and $u(t) <| u(t) |\leq \|U(t)\|$ into (2.236) yields:

$$u(t) \leq \beta(\| V(x_0) \|,t), \quad \forall t \geq 0, \quad \forall \| V(x_0) \| < c, \beta \in \mathcal{KL}. \tag{2.237}$$

Then, using Lemma 2.2, there exists $\psi \in$ such that: $\| V(x_0) \| \leq \psi(\| x_0 \|)$. Replacing this into (2.237) yields:

$$u(t) \leq \beta(\psi(\| \underline{x}_0 \|),t), \quad \forall t \geq 0, \quad \forall \| \underline{V}(\underline{x}_0) \| < c, \beta \in \mathcal{KL}. \tag{2.238}$$

Finally, substituting (2.231) and (2.229) for $i = 0$ yields:

$$\alpha_1(\| x(t) \|) \le V(x(t)) \le u(t) \le \beta(\psi(\| x_0 \|), t)$$
$$\| x(t) \| \le \alpha_1^{-1}[\beta(\psi(\| x_0 \|), t)], \tag{2.239}$$

which implies asymptotic stability of the zero equilibrium state of $\dot{x} = f(x)$. ■

**Example 2.23:**

Consider the following nonlinear system:

$$\dot{x} = -(1 + x^2)x.$$

Let: $V(x) \triangleq x^2$, then,

$$\dot{V}(x) = -2x\dot{x} = -2x(1 + x^2)x = -2x^2 - 2x^4 \le -2x^2$$

Thus: $\dot{V}(x) \le -2V(x)$.

Now, let the cosystem be as follows:

$$\dot{u} = -2u,$$

$$\text{with } u(0) = x^2(0),$$

$$\text{thus: } u(t) = x^2(0)e^{-2t} \le x^2(0), \ \forall t \ge 0.$$

Therefore, the condition (a) of Theorem 2.15, that is, (2.233) is satisfied, which implies asymptotic stability of the zero equilibrium state of the system. ■

**Remark 2.2 [m2]:**

Theorems 2.14 and 2.15 use comparison lemma (Lemma 2.1) to obtain some stability property for the zero equilibrium state of $\dot{x} = f(x)$, using m higher-order derivatives of an LF candidate. However, they require some stability property for a cosystem such as (2.226). The requirement for stability of a new nonlinear cosystem is a restriction for this method. The idea is to use an LTI cosystem, where stability analysis of the cosystem is not complicated. ■

However, in the case of LTI cosystems, asymptotic stability of the zero equilibrium state contradicts the class requirement of cosystem, because an LTI cosystem is of the following form:

$$\begin{bmatrix} \dot{u}_1 \\ \dot{u}_2 \\ \vdots \\ \dot{u}_m \end{bmatrix} = \begin{bmatrix} 0 & 1 & \cdots & 0 \\ 0 & 0 & \ddots & \vdots \\ 0 & 0 & \cdots & 1 \\ -a_0 & -a_1 & \cdots & -a_{m-1} \end{bmatrix} \begin{bmatrix} u_1 \\ u_2 \\ \vdots \\ u_m \end{bmatrix} \tag{2.240}$$

Using Remark 2.1, the class condition for (2.240) implies $-a_i \geq 0$ for $i = 0,1,\ldots, m-2$, thus, the characteristic equation $s^m + a_{m-1}s^{m-1} + \cdots + a_1 s + a_0 = 0$ is not Hurwitz. Therefore, Theorems 2.14 and 2.15 could not be implemented using LTI cosystems. However, the next theorem will use an LTI cosystem instead of class condition.

---

**Theorem 2.16 [m2]:**

Consider the nonlinear system (2.199) and a $C^1$ m-vector function $V(x)$ of the form:

$$V(x)=[V_1(x),V_2(x),\ldots,V_m(x)]^T \tag{2.241}$$

whose derivative $\dot{V}(x)$ along the solutions of (2.199) satisfies the following controllable canonical differential inequality form:

$$\begin{bmatrix} \dot{V}_1(x) \\ \dot{V}_2(x) \\ \vdots \\ \dot{V}_m(x) \end{bmatrix} \leq \begin{bmatrix} 0 & 1 & \cdots & 0 \\ 0 & 0 & \ddots & \vdots \\ 0 & 0 & \cdots & 1 \\ -a_0 & -a_1 & \cdots & -a_{m-1} \end{bmatrix} \begin{bmatrix} V_1(x) \\ V_2(x) \\ \vdots \\ V_m(x) \end{bmatrix} \tag{2.242}$$

Moreover, assume $V(0) = 0$ and all the roots of the following characteristic equation:

$$s^m + a_{m-1}s^{m-1} + \cdots + a_1 s + a_0 = 0, \tag{2.243}$$

of the cosystem (2.240) are negative real numbers. Therefore:

(i) If $V_1(x)$ is LPDF then, the zero equilibrium state of (2.199) is asymptotically stable.
(ii) If $V_1(x)$ is PDF and Radially Unbounded (RU), then the zero equilibrium state of (2.199) is globally asymptotically stable. ■

Before proving this theorem let us give a useful definition.

---

**Definition 2.4 [m2]:**

A linear system is said to be externally positive if for any positive input, the system output is also positive. It is said to be positive, if all the states remain positive as well. ■

**Proof of Theorem 2.16:**

The differential inequality (2.242) could be considered as the following LTI cosystem:

$$\begin{bmatrix} \dot{V}_1(x(t)) \\ \dot{V}_2(x(t)) \\ \vdots \\ \dot{V}_m(x(t)) \end{bmatrix} = \begin{bmatrix} 0 & 1 & \cdots & 0 \\ 0 & 0 & \ddots & \vdots \\ 0 & 0 & \cdots & 1 \\ -a_0 & -a_1 & \cdots & -a_{m-1} \end{bmatrix} \begin{bmatrix} V_1(x(t)) \\ V_2(x(t)) \\ \vdots \\ V_m(x(t)) \end{bmatrix} + \begin{bmatrix} d_1(x(t)) \\ d_2(x(t)) \\ \vdots \\ d_m(x(t)) \end{bmatrix}, \tag{2.244}$$

$$y(t) = V_1(x(t)),$$

with negative inputs (compared to 2.242), that is,

$$d_i(x(t)) \leq 0, \quad i = 1,2,\ldots,m, \tag{2.245}$$

and an output $y(t)$. By the superposition principle for the LTI systems, one has the following:

$$V_1(x(t)) \triangleq y(t) = y_0(t) + y_d(t), \tag{2.246}$$

where $y_0(t)$ is the response to the initial conditions $V_i(x_0)$ and $y_d(t)$ is the response to negative inputs $d_i(x(t))$. It is not difficult to verify that the transfer function $H_i^m(s)$ from each input $D_i(s)$ of (2.444) to the output $Y(s)$ is given by:

$$\frac{Y(s)}{D_i(s)} \triangleq h_i^m(s) = \frac{a_i + a_{i+1}s + a_{i+2}s^2 + \cdots + a_{m-1}s^{m-1-i} + s^{m-i}}{a_0 + a_1 s + a_2 s^2 + \cdots + a_{m-1}s^{m-1} + s^m} \triangleq \frac{{}^{m}_{i}}{{}^{m}_{0}}, \tag{2.247}$$

$$i = 1,2,\ldots,m.$$

All poles of the controllable canonical form of system (2.244) have negative real values. Using this assumption, this system is shown to be externally positive.

Let us recall some facts: First, an LTI system with a transfer function $h(s)$ is externally positive iff $h(t) > 0$ for all $t > 0$. Second, the summation or the multiplication of any collection of externally positive LTI systems is also an externally positive system.

The external positiveness of $h_i^m(s)$ in (2.247) for $i = 1,2,\ldots,m$, is proved by mathematical induction on $m \in N$. Note that $h_1^1(s) = 1/(a_0 + s)$, that is, $h_1^1(t) = e^{-a_0 t}u_{-1}(t)$[17] is externally positive. Thus, our claim is true in the case of the one-dimensional system ($m = 1$). Assume this claim is true for $m \in N$, then it is shown that the claim is true for $m + 1$. Assume the denominator of $h_i^{m+1}(s)$ is given by:

$${}^{m+1}_{0} = (a + s)(a_0 + a_1 s + a_2 s^2 + \cdots + a_{m-1}s^{m-1} + s^m) \tag{2.248}$$

where $(-a)$ and all its other poles are negative reals. Thus, $h_{m+1}^{m+1}(s) = 1/(a+s)(a_0 + a_1 s \cdots + a_{m-1}s^{m-1} + s^m)$ is the product of $m$ + 1 externally positive first-order systems and hence it is externally positive.

The relation (2.248) could be written in the following form:

$$\begin{aligned} {}_{0}^{m+1} &= aa_0 + (aa_1 + a_0)s + (aa_2 + a_1)s^2 + \cdots \\ &+ (aa_{m-1} + a_{m-2})s^{m-1} + (a + a_{m-1})s^m + s^{m+1}. \end{aligned} \tag{2.249}$$

Using this form and comparing it with (2.247), the numerator of $h_i^{m+1}(s)$ is given by:

$$\begin{aligned} {}_{i}^{m+1} &= (aa_i + a_{i-1}) + (aa_{i+1} + a_i)s + \cdots + (aa_{m-1} + a_{m-2})s^{m-1-i} \\ &+ (a + a_{m-1})s^{m-i} + s^{m+1-i}. \end{aligned}$$

Some rearranging yields:

$${}_{i}^{m+1} = a_{i-1} + (a+s)(a_i + a_{i+1}s + \cdots + a_{m-1}s^{m-1-i} + s^{m-i}). \tag{2.250}$$

Then dividing (2.250) by (2.249) and using (2.247) yields:

$$h_i^{m+1}(s) = \frac{{}_{i}^{m+1}}{{}_{0}^{m+1}} = \frac{a_{i-1}}{{}_{0}^{m+1}} + \frac{{}_{i}^{m}}{{}_{0}^{m}} = a_{i-1}h_{m+1}^{m+1}(s) + h_i^m(s) \tag{2.251}$$

where $h_{m+1}^{m+1}(s)$ and $h_i^m(s)$ are both externally positive transfer functions of degrees $m$ + 1 and $m$, respectively. Thus, $h_i^{m+1}(s)$ is also externally positive, and the mathematical induction is completed.

The cosystem (2.244) is an externally positive system, while it is linear with negative inputs, $d_i(x(t))$, implying $y_d(t) \le 0$ in (2.246). Using this relation yields $V_1(x(t)) = y(t) \le y_0(t)$, where $y_0(t)$ is exponentially stable and depends only on $V_1(x_0)$. Thus, there are constants $a$, $b > 0$ such that:

$$V_1(x(t)) \le y_0(t) \le \mid y_0(t) \mid \le a \left\| V_1(x_0) \right\| \exp[-b(t - t_0)] \ \ \forall t \ge t_0. \tag{2.252}$$

On the other hand, $V(0) = 0$ and $V(x_0)$ is $C^1$. Similarly to part (ii) of Lemma 2.2, there exists some $\psi \in$ such that $\| V(x_0) \| \le \psi(\| x_0 \|)$. Substituting this in (2.252) yields:

$$V_1(x(t)) \le a\psi(\| x_0 \|)\exp[-b(t - t_0)], \quad \forall t \ge t_0. \tag{2.253}$$

Part (i) $V_1(x)$ is LPDF, thus for some $\phi_1 \in \mathcal{K}$, one has:

$$\phi_1(\| x(t) \|) \le V_1(x(t)) \le a\psi(\| x_0 \|)\exp[-b(t - t_0)], \quad \forall t \ge t_0. \tag{2.254}$$

If $\| x_0 \| < c$ is chosen sufficiently small, then all the inequalities in (2.254) are preserved; moreover the following is obtained:

$$\| x(t) \| \leq \phi_1^{-1}\{a\psi(\| x_0 \|)\exp[-b(t-t_0)]\} \triangleq \beta(\| x_0 \|, t-t_0),$$

$$\forall t \geq t_0, \forall \| x_0 \| < c. \tag{2.255}$$

Note that $\phi_1^{-1} \in \mathcal{K}$ since $\phi_1 \in \mathcal{K}$. Moreover $\beta \in \mathcal{L}$, this implies that the zero equilibrium state of (2.199) is asymptotically stable.

Part (ii) Since $V_1(x)$ is PDF and RU, the relations (2.254) and (2.255) are satisfied globally for some $\phi_1 \in \mathcal{K}_\infty$, then the zero equilibrium state is globally asymptotically stable. ■

---

**Corollary 2.1 [m2]:**

Consider the nonlinear system (2.199) and let $V(x)$ be smooth enough such that its higher-order derivatives, that is, $\dot{V}(x), \ddot{V}(x), \ldots$ and $V^{(m)}(x)$ are well-defined and satisfy the following differential inequality:

$$V^{(m)}(x) + a_{m-1}V^{(m-1)}(x) + \cdots + a_1\dot{V}(x) + a_0V(x) \leq 0 \tag{2.256}$$

Also, the characteristic Equation (2.256) has only negative real roots.

(i) If $V(x)$ is LPDF, then the zero equilibrium state of (2.199) is asymptotically stable.

(ii) If $V(x)$ is PDF and RU then, the zero equilibrium state of (2.199) is globally asymptotically stable.

**Proof:**

The conditions of this corollary are the special case of the conditions of Theorem 2.16, defining $V_i(x) \triangleq V^{(i-1)}(x)$ for $i = 1,2,\ldots,m$. ■

**Remark 2.3:**

Theorem 2.16 or Corollary 2.1 could be implemented by the following approach: Consider a $C^1$ LF candidate $V_1(x)$, if $\dot{V}_1(x)$ is not negative definite, so using this LF we may not conclude the asymptotic stability of the zero equilibrium state of (2.199) using the Lyapunov direct method. In Corollary 2.1, the higher-order time derivatives, $V_i(x) = V_1^{(i-1)}(x), i = 2,3,\ldots,m$, are used in the stability analysis, to compensate for the nonnegative definiteness of $\dot{V}_1(x)$. However, if the nonlinear system (2.199) and/or the function $V_1(x)$ are not smooth enough, then the higher-order derivatives $V_1^{(i-1)}(x)$ are not well-defined. In this case, Theorem 2.16 proposes that a $C^1$ functions $V_1(x)$ to be found iteratively, instead of $V_1^{(i-1)}(x)$ as follows:

Compute $\dot{V}_1(x)$, if it is not smooth enough, then choose an upper bounding $C^1$ function $V_{i+1}(x) \geq \dot{V}_i(x)$ (The equality is permitted for the smooth cases). Repeat this procedure for $i$ = 1, 2, …, $m$–1, until the inequality of the last line in (2.242) is satisfied for some $m$. ■

The constraints on the characteristic equation (2.243) are the main restrictions of Theorem 2.16, and are relaxed in what follows. For example, the conclusions of the theorem would be true, if the characteristic equation (2.243) is Hurwitz [m4]. However, proof of Theorem 2.16 fails in this case, because the LTI cosystem (2.244) with complex conjugated roots is not externally positive. For more details of this case refer to Meigoli and Nikravesh [m4]. However, the Hurwitz condition is not considered a characteristic equation (2.243), because a more relaxed condition will be followed.

Let us make some manipulations on (2.242). Starting from the last row of (2.242) and substituting from previous rows of (2.242) into it, yields:

$$0 \geq \dot{V}_m(x) + \sum_{i=1}^{m-1} a_i V_{i+1}(x) + a_0 V_1(x) \geq \dot{V}_m(x) + \sum_{i=1}^{m-1} a_i \dot{V}_i(x) + a_0 V_1(x).$$

Thus, the following relation is obtained:

$$\begin{bmatrix} 1 & 0 & 0 & 0 & 0 \\ 0 & 1 & 0 & 0 & 0 \\ 0 & 0 & \ddots & 0 & 0 \\ 0 & 0 & 0 & 1 & 0 \\ a_1 & a_2 & \cdots & a_{m-1} & 1 \end{bmatrix} \begin{bmatrix} \dot{V}_1 \\ \dot{V}_2 \\ \vdots \\ \dot{V}_{m-1} \\ \dot{V}_m \end{bmatrix} \leq \begin{bmatrix} V_2 \\ V_3 \\ \vdots \\ V_m \\ -a_0 V_1(\underline{x}) \end{bmatrix}. \quad (2.257)$$

Note that (2.242) implies (2.257), but the converse is not true, therefore (2.257) is more relaxed. The following theorem considers a more generalized version of (2.257).

---

**Theorem 2.17 [m3,m5]:**

Let an m-vector $C^1$ $V(x) = [V_1(x), V_2(x), \ldots, V_m(x)]^T$ satisfy $V(0) = 0$.

(i) If the first component $V_1(x)$ of $V(x)$ is PDF and RU, that is,

$$V_1(x) \geq \phi_1(\| x \|) \quad \forall x \in R^n \quad , \quad \phi_1 \in \mathcal{K}_\infty \tag{2.258}$$

and along the trajectories of (2.199), the following differential inequality for $\phi_2 \in \mathcal{K}$ holds:

$$\begin{bmatrix} a_{11} & 0 & 0 & \cdots & 0 \\ a_{21} & a_{22} & 0 & 0 & \vdots \\ \vdots & a_{ij} & \ddots & 0 & 0 \\ a_{m-1,1} & \cdots & & a_{m-1,m-1} & 0 \\ a_{m1} & \cdots & & a_{m,m-1} & a_{mm} \end{bmatrix} \begin{bmatrix} \dot{V}_1 \\ \dot{V}_2 \\ \vdots \\ \dot{V}_{m-1} \\ \dot{V}_m \end{bmatrix} \leq \begin{bmatrix} V_2 \\ V_3 \\ \vdots \\ V_m \\ -\phi_2(\| \underline{x} \|) \end{bmatrix} \tag{2.259}$$

and a lower triangular matrix $A = [a_{ij}]_{m \times m}$ with the following properties:

$$a_{ij} \begin{cases} = 0 \quad , \quad \mathit{if}\ i < j, \\ > 0 \quad , \quad \mathit{if}\ i = j, \\ \geq 0 \quad , \quad \mathit{if}\ i > j, \end{cases} \tag{2.260}$$

exists, then the zero equilibrium state of (2.199) is globally asymptotically stable.

(ii) If the above conditions hold only locally, that is, for $\|x\| < r$ for some $r > 0$ then, the zero equilibrium state of (2.199) is asymptotically stable.

The proof of this theorem is deferred until the time-varying version of the theorem is introduced in the next chapter. Two corollaries for this theorem will be introduced. ■

---

**Corollary 2.2 [m3,m5]:**

Let the nonlinear system $\dot{x} = f(x)$ and $V(x)$ be smooth enough such that the higher-order derivatives $\dot{V}(x), \ddot{V}(x), \ldots$ and $V^{(m)}(x)$ are well-defined.

(i) If $V(x)$ is PDF and RU and,

$$\sum_{i=1}^{m} a_i V^{(i)}(\underline{x}) \leq -\phi_2(\| \underline{x} \|), \quad \phi_2 \in \mathcal{K}_\infty \tag{2.261}$$

for,

$$\begin{cases} a_i \geq 0, \quad i = 1, 2, \ldots, m-1, \\ a_m > 0, \end{cases} \tag{2.262}$$

then, the zero equilibrium state of $\dot{x} = f(x)$ is globally asymptotically stable.

(ii) If the above conditions hold only locally, that is, for $\|x\| < r$ for some $r > 0$, then, the zero equilibrium state of $\dot{x} = f(x)$ is asymptotically stable.

**Proof:**

This corollary essentially generalizes Theorem 2.13 for the use of higher-order derivatives of an LF candidate up to $m \in N$ order. The conditions of this corollary is a special case of Theorem 2.17 because defining $V_i(x) \triangleq V^{(i-1)}(x)$ for $i = 1, 2, \ldots, m$, and using (1-63) implies:

$$\begin{bmatrix} 1 & 0 & 0 & \cdots & 0 \\ 0 & 1 & 0 & 0 & \vdots \\ 0 & 0 & \ddots & 0 & 0 \\ 0 & 0 & 0 & 1 & 0 \\ a_1 & \cdots & & a_{m-1} & a_m \end{bmatrix} \begin{bmatrix} \dot{V}_1 \\ \dot{V}_2 \\ \vdots \\ \dot{V}_{m-1} \\ \dot{V}_m \end{bmatrix} \leq \begin{bmatrix} V_2 \\ V_3 \\ \vdots \\ V_m \\ -\phi_2(\| \underline{x} \|) \end{bmatrix} \tag{2.263}$$

which is the special case of (2.259). On the other hand, implementing Lemma 2.2 yields $V(0) = 0$. ■

---

**Corollary 2.3 [m3,m5]:**

All conclusions of Theorem 2.16 hold, if the characteristic equation (2.243) has only the following conditions:

$$\begin{cases} a_j \geq 0 & , j = 1, 2, \ldots, m-1, \\ a_0 > 0. \end{cases} \tag{2.264}$$

■

**Proof:**

It has been shown that (2.242) implies (2.257). Moreover substituting $V_1(x) \geq \phi_1(\| x \|)$ into (2.257) implies (2.263) for $a_m = 1$ and $\phi_2 \triangleq a_0 \phi_1$. Therefore, conditions of Theorem 2.16 with (2.264) imply corresponding conditions of Theorem 2.17. ■

**Example 2.24 [m3,m5]:**

Consider the following LTI system together with an arbitrary quadratic PDF $V(x)$:

$$\dot{x} = Ax, \quad V(x) = x^T P_0 x, \quad P_0 > 0. \tag{2.265}$$

Then, the higher-order derivatives $V^{(i)}(x)$ can be computed by the following iterating equation:

$$V^{(i)}(x) = x^T P_i x, \quad P_i = P_{i-1}A + A^T P_{i-1}, \quad i = 1, 2, \ldots \tag{2.266}$$

As a numerical example, consider the following system and a positive definite matrix $P_0$:

$$\begin{bmatrix} \dot{x}_1 \\ \dot{x}_2 \\ \dot{x}_3 \end{bmatrix} = \begin{bmatrix} 0 & 1 & 0 \\ 0 & 0 & 1 \\ -1 & -2.2 & -2.2 \end{bmatrix} \begin{bmatrix} x_1 \\ x_2 \\ x_3 \end{bmatrix}, \quad P_0 = I_{3\times 3} \tag{2.267}$$

Iterating (2.266) for $i = 1,2,\ldots,5$ yields:

$$P_1 = \begin{bmatrix} 0 & 1 & -1 \\ 1 & 0 & -1.2 \\ -1 & -1.2 & -4.4 \end{bmatrix}, P_2 = \begin{bmatrix} 2 & 3.4 & 7.6 \\ 3.4 & 7.28 & 11.32 \\ 7.6 & 11.32 & 16.96 \end{bmatrix},$$

$$P_3 = \begin{bmatrix} -15.2 & -26.04 & -30.28 \\ -26.04 & -43.008 & -47.336 \\ -30.28 & -47.336 & -51.984 \end{bmatrix}, P_4 = \begin{bmatrix} 60.56 & 98.752 & 92.56 \\ 98.752 & 156.1984 & 145.216 \\ 92.56 & 145.216 & 134.0576 \end{bmatrix},$$

$$P_5 = \begin{bmatrix} -185.12 & -288.288 & -238.9376 \\ -288.288 & -441.4464 & -365.64352 \\ -238.9376 & -365.64352 & -299.42144 \end{bmatrix}. \tag{2.268}$$

Then the following linear matrix inequality for the unknown parameters $a_i$ is considered:

$$\sum_{i=1}^{5} a_i P_i = -Q < 0, \quad a_i \geq 0 \quad i = 1,\ldots,5 \tag{2.269}$$

The following coefficients are computed by solving (2.269) numerically:

$$a_1 = 1, a_2 = 0, a_3 = 1.7, a_4 = 1.7, a_5 = 0.5 \tag{2.270}$$

The corresponding $Q$ is the following positive definite matrix with positive eigenvalues:

$$Q = \begin{bmatrix} 15.448 & 19.5336 & 14.5928 \\ 19.5336 & 28.2995 & 17.6258 \\ 14.5928 & 17.6258 & 14.5856 \end{bmatrix}, \tag{2.271}$$

$$\lambda(Q) \in \{0.2156, \quad 2.85, \quad 55.2669\}.$$

Using (2.269), the following relation can be found:

$$\begin{aligned} &a_5 V^{(5)}(x) + a_4 V^{(4)}(x) + a_3 V^{(3)}(x) + a_1 \dot{V}(x) \\ &= -x^T Q x \leq -0.2156 \| x \|^2 \triangleq -\phi_2(\| x \|). \end{aligned} \tag{2.272}$$

Thus, using Corollary 2.2 the global asymptotic stability of the zero equilibrium state of (2.267) is proved. Note that the characteristic equation of (2.267), that is, $_1 = s^3 + 2.2s^2 + 2.2s + 1 = (s+1)[(s+0.6)^2 + 0.64] = 0$ is Hurwitz. ■

Also note that the system of (2.267) is a linear system, where its stability could be verified using (2.12) with Q = 2I > 0. This implies P > 0 which in turn implies the globally asymptotic stability of the zero equilibrium state of (2.267).

**Example 2.25 [m3,m4]:**

Consider a nonlinear homogeneous system $\dot{x} = f(x)$, that is, $f(kx) = kf(x)$, $\forall k \geq 0$. Such a system has a radial symmetry, which may help us in obtaining (2.256). To demonstrate this, let $V(x)$ be homogeneous of degree $2q$, that is, $V(kx) = k^{2q}V(x)$. In this case all the higher-order time derivatives $V^{(i)}(x) \triangleq [\partial V^{(i-1)}(x)/\partial x]^T f(x)$ are also homogeneous of degree $2q$. Using the radial symmetrical property, the inequality (2.256) is equivalent to:

$$\sum_{i=0}^{m} a_i V^{(i)}(x) \leq 0, \quad \forall \|x\| = 1 \tag{2.273}$$

Since the unit sphere $\|x\| = 1$ is a compact set, the inequality (2.273) could be converted to a finite dimensional Linear Matrix Inequality (LMI), using finite element methods.

For a numerical example, consider the following continuous nonlinear dynamic system with a zero equilibrium state:

$$\begin{cases} \dot{x}_1 = x_2 \\ \dot{x}_2 = -ax_1 - x_2\left(b + x_2/\sqrt{x_1^2 + x_2^2}\right) \quad , x_1 \neq 0,\ x_2 \neq 0. \end{cases} \tag{2.274}$$

This is obviously a nonlinear homogeneous system. For simplicity, the dynamic equation (2.274) is rewritten in the polar coordinate, that is, $x_1 = r\cos\theta \triangleq rC_\theta$ and $x_2 = r\sin\theta \triangleq rS_\theta$. Differentiating with respect to time and finding $\dot{r}$ and $\dot{\theta}$ yields the following relations:

$$\begin{cases} \dot{r} = -rS_\theta[S_\theta^2 + bS_\theta + C_\theta(a-1)], \\ \dot{\theta} = -S_\theta^2 - aC_\theta^2 - (b + S_\theta)S_\theta C_\theta. \end{cases} \tag{2.275}$$

Using the LF candidate; $V(x) = x_1^2 + x_2^2 = r^2$, yields:

$$\dot{V}(x) = 2r\dot{r} = -2r^2 S_\theta[S_\theta^2 + bS_\theta + C_\theta(a-1)] \tag{2.276}$$

Choosing $a = 0.1$ and $b = 1.2$, makes it obvious that $\dot{V}(x)$ is sign indefinite. Using (2.275) and (2.276) (and implementing the MAPLE software) yield:

$$\ddot{V}(x) = \dot{r}\,\partial\dot{V}(x)/\partial r + \dot{\theta}\,\partial\dot{V}(x)/\partial\theta = r^2[11.56 + 0.6C_\theta - 13.74C_\theta^2 - 2.4C_\theta^3$$

$$+1.8C_\theta^5 + 2C_\theta^6 + S_\theta(9.6 - 1.68C_\theta - 7.2C_\theta^2 - 2.4C_\theta^4)],$$

$$\dddot{V}(x) = r^2[-48.624 - 5.76C_\theta + 58.392C_\theta^2 - 0.48C_\theta^3 - 7.2C_\theta^4 + 19.2C_\theta^5 + 14.4C_\theta^6$$

$$-12.96C_\theta^7 - 16.8C_\theta^8 + S_\theta(-45.56 - 5.256C_\theta + 27.98C_\theta^2 - 1.8C_\theta^3 + 14.04C_\theta^4$$

$$+19.8C_\theta^5 + 11.78C_\theta^6 - 12.6C_\theta^7 - 8C_\theta^8)].$$

Then for obtaining (2.273) for $m = 3$, it is sufficient to solve the inequality $\sum_{i=0}^{3} a_i V^{(i)}(x) \le 0$, $(\|x\| = r = 1)$ for the $a_i$ parameters.

Implementing the finite element method on the unit circle, that is, considering $(C_\theta, S_\theta)$ for $\theta_k = k$ and $\Delta$»1, $k = 0,1,\ldots,[2\pi/\Delta]$, we have a finite dimensional LMI in the following form:

$$\sum_{i=0}^{3} a_i V^{(i)}(\theta_k) \le 0, \quad k = 0,1,\ldots,[2\pi/\ ],$$
$$a_i > 0, \quad i = 0,1,2,3. \tag{2.277}$$

Using $\Delta = 0.01$, the coefficients of this LMI are obtained as follows: $\dddot{V}(x) + 1.237\ddot{V}(x) + 0.526\dot{V}(x) + 0.0263V(x) \le 0$.

Thus, the conditions of Corollary 2.3 are satisfied and the zero equilibrium state of (2.274) is globally asymptotically stable. ■

### 2.4.5 Stability Analysis of Nonlinear Homogeneous Systems

In this subsection, the application of higher-order derivatives of Lyapunov functions in stability analysis of nonlinear homogeneous systems is considered [m6]. First, some preliminary materials and definitions are provided.

#### 2.4.5.1 Homogeneity

An autonomous nonlinear system $\dot{x} = f(x)$ and a function $V(x): R^n \to R$ are said to be standard homogeneous of order p if $f(\alpha x) = \alpha^p f(x)$ and $V(\alpha x) = \alpha^p V(x)$, respectively, with a nonnegative variable $\alpha \ge 0$. The standard homogeneity is defined with respect to $\alpha x = (\alpha x_1, \alpha x_2, \ldots, \alpha x_n)^T$, which is clearly a linear mapping. Also, there exists a generalized definition for homogeneity as follows:

For a sequence of positive weights $r = (r_1, \ldots, r_n)$, $r_i \ge 1$ and a nonnegative variable $\alpha \ge 0$, a dilation is defined as a linear mapping $\ _\alpha^r(x) \triangleq (\alpha^{r_1} x_1, \alpha^{r_2} x_2, \ldots, \alpha^{r_n} x_n)^T$. Then, the $V(x)$ function and the $f(x)$ vector field are defined to be homogeneous of order p with respect to $\ _\alpha^r$ dilation, if $V(\ _\alpha^r x) = \alpha^p V(x)$ (for scalar cases) and $f(\ _\alpha^r x) = \alpha^p\ _\alpha^r f(x)$ (for vector cases), respectively. In this case we briefly define the $V(x)$ and $f(x)$ to be

Δ-homogeneous of order p and symbolize them with $V \in H_p$ and $f \in n_p$. For example, consider the following nonlinear system:

$$\dot{x}_1 = x_2^2 \triangleq f_1(x_1, x_2)$$

$$\dot{x}_2 = x_1^{1/2} + x_2 \triangleq f_2(x_1, x_2)$$

If r is chosen to be r = (2,1), then:

$${}_{\alpha}^{r}x = (\alpha^2 x_1, \alpha^1 x_2)$$

and, thus:

$${}_{\alpha}^{r}f(x) = (\alpha^{r_1} f_1, \alpha^{r_2} f_2) = (\alpha^2 (x_2^2), \alpha^1 (x_1^{1/2} + x_2))$$

$$= (\alpha^2 f_1, \alpha((\alpha^2 x_1)^{1/2} + x_2))$$

The special weights $r = (1,1,\ldots,1)$ are referred to, as standard weights. Thus, using the standard dilation ${}_{\alpha}^{r}x = \alpha x$, a system $f(x)$ is Δ-homogeneous of order p iff it is standard homogeneous of order $p + 1$, because $f(\alpha x) = \alpha^p \; {}_{\alpha}^{r}f(x) = \alpha^{p+1} f(x)$.

Any continuous $\rho : R^n \to R_+$function, which is PDF and $\rho \in H_1$ is called a Δ-homogeneous norm. The $\rho(x)$ is not a real norm, because it does not satisfy the triangular inequality or the linear property. A Δ-homogeneous sphere is defined as:

$$S^{n-1} = \{u \in R^n \mid \rho(u) = 1\} \tag{2.278}$$

The Δ-homogeneous ray is defined as a set $\{\; {}_{\alpha}^{r}x : \forall \alpha \geq 0\}$ for a given $x \in R^n$. It could be shown that every Δ-homogeneous ray has exactly one cross point with a given Δ-homogeneous sphere $S^{n-1}$.

In the following, let the nonlinear system (2.199) be Δ-homogeneous of order $k \geq 0$, that is, $f \in n_k$. The following lemma states the main geometric property of Δ-homogeneous systems.

---

**Lemma 2.3 [m6]:**

Let the system of (2.199) be a given Δ- homogeneous system of order k, and $(t, x_0)$ denote a trajectory of (2.199) starting from any initial state $x_0$ and parameterized by t. Then, applying ${}_{\alpha}^{r}$ on $(t, x_0)$ leads to a new trajectory of (2.199), that is,

$${}_{\alpha}^{r}\;(t, x_0) = \;(t/\alpha^k,\; {}_{\alpha}^{r}x_0), \quad \forall \alpha \in R^+, \quad \forall x_0 \in R^n \tag{2.279}$$ ■

This lemma means that the solution $(t, x_0)$ of (2.199) system and the dilation ${}_{\alpha}^{r}$ commute with each other. This property is called the radial symmetry of trajectories

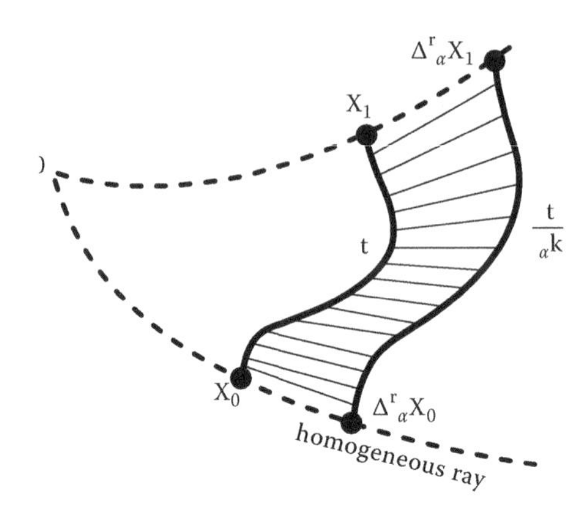


**FIGURE 2.7** Radial symmetry in trajectories of a Δ-homogeneous system.

and it is shown in Figure 2.7. As a consequence of this lemma, note the following fact:

If zero equilibrium state of a homogeneous system (2.199) is asymptotically stable, then it is globally asymptotically stable.

#### 2.4.5.2 Application of Higher-Order Derivatives of Lyapunov Functions

Using some suitable Δ-homogeneous LF in the stability analysis of a given Δ-homogeneous vector field is a usual task in the literature [m6]. In the following, we concentrate on the applications of the higher-order time derivatives of Δ-homogeneous LFs to stability analysis of Δ-homogeneous systems.

All higher-order derivatives of a Δ-homogeneous function $V(x)$ along the solutions of $\dot{x} = f(x)$ a Δ-homogeneous system (2.199), are also Δ-homogeneous functions. This important fact is summarized in the following lemma.

---

**Lemma 2.4 [m6]:**

If $V(x) \in H_p$(homogeneous of order p) and $f(x) \in n_k$ with respect to some dilation $^r_\alpha$, then the scalar multiplication of these two satisfies $V(x) \cdot f(x) \in n_{p+k}$, and the total time derivative of $V(x)$ along the solutions of $\dot{x} = f(x)$, satisfies $\dot{V}(x) \in H_{p+k}$. Therefore, by induction $[V(x)]^i .f(x) \in n_{pi+k}$ and $V^{(i)}(x) \in H_{p+ki}$ for $i = 1,2,\ldots$. ■

Note that in Example 2.20, $V(x) \in H_2$ and $f(x) \in n_0$ with respect to the standard dilation. Thus, $V^{(i)}(x) \in H_2$ for $i = 1,2,\ldots$. Therefore, using Lemma 2.4 implies $\sum_{i=1}^{m} a_i V^{(i)}(x) \in H_2$ for $m \in N$. Thus, one could easily obtain the sign of (2.277) using polar coordinates. However, the following remark shows some difficulties for homogeneous nonlinear systems of order $k > 0$.

**Remark 2.4 [m6]:**

If a nonlinear system is homogeneous of order $k > 0$ then, the higher-order derivatives of a homogeneous LF $V(x)$ are homogeneous of different orders and one cannot easily determine the sign of their linear combinations. Moreover, in this case, if $\sum_{i=1}^{m} a_i V^{(i)}(x) < 0, \quad \forall x \neq 0$ and $a_1 > 0$ then, $\dot{V}(x) \leq 0$ in some region very close to the origin. This is true because, for $k > 0$, the first derivative term dominates the other derivatives' terms in a very small neighborhood of zero (see Lemma 2.4) and the Lyapunov direct method would be useful for this case. Therefore, Theorem 2.17 could not directly be used for stability analysis of the homogeneous nonlinear systems of order $k > 0$. ■

It was shown in the previous remark that Theorem 2.17 is not useful for stability analysis of nonlinear homogeneous systems of order $k > 0$. One needs to slightly modify Theorem 2.17 in order to make it useable for stability analysis of nonlinear homogeneous systems of arbitrary order.

Let $f(x) \in n_k$ for some $k > 0$ in (2.199) and $\rho(x)$ be a given homogeneous norm with respect to a given dilation $\,^{r}_{\alpha}$. Define the following nonlinear system:

$$\dot{x} = \tilde{f}(x) = \begin{cases} f(x)/\rho^k(x) & , \quad x \neq 0, \\ 0 & , \quad x = 0. \end{cases} \tag{2.280}$$

It is clear that $\tilde{f} \in n_0$ and $\tilde{f}(x)$ are continuous at zero.

---

**Lemma 2.5 [m6]:**

The zero equilibrium state of (2.199) is asymptotically stable iff the zero equilibrium state of (2.280) is asymptotically stable. ■

**Proof:**

It is clear from the definition of homogeneity that any nonlinear homogeneous system of nonnegative order such as (2.199) and (2.280) has an equilibrium state at the origin. Also the nonlinear system (2.199) does not have any nonzero equilibrium state iff the system (2.280) does not have any either, since $\rho(x) > 0$ for $x \neq 0$. The phase portraits of both systems coincide, but with different velocities at each $x \in R^n$. Since for $\rho^k(x) > 0$ the phase portraits of the two systems are equivalent, then the asymptotical stability property of zero equilibrium state for both systems are equivalent. ■

Now consider a given class $C^1$ function $V(x)$, let us compare the time derivative $\dot{V}(x)$ along the solutions of (2.199) and (2.280) at each point $x$. This is simply done by using (2.217), (2.199), and (2.280). Let t and $\tilde{t}$ be the time variables in (2.199) and (2.280), respectively, and thus:

$$dV(x)/d\tilde{t} = [\partial V(x)/\partial x]^T \tilde{f}(x) = [\partial V(x)/\partial x]^T f(x)/\rho^k(x)$$
$$= \frac{dV(x)/dt}{\rho^k(x)}. \tag{2.281}$$

Both systems (2.199) and (2.280) are equivalent using the same state vector $x$ and the variable timescaling (depending on state), because:

$$dx/d\tilde{t} = \tilde{f}(x) = f(x)/\rho^k(x) = (dx/dt)/\rho^k(x) \quad d\tilde{t} = \rho^k(x)dt \tag{2.282}$$

The relation (2.282) shows the time scaling between the two systems. It depends on the homogeneous norm of the state vector x. Also, (2.282) gives a new interpretation of (2.281).

Note that Theorem 2.17 could be applied for stability analysis of the nonlinear system (2.199) indirectly through stability analysis of the nonlinear system (2.280) (see Lemma 2.5).

The following theorem addresses the stability analysis of (2.199), which uses (2.282) and Lemma 2.5, in its proof.

---

**Theorem 2.18 [m6]:**

Let the nonlinear system (2.199) be Δ-homogeneous of order k, that is, $f \in n_k$. If there exist $V_1(x), V_2(x), \ldots$ and $V_{m+1}(x)$ functions of class $C^1$ such that the following conditions are satisfied:

(i) $V_i(x) \in H_p$ for $i = 1,\ldots,m+1$ and for some $p \in N$.
(ii) $V_1(x)$ is PDF and $V_{m+1}(x)$ is NDF.
(iii) For the time derivatives along the solutions of (2.199):

$$\frac{1}{\rho^k(x)}\begin{bmatrix} a_{11} & 0 & 0 & \cdots & 0 \\ a_{21} & a_{22} & 0 & 0 & \vdots \\ \vdots & a_{ij} & \ddots & 0 & 0 \\ a_{m-1,1} & \cdots & & a_{m-1,m-1} & 0 \\ a_{m1} & \cdots & & a_{m,m-1} & a_{mm} \end{bmatrix}\begin{bmatrix} \dot{V}_1 \\ \dot{V}_2 \\ \vdots \\ \dot{V}_{m-1} \\ \dot{V}_m \end{bmatrix} \le \begin{bmatrix} V_2 \\ V_3 \\ \vdots \\ V_m \\ V_{m+1} \end{bmatrix} \tag{2.283}$$

$\forall x \in R^n$,

where $A = [a_{ij}]_{m\times m}$ is a matrix with the property (2.260), then the zero equilibrium state of (2.199) is asymptotically stable. ■

**Proof:**

Using (2.281) yields $(dV_i/dt)/\rho^k(x) = dV_i/d\tilde{t}$ for the time derivatives of each $V_i(x)$ along the solutions of (2.199) and (2.280). Thus, using $(dV_i/dt)/\rho^k(x) = dV_i/d\tilde{t}$ and (2.283) yields the following relation, component-wise:

$$A\,dV(x)/d\tilde{t} \le [V_2(x),\ldots,V_m(x),V_{m+1}(x)]^T. \tag{2.284}$$

Using (2.284), the conditions of Theorem 2.17 are satisfied for asymptotical stability of the zero equilibrium state of (2.280). Using Lemma 2.5 yields asymptotical stability of the zero equilibrium state of (2.199). ■

Using the Δ-homogeneous sphere $S^{n-1}$ defined in (2.278) and the radial symmetry of Δ-homogeneous functions, the conditions of Theorem 2.18 could be stated simpler as follows:

---

**Corollary 2.4 [m6]:**

Let the nonlinear system (2.199) be Δ-homogeneous of order $k \ge 0$, that is, $f \in n_k$, $\rho(x)$ be a Δ-homogeneous norm, and $S^{n-1}$ be the corresponding Δ-homogeneous sphere using (2.278). If there exist $V_1(x), V_2(x),\ldots,V_{m+1}(x)$ functions of class $C^1$, such that the following conditions are satisfied:

(i) $V_i \in H_p$ for $i = 1,\ldots,m+1$ and for some $p \in N$.
(ii) $V_1(u) > 0$ and $V_{m+1}(u) < 0$, $\forall u \in S^{n-1}$.
(iii) For the time derivatives along the solutions of (2.199):

$$\begin{bmatrix} a_{11} & 0 & 0 & \cdots & 0 \\ a_{21} & a_{22} & 0 & 0 & \vdots \\ \vdots & a_{ij} & \ddots & 0 & 0 \\ a_{m-1,1} & \cdots & & a_{m-1,m-1} & 0 \\ a_{m1} & \cdots & & a_{m,m-1} & a_{mm} \end{bmatrix} \begin{bmatrix} \dot{V}_1 \\ \dot{V}_2 \\ \vdots \\ \dot{V}_{m-1} \\ \dot{V}_m \end{bmatrix} \le \begin{bmatrix} V_2 \\ V_3 \\ \vdots \\ V_m \\ V_{m+1} \end{bmatrix}, \forall u \in S^{n-1}, \tag{2.285}$$

where $A = [a_{ij}]_{m\times m}$ is a matrix with the property given in (2.260), then the zero equilibrium state of (2.199) is asymptotically stable.

**Proof:**

Let $w_1(x) \in H_p$ and $w_2(x) \in H_p$ be two given Δ-homogeneous functions. Then the radial symmetry of these functions yields the following conclusion:

$$\left(w_1(u) \le w_2(u),\quad \forall u \in S^{n-1}\right) \qquad \left(w_1(x) \le w_2(x),\quad \forall x \in R^n\right),(2.4-88) \tag{2.286}$$

Because $w_1(0) = w_2(0) = 0$. On the other hand, $\forall x \neq 0, u = {}^{r}_{(\rho(x))^{-1}}(x) \in S^{n-1}$. Thus, implementing $w_i(x) \in H_p$ for $i$ = 1,2 yields,

$$w_1(u) \leq w_2(u) \qquad w_1\left( {}^{r}_{(\rho(x))^{-1}}(x)\right) \leq w_2\left( {}^{r}_{(\rho(x))^{-1}}(x)\right)$$

$$\frac{w_1(x)}{(\rho(x))^p} \leq \frac{w_2(x)}{(\rho(x))^p} \qquad w_1(x) \leq w_2(x).$$

Now having (2.286) yields that the relative condition in Corollary 2.4 implies the similar condition of Theorem 2.18 and vice versa. Thus, (2.285) implies (2.283).

Each row of (2.283) is of the following form.

$$\sum_{j=1}^{i} a_{ij}\left(\frac{\dot{\mathrm{V}}_j(x)}{\rho^k(x)}\right) \leq \mathrm{V}_{i+1}(x), \quad \forall x \in R^n \quad ,i = 1,2,\ldots,m, \tag{2.287}$$

Implementing $\mathrm{V}_j \in H_p$ and $\mathrm{f} \in n_k$ yields $\dot{\mathrm{V}}_j(x) \in H_{p+k}$ for $j$ = 1, …, $m$. Dividing each term in (2.287) by $\rho^k(x)$ implies that the terms of (2.287) are Δ-homogeneous function of order p. The restriction of (2.287) on the Δ-homogeneous sphere (2.278) is the following:

$$\sum_{j=1}^{i} a_{ij}\dot{\mathrm{V}}_j(u) \leq \mathrm{V}_{i+1}(u), \quad \forall \rho(u) = 1 \quad ,i = 1,2,\ldots,m. \tag{2.288}$$

A similar result as in (2.286) would be obtained if one applies the same procedure of the proof of (2.286) (in Corollary 2.4) to conclude:

(2.288) ⇒ (2.287), therefore (2.285) ⇒ (2.283).

■

#### 2.4.5.3 Polynomial Δ-Homogeneous Systems of Order k = 0

In the case of Δ-homogeneous polynomial systems of order $k = 0$, the use of higher-order derivatives of Δ-homogeneous LF candidates, leads to a necessary and sufficient condition of asymptotical stability of the zero equilibrium state of these systems. The following example demonstrates this concept.

**Example 2.26 [m3]:**

A nonlinear system is given below:

$$\begin{cases} \dot{x}_1 = -x_1, \\ \dot{x}_2 = x_1^2 - x_3, \\ \dot{x}_3 = -2x_1^2 + x_2 - x_3. \end{cases} \tag{2.289}$$

It is Δ-homogeneous of order $k = 0$ with respect to $_{\alpha}^{r}x = (\alpha x_1, \alpha^2 x_2, \alpha^2 x_3)$, because:

$$f(\,_{\alpha}^{r}x) = f\begin{bmatrix} \alpha x_1 \\ \alpha^2 x_2 \\ \alpha^2 x_3 \end{bmatrix} = \begin{bmatrix} -(\alpha x_1) \\ (\alpha x_1)^2 - (\alpha^2 x_3) \\ -2(\alpha x_1)^2 + (\alpha^2 x_2) - (\alpha^2 x_3) \end{bmatrix}$$

$$= \begin{bmatrix} \alpha(-x_1) \\ \alpha^2(x_1^2 - x_3) \\ \alpha^2(-2x_1^2 + x_2 - x_3) \end{bmatrix} = \,_{\alpha}^{r}f(x). \tag{2.290}$$

The stability of the zero equilibrium state of this system is analyzed using the higher-order derivatives of LF candidate $V(x) = (x_1^4 + x_2^2 + x_3^2) \in H_4$. Iterating (2.217) for $i = 1,2,3$ yields:

$$\begin{cases} \dot{V}(x) = -4x_1^4 + 2x_1^2x_2 - 4x_1^2x_3 - 2x_3^2, \\ \ddot{V}(x) = 26x_1^4 - 8x_1^2x_2 + 18x_1^2x_3 - 4x_2x_3 + 4x_3^2, \\ \dddot{V}(x) = -148x_1^4 + 42x_1^2x_2 - 66x_1^2x_3 - 4x_2^2 + 12x_2x_3 - 4x_3^2. \end{cases} \tag{2.291}$$

Clearly $\dot{V}(x)$ is not NDF. Using Lemma 2.4 yields $V^{(i)}(x) \in H_4$ for $i = 1,2,3,\ldots$; for example,

$$\begin{aligned} \dot{V}(\,_{\alpha}^{r}x) &= -4(\alpha x_1)^4 + 2(\alpha x_1)^2(\alpha^2 x_2) - 4(\alpha x_1)^2(\alpha^2 x_3) \\ &- 2(\alpha^2 x_3)^2 = \alpha^4(-4x_1^4 + 2x_1^2x_2 - 4x_1^2x_3 - 2x_3^2) = \alpha^4\dot{V}(x). \end{aligned} \tag{2.292}$$

All the polynomial Δ-homogeneous functions of order $p = 4$ are linear combinations of the following independent terms:

$$\{x_1^4\,,\, x_1^2x_2\,,\, x_1^2x_3\,,\, x_2^2\,,\, x_2x_3\,,\, x_3^2\} \tag{2.293}$$

Thus, the $V(x), \dot{V}(x), \ldots$ and $V^{(6)}(x)$ must be linearly dependent functions. It can be shown that:

$$[V(x)\;\; \dot{V}(x)\;\; \ddot{V}(x) \ldots V^{(6)}(x)] = [x_1^4\;\; x_1^2x_2\;\; x_1^2x_3\;\; x_2^2\;\; x_2x_3\;\; x_3^2]M\,,$$

where:

$$M = \left[\begin{array}{cccccc|c} 1 & -4 & 26 & -148 & 766 & -3614 & 15784 \\ 0 & 2 & -8 & 42 & -182 & 596 & -1542 \\ 0 & -4 & 18 & -66 & 184 & -366 & 402 \\ 1 & 0 & 0 & -4 & 12 & -12 & -20 \\ 0 & 0 & -4 & 12 & -12 & -20 & 84 \\ 1 & -2 & 4 & -4 & -4 & 20 & -20 \end{array}\right]. \tag{2.294}$$

The left $6 \times 6$ submatrix of M as shown in (2.294) is nonsingular, hence: $V(x), \dot{V}(x), \ldots$ and $V^{(5)}(x)$ functions are linearly independent. Substituting (2.294) into $\sum_{i=0}^{6} a_i V^{(i)}(x) = [V(x)\ \dot{V}(x)\ \ldots V^{(6)}(x)]\mathbf{a} = 0$, yields $Ma = 0$. Solving $Ma = 0$ for the $a_i$ parameters implies:

$$\sum_{i=0}^{6} a_i V^{(i)}(x) = V^{(6)}(x) + 12V^{(5)}(x) + 60V^{(4)}(x) + 167\dddot{V}(x) + 282\ddot{V}(x) + 276\dot{V}(x) + 112V(x) \equiv 0. \tag{2.295}$$

The relation (2.295) can be written in the following form:

$$\sum_{i=1}^{6} a_i V^{(i)}(x) = -a_0 V(x) < 0, \quad \forall x \neq 0 \quad , a_i > 0. \tag{2.296}$$

Then, using Corollary 2.2 for (2.296) implies the asymptotical stability of the zero equilibrium state of (2.289). ■

The procedure used in Example 2.21 could be implemented for all polynomial Δ-homogeneous systems of order $k = 0$. In fact, we have the following theorem:

---

**Theorem 2.19 [m3,m6]:**

Consider a nonlinear polynomial system $\dot{x} = f(x)$, and a PDF polynomial $V(x)$. If $f(x) \in n_0$ and $V(x) \in H_p$ with respect to some dilation $\Delta_\alpha^r$, then:

(i) $V^{(i)}(x) \in H_p$, for $i = 1, 2, \ldots$ .
There exist m∈N such that $V(x), \dot{V}(x), \ldots$ and $V^{(m-1)}(x)$ are linearly independent, but $V(x), \dot{V}(x), \ldots$ and $V^{(m)}(x)$ are linearly dependent and we have:

$$a_m V^{(m)}(x) + a_{m-1} V^{(m-1)}(x) + \cdots + a_1 \dot{V}(x) + a_0 V(x) \equiv 0 \tag{2.297}$$

(ii) The following conditions are equivalent:
(a) The zero equilibrium state of $\dot{x} = f(x)$ is asymptotically stable.
(b) The characteristic polynomial of (2.297) is Hurwitz.
(c) All the coefficients of the characteristic polynomial of (2.297) are positive.

**Proof:**

Part (i) It is a consequence of Lemma 2.4.
Part (ii) Note that the maximal number of linearly independent Δ-homogeneous polynomial functions $V_i(x) \in H_p$, is equal to the number of different terms $x_1^{p_1} x_2^{p_2} \cdots x_n^{p_n}$ such that:

$$\sum_{i=1}^{n} r_i p_i = p, \quad p_i \geq 0, i = 1, 2, \ldots, n. \tag{2.298}$$

Thus, there exists $m \in N$ with the properties given in part (ii).

Part (iii) It is clear that (b) implies (c). Also (2.297) could be written as: $\sum_{i=1}^{m} a_i V^{(i)}(x) = -a_0 V(x) < 0, \quad \forall x \neq 0$ and for $a_i > 0$.

Thus, using Corollary 2.2, yields (c)⇒(a).
The conclusion of (a)⇒(b) will be shown as follows.

Defining $\mathrm{V}(x) \triangleq [V(x) \quad \dot{V}(x) \cdots V^{(m-1)}(x)]^T$, we have $\mathrm{V}(0) = 0$ as a consequence of properties of zero equilibrium state. Let the zero equilibrium state of $\dot{x} = f(x)$ be asymptotically stable, thus $x(t) \to 0$ as:

$$t \to +\infty, \forall x_0 \in R^n, \quad \mathrm{V}(x(0)) = \mathrm{V}(x_0) \quad \lim_{t \to 0} \mathrm{V}(x(t)) = \mathrm{V}(0) = 0. \quad (2.299)$$

Now consider the following LTI cosystem:

$$a_m y^{(m)}(t) + a_{m-1} y^{(m-1)}(t) + \cdots + a_1 \dot{y}(t) + a_0 y(t) = 0. \quad (2.300)$$

with the state vector $Y(t) \triangleq [y(t) \quad \dot{y}(t) \cdots y^{(m-1)}(t)]^T$. Using (2.297) implies that the trajectories of $\mathrm{V}(x(t)) \quad R^m$ are subsets of the trajectories of (2.300). Thus, (2.299) yields:

$$\forall x_0 \in R^n \quad , \quad Y(0) = \mathrm{V}(x_0) \qquad \lim_{t \to +\infty} Y(t) = 0. \quad (2.301)$$

Since the $V(x), \dot{V}(x), \ldots$ and $V^{(m-1)}(x)$ functions are linearly independent, there exist m points $x^1, x^2, \ldots$ and $x^m \in R^n$ such that $V(x^1), V(x^2), \ldots$ and $V(x^m)$ are linearly independent. On the other hand, let $V(x^1), V(x^2), \ldots$ and $V(x^k)$ for some $k < m$ be the maximum number of such vectors, thus: $V(x) \in \mathrm{span}\{V(x^1), V(x^2), \ldots, V(x^k)\}, \forall x$. Choosing a nonzero vector $\alpha \in R^m$ normal to this subspace yields $\alpha^T V(x) = 0, \forall x \in R^n$. This contradicts the linear independence of $V(x), \dot{V}(x), \ldots, V^{(m-1)}(x)$. Now, substitute the above m linearly independent vectors into (2.301), therefore:

$$\forall i = 1, 2, \ldots, m, \quad Y(0) = Y^i \triangleq V(x^i) \qquad \lim_{t \to +\infty} Y(t) = 0 \quad (2.302)$$

Hence, there exist m independent vectors $Y^1, Y^2, \ldots, Y^m \in R^m$ as initial states for the LTI cosystem (2.300), converging to zero. This result yields the asymptotical stability of the zero equilibrium state of (2.300), which is equivalent to the Hurwitz condition for the characteristic equation of (2.300). ■

#### 2.4.5.4 The Δ-Homogeneous Polar Coordinate

Although Theorem 2.18 is applicable to homogeneous systems of any arbitrary order, some designing tools are required to find the useful $V_i(x)$ functions for a given nonlinear system. In Example 2.25, the polar coordinate is used. The usual polar coordinate is useful only for standard nonlinear homogeneous systems, but it is useless in the case of general homogeneity. Here, a new polar coordinate with respect to the given weights $r = (r_1, r_2)$ for $n = 2$ is introduced.

A Δ-homogeneous p-norm is defined by $\| \cdot \|_{,p} \triangleq \left( \sum_{i=1}^{n} | x_i |^{p/r_i} \right)^{1/p}$ for some $p \geq (2 \max r_i)$. It is clear that $\| \cdot \|_{,p} \in H_1$. Using (2.278) the corresponding Δ-homogeneous sphere is defined as $S^{n-1} \triangleq \{ x \in R^n : \| x \|_{,p} = 1 \}$. So each pair of $x = [x_1, x_2]^T$ is mapped to a pair of (ρ,θ) as Δ-polar coordinate. Considering a given Δ-homogeneous norm $\| \cdot \|_{,p}$, let us define:

$$
\begin{cases} x_1 = \rho^{r_1} C_\theta^{2r_1/p}, \\ x_2 = \rho^{r_2} S_\theta^{2r_2/p}. \end{cases} \tag{2.303}
$$

Defining $u_\theta \triangleq [C_\theta^{2r_1/p}, S_\theta^{2r_2/p}]^T$ yields:

$$
\| u_\theta \|_{,p} = \sqrt[p]{(C_\theta^{2r_1/p})^{p/r_1} + (S_\theta^{2r_2/p})^{p/r_2}} = 1
$$

and $x = {}^{r}_{\rho} u_\theta$, thus, $\| x \|_{,p} = \rho$ and $\rho(x) \in H_1$. Moreover, each $V(x) \in H_p$ and $f(x) \in n_k$ could be decomposed as:

$$
\begin{aligned} V(x) &= \rho^p V(u_\theta), \\ f(x) &= \rho^k \; {}^{r}_{\rho} f(u_\theta), \end{aligned} \tag{2.304}
$$

The decomposition of ρ and θ are very important and will be used in this subsection. Differentiating (2.303) with respect to time and solving for $\dot{\rho}$ and $\dot{\theta}$ yields:

$$
\begin{bmatrix} \dot{\rho} \\ \dot{\theta} \end{bmatrix} = \begin{bmatrix} \rho & 0 \\ 0 & 1 \end{bmatrix} \begin{bmatrix} \frac{1}{r_1} C_\theta & \frac{1}{r_2} S_\theta \\ -\frac{p}{2r_1} S_\theta & \frac{p}{2r_2} C_\theta \end{bmatrix} \begin{bmatrix} C_\theta^{(1-2r_1/p)} & 0 \\ 0 & S_\theta^{(1-2r_2/p)} \end{bmatrix} \times \; {}^{r}_{\rho^{-1}} \begin{bmatrix} \dot{x}_1 \\ \dot{x}_2 \end{bmatrix}. \tag{2.305}
$$

Using (2.199) and (2.304), one can write:

$$
{}^{r}_{\rho^{-1}} \dot{x} = {}^{r}_{\rho^{-1}} f(x) = \rho^k \; {}^{r}_{\rho\rho^{-1}} f(u_\theta) = \rho^k f(u_\theta).
$$

Substituting this relation into (2.305) yields:

$$
\begin{bmatrix} \dot{\rho} \\ \dot{\theta} \end{bmatrix} = \begin{bmatrix} \rho^{k+1} & 0 \\ 0 & \rho^k \end{bmatrix} \begin{bmatrix} \frac{1}{r_1} C_\theta & \frac{1}{r_2} S_\theta \\ -\frac{p}{2r_1} S_\theta & \frac{p}{2r_2} C_\theta \end{bmatrix} \begin{bmatrix} C_\theta^{(1-2r_1/p)} & 0 \\ 0 & S_\theta^{(1-2r_2/p)} \end{bmatrix} f(u_\theta). \tag{2.306}
$$

The last equation is the Δ-polar differential equation of the nonlinear system with $f(x) \in n_k$. The extension of Δ-polar coordinates to $n > 2$ is straightforward.

An important question arises here: How to use the Δ-polar coordinates to implement Theorem 2.18? The answer would be:

Let $f(x) \in n_k$. In our procedure of implementing Theorem 2.18, the equation (2.283) is constructed one row after another. Assume one is at the ith iteration, that is, $V_j(x)$ for $j = 1, 2, \ldots, i$ are previously defined and one wants to find $V_{i+1}(x)$ and $a_{ij}$ for $j = 1, \ldots, i$ and construct the $i$th row of (2.283), that is, (2.287) or equivalently:

$$\sum_{j=1}^{i} a_{ij} \left( \frac{\dot{V}_j(x)}{\rho^{k+p}} \right) \leq \frac{V_{i+1}(x)}{\rho^p}. \tag{2.307}$$

According to the assumption $V_j(x) \in H_p$, $\dot{V}_j(x) \in H_{k+p}$ and $\rho(x) \in H_1$, therefore each term in (2.307) is a Δ-homogeneous function of order 0, independent of ρ, that is, (2.307) could be written as follows:

$$\sum_{j=1}^{i} a_{ij} \dot{V}_j(u_\theta) \leq V_{i+1}(u_\theta), \quad \forall x = {}_\rho^r u_\theta, \quad 0 \leq \theta \leq 2\pi \tag{2.308}$$

Using numerical methods for plotting the known functions $\dot{V}_j(u_\theta)$ versus θ, one can find a linear combination of these functions together with the upper bound $V_{i+1}(u_\theta) \in H_0$ such that (2.308) is satisfied.

#### 2.4.5.5 Numerical Examples

**Example 2.27 [m6]:**

The nonlinear system:

$$\begin{bmatrix} \dot{x}_1 \\ \dot{x}_2 \end{bmatrix} = \begin{bmatrix} b_{11} & b_{12} \\ b_{21} & b_{22} \end{bmatrix} \begin{bmatrix} x_1^3 \\ x_2^3 \end{bmatrix} \tag{2.309}$$

is standard homogeneous of order 3, that is, $f \in n_3$. Similar to Example 2.25 one may change (2.309) to the Δ-polar differential equations of the form (2.306). Substituting $u_\theta = [C_\theta, S_\theta]^T$, $k = 2$, $r_1 = r_2 = 1$, and $p = 2$ into (2.306) yields:

$$\begin{bmatrix} \dot{\rho} \\ \dot{\theta} \end{bmatrix} = \begin{bmatrix} \rho^3 & 0 \\ 0 & \rho^2 \end{bmatrix} \begin{bmatrix} C_\theta & S_\theta \\ -S_\theta & C_\theta \end{bmatrix} \begin{bmatrix} b_{11} & b_{12} \\ b_{21} & b_{22} \end{bmatrix} \begin{bmatrix} C_\theta^3 \\ S_\theta^3 \end{bmatrix} \tag{2.310}$$

Let $||x||^k = \rho^2$ be applied in Theorem 2.18 for this example. For stability analysis of (2.310), starting with $V_1(x) = x_1^2 + x_2^2 = \rho^2$, the following special form of (2.283) is used:

$$\frac{1}{\rho^2} \begin{bmatrix} 1 & & & & \\ 0 & 1 & & & \\ \vdots & 0 & \ddots & & \\ 0 & \cdots & 0 & 1 & \\ a_1 & \cdots & & a_{m-1} & a_m \end{bmatrix} \begin{bmatrix} \dot{V}_1 \\ \dot{V}_2 \\ \vdots \\ \dot{V}_{m-1} \\ \dot{V}_m \end{bmatrix} = \begin{bmatrix} V_2 \\ V_3 \\ \vdots \\ V_m \\ V_{m+1} \end{bmatrix}. \tag{2.311}$$

Hence, for $i = 1,2,\ldots,m-1$:

$$V_{i+1}(x)=\frac{\dot{V}_i(x)}{\rho^2}=\begin{bmatrix} \frac{\partial V_i(x)}{\partial \rho} & \frac{\partial V_i(x)}{\partial \theta} \end{bmatrix}\begin{bmatrix} \dot{\rho} \\ \dot{\theta} \end{bmatrix}\frac{1}{\rho^2} \tag{2.312}$$

Substituting (2.310) into (2.312) yields:

$$V_{i+1}(x)=\frac{\dot{V}_i(x)}{\rho^2}=\begin{bmatrix} \rho\frac{\partial V_i(x)}{\partial \rho} & \frac{\partial V_i(x)}{\partial \theta} \end{bmatrix}\begin{bmatrix} C_\theta & S_\theta \\ -S_\theta & C_\theta \end{bmatrix}$$

$$\times\begin{bmatrix} b_{11} & b_{12} \\ b_{21} & b_{22} \end{bmatrix}\begin{bmatrix} C_\theta^3 \\ S_\theta^3 \end{bmatrix}, \quad i=1,2,\ldots,m-1. \tag{2.313}$$

Cahoosing $B=\begin{bmatrix} -0.2 & -1 \\ 1 & -1 \end{bmatrix}$, all the higher-order derivatives are of class $C_1$, well defined and belong to $H_2$, for example,

$$V_2(x)=\dot{V}_1(x)/\rho^2=2\rho^2\begin{bmatrix} C_\theta & S_\theta \end{bmatrix}\begin{bmatrix} -0.2 & -1 \\ 1 & -1 \end{bmatrix}\begin{bmatrix} C_\theta^3 \\ S_\theta^3 \end{bmatrix},$$

$$V_3(x)=\dot{V}_2(x)/\rho^2=\rho^2\big(19.2C_\theta^7S_\theta+12C_\theta^3S_\theta-28.8C_\theta^5S_\theta-2C_\theta S_\theta$$
$$+6+30.4C_\theta^4-24C_\theta^2-20.48C_\theta^6+10.24C_\theta^8\big),$$

$$\dot{V}_3(x)=\rho^4\big(38C_\theta S_\theta-69.12C_\theta^9S_\theta+175.68C_\theta^5S_\theta$$
$$+15.36C_\theta^{11}S_\theta-135.2C_\theta^3S_\theta-53.12C_\theta^7S_\theta$$
$$-10+296C_\theta^4-16C_\theta^2+304.128C_\theta^{12}$$
$$+1215.616C_\theta^8-861.44C_\theta^6-928.768C_\theta^{10}\big). \tag{3.314}$$

In spite of $V_1(x)$ being PDF, $\dot{V}_1(x)$ is not negative definite, and thus the Lyapunov direct method fails to prove asymptotical stability of the zero equilibrium state of (2.309). Substitute $\rho = 1$ into (2.314) for computing $\dot{V}_j(u_\theta)$, $j=1,2,3,\ldots$, if one considers the following points on the $\Delta$-homogeneous sphere:

$$u_{\theta_s}=[C_{\theta_s}^{2r_1/p}, S_{\theta_s}^{2r_2/p}]^T \in S^{n-1};$$
$$\theta_s = sT, \quad s=0,1,\ldots,[2\pi/T], \quad T\ll 1, \tag{2.315}$$

and solves the following LMI for their $a_j$ coefficients:

$$\begin{cases} \sum_{j=1}^{3} a_j \dot{V}_j(u_{\theta_s}) < 0, & \theta_s = sT, \quad s = 0,1,\ldots,[2\pi / T], \\ a_j \geq 0, & j = 1,2,3, \end{cases} \tag{2.316}$$

then $\dot{V}_1(u_\theta) + 0.1\dot{V}_3(u_\theta) < 0$ would be obtained for all $0 \leq \theta \leq 2\pi$ and thus $\frac{1}{\rho^2}[\dot{V}_1(x) + 0.1\dot{V}_3(x)] \triangleq V_4(x) \in H_2$ is NDF. Then for $m$ = 3, the relation (2.311) (and thus [2.283]) is satisfied, and then the zero equilibrium state of (2.309) is asymptotically stable. ■

## Example 2.28 [m6]:

The nonlinear system:

$$\begin{bmatrix} \dot{x}_1 \\ \dot{x}_2 \end{bmatrix} = \begin{bmatrix} 1 & 0 \\ 0 & x_1^2 \end{bmatrix} \begin{bmatrix} b_{11} & b_{12} \\ b_{21} & b_{22} \end{bmatrix} \begin{bmatrix} x_1^3 \\ x_2 \end{bmatrix} = \begin{bmatrix} b_{11}x_1^3 + b_{12}x_2 \\ b_{21}x_1^5 + b_{22}x_1^2 x_2 \end{bmatrix} \tag{2.317}$$

is $\Delta$-homogeneous of order 2 with the weights $r = (r_1, r_2) = (1,3)$, that is, $f \in n_2$, since:

$$f({}_{\alpha}^{r}x) = \begin{bmatrix} b_{11}(\alpha x_1)^3 + b_{12}(\alpha^3 x_2) \\ b_{21}(\alpha x_1)^5 + b_{22}(\alpha x_1)^2(\alpha^3 x_2) \end{bmatrix}$$

$$= \alpha^2 \begin{bmatrix} \alpha & 0 \\ 0 & \alpha^3 \end{bmatrix} \begin{bmatrix} b_{11}x_1^3 + b_{12}x_2 \\ b_{21}x_1^5 + b_{22}x_1^2 x_2 \end{bmatrix} = \alpha^2 \; {}_{\alpha}^{r}f(x).$$

We use the $\Delta$-polar coordinate $(x_1, x_2) \triangleq (\rho\sqrt[3]{\cos\theta}, \rho^3 \sin\theta)$ and the $\Delta$-homogeneous norm $\|x\|_{,6} \triangleq \sqrt[6]{x_1^6 + x_2^2} = \rho$ for this system. Substituting: $(r_1, r_2) = (1,3)$, $p$ = 6, $u_\theta = [\sqrt[3]{C_\theta}\ , S_\theta]^T$ and $k$ = 2 into (2.306), we obtain the $\Delta$-polar differential equation as follows:

$$\begin{bmatrix} \dot{\rho} \\ \dot{\theta} \end{bmatrix} = C_\theta^{-2/3} \begin{bmatrix} \rho^3 & 0 \\ 0 & \rho^2 \end{bmatrix} \begin{bmatrix} C_\theta & \frac{1}{3}S_\theta \\ -3S_\theta & C_\theta \end{bmatrix} \begin{bmatrix} b_{11} & b_{12} \\ b_{21} & b_{22} \end{bmatrix} \begin{bmatrix} C_\theta \\ S_\theta \end{bmatrix} \tag{2.318}$$

The PDF $V_1(x) \triangleq \|x\|_{,6}^6 = x_1^6 + x_2^2 = \rho^6$ and Equation (2.311) are used for stability analysis of the zero equilibrium state of (2.317) using Theorem 2.18. Substituting (2.318) into (2.312) yields:

$$V_{i+1}(x) = \frac{\dot{V}_i(x)}{\rho^2} = C_\theta^{2/3} \begin{bmatrix} \rho \dfrac{\partial V_i(x)}{\partial \rho} & \dfrac{\partial V_i(x)}{\partial \theta} \end{bmatrix} \begin{bmatrix} C_\theta & \frac{1}{3}S_\theta \\ -3S_\theta & C_\theta \end{bmatrix}$$
$$\times \begin{bmatrix} b_{11} & b_{12} \\ b_{21} & b_{22} \end{bmatrix} \begin{bmatrix} C_\theta \\ S_\theta \end{bmatrix}, \quad \forall\, i = 1,2,\ldots,m-1. \tag{2.319}$$

All the higher-order derivatives are of class $C^0$, well defined and belong to $H_6$. Using the numerical value for $B = \begin{bmatrix} 0 & -1 \\ 2 & -1 \end{bmatrix}$ and iterating (2.319) one has:

$$V_2(x) = \dot{V}_1(x)/\rho^2 = -2\rho^6 C_\theta^{(2/3)} S_\theta (C_\theta + S_\theta),$$

$$V_3(x) = \dot{V}_2(x)/\rho^2 = \tfrac{2}{3}\rho^6 C_\theta^{(1/3)} \left(6S_\theta + 4C_\theta^4 S_\theta - 19C_\theta^2 S_\theta - 25C_\theta^3 + 19C_\theta\right),$$

$\dot{V}_1(x)$ is not negative definite, therefore the Lyapunov direct method fails to prove asymptotic stability of the zero equilibrium state of (2.317).

$$\dot{V}_3(x) = -\tfrac{2}{9}\rho^8 \left(258C_\theta S_\theta - 4C_\theta^5 S_\theta - 705C_\theta^3 S_\theta + 40C_\theta^7 S_\theta\right.$$
$$\left. - 403C_\theta^2 + 40C_\theta^8 - 396C_\theta^6 + 795C_\theta^4 + 18\right), \quad (2.320)$$

Letting $m = 3$ and following the same procedure of the previous example, one has $\dot{V}_1(u_\theta) + \frac{1}{60}\dot{V}_3(u_\theta) < 0$ for all $0 \le \theta \le 2\pi$, then $\frac{1}{\rho^2}[\dot{V}_1(x) + \frac{1}{60}\dot{V}_3(x)] \triangleq V_4(x) \in H_6$ is NDF, and thus the zero equilibrium state of (2.317) is asymptotically stable.[18] ■

## 2.5 NEW STABILITY THEOREMS

In this section a method for stability analysis of nonlinear systems is proposed. Although, this method uses Lyapunov stability definitions, it does not use the Lyapunov function properties for its analysis, and therefore, it may represent sufficient as well as necessary conditions (in some view) for stability analysis of nonlinear systems. Also, this method is suitable for recognizing the existence of an asymptotically stable limit cycle in nonlinear systems [f2].

### 2.5.1 Fathabadi–Nikravesh's Method [f1,f2:]

#### 2.5.1.1 Low-Order Systems

Consider the following second-order nonlinear autonomous system:

$$\dot{x}_1 = f_1(x_1, x_2),$$
$$\dot{x}_2 = f_2(x_1, x_2), \quad (2.321)$$

---

**Definition 2.5:**

The closed curves with the algebraic relation of the form $u(x_1, x_2) = c_1$ are called *equipotential curves*, because they are the response of the following Pfaffian differential equation:

$$\sum_{i=1}^{2} F_i . dx_i = 0. \quad (2.322)$$

where

$$F_1 \triangleq \frac{\partial u(x_1,x_2)}{\partial x_1} \quad \text{and} \quad F_2 \triangleq \frac{\partial u(x_1,x_2)}{\partial x_2}.$$

■

If $u(x_1,x_2)=c_1$ are equipotential curves, we have:

$$\frac{\partial u(x_1,x_2)}{\partial x_1}dx_1 + \frac{\partial u(x_1,x_2)}{\partial x_2}dx_2 = 0, \tag{2.323}$$

so the following state equation for the dynamic movement on the equipotential curves is obtained:

$$\begin{aligned} \dot{x}_1 &= \frac{\partial u(x_1,x_2)}{\partial x_2}, \\ \dot{x}_2 &= -\frac{\partial u(x_1,x_2)}{\partial x_1}, \end{aligned} \tag{2.324}$$

From the above state equation one concludes that:

$$\mathrm{V}_u = \frac{\partial u(x_1,x_2)}{\partial x_2}u_{x1} + \left(-\frac{\partial u(x_1,x_2)}{\partial x_1}\right)u_{x2}, \tag{2.325}$$

where $V_u$ is the velocity vector of the movement on the curves. Also $u_{x_1}$ and $u_{x_2}$are the unit vectors of $x_1$ and $x_2$ axes, respectively. The gradient vector $\nabla u$, which is perpendicular to the equipotential curves is given by:

$$u = \frac{\partial u(x_1,x_2)}{\partial x_1}u_{x_1} + \frac{\partial u(x_1,x_2)}{\partial x_2}u_{x2}. \tag{2.326}$$

Also, the velocity vector of the system (2.321) would be:

$$\dot{x} = f_1(x_1,x_2)\,u_{x_1} + f_2(x_1,x_2)\,u_{x_2}. \tag{2.327}$$

Now, let us present the new stability theorem:

---

**Theorem 2.20:**

Assume that there are clockwise closed curves $u(x_1,x_2)=c_i$ that encloses $x_e=0$, then the equilibrium state of system (2.321) is asymptotically stable if and only if (iff) the following inequality holds [f1]:

$$u \times \dot{x} < 0, \tag{2.328}$$

otherwise, the equilibrium state would be unstable. If the "<" sign in (2.328) is replaced with "≤" then, only the stability will be the result. ■

**Proof:**

If (2.321) is locally asymptotically stable, then all solutions starting in some $\Omega$ neighborhood of the equilibrium state will eventually approach $x_e = 0$. The largest such $\Omega$ will be the attraction region. In (2.328), $\dot{x}$ which is given by (2.327) is the velocity vector of dynamic system (2.321). Since $x_e = 0$ is asymptotically stable, then the system's trajectories do not diverge from the equilibrium state, therefore the equipotential curves should be closed curves, otherwise in the open parts of these curves, the trajectories could escape from $x_e$. It is necessary that the vector product of the gradient vector of equipotential curves and the system's velocity vector be negative in order for the trajectories to enter the closed equipotential curves. This proves the necessary part of the theorem.

The closeness of $u(x_1, x_2) = c_i$ curves is sufficient for the trajectories to converge $x_e = 0$, since the curves shrink to the equilibrium state. This completes the rigors proof of this theorem. ■

Note that if $u(x_1, x_2) = c_i$ are radially unbounded, the stability properties would be global.

It is obvious that if (2.328) replaced by $u \times \dot{x} \leq 0$ (but not always zero) then, only the stability will be the result. Also, note that since $u(x_1, x_2) = c_i$ would be any closed curves around the $x_e = 0$, one can choose (2.324) arbitrarily, for example,

$$\begin{cases} \dot{x}_1 = g_1(x_1, x_2), \\ \dot{x}_2 = g_2(x_1, x_2), \end{cases} \tag{2.329}$$

could be considered as an auxiliary system, as long as the $u(x_1, x_2) = c_i$ obtained from (2.330) are closed curves. The closedness happens if for all $x_1, x_2 \in R$, the constant $c_i$ in $u(x_1, x_2) = c_i$ are either all positive or all negative. If the equal sign holds for both positive as well as negative values of $c_i$, then the corresponding curve is not closed.

The above method will guide us to the following algorithm:

---

**Algorithm 2.1:**

Choose again the 2.329 auxilliary system as follows:

$$\begin{cases} \dot{x}_1 = g_1(x_1, x_2), \\ \dot{x}_2 = g_2(x_1, x_2), \end{cases}$$

such that the equilibrium state ($x_e = 0$) of the system (2.321) is enclosed by the response of the following Pfaffian differential equation:

$$g_1(x_1, x_2)dx_1 + g_2(x_1, x_2)dx_2 = 0. \tag{2.330}$$

And:

$$g_1(x_1,x_2)f_1(x_1,x_2)+g_2(x_1,x_2)f_2(x_1,x_2)<0, \tag{2.331}$$

then the equilibrium state of the system (2.321) is asymptotically stable [f1]. ■

**Example 2.29:**

Consider the following system:

$$\begin{aligned}\dot{x}_1 &= x_1(x_1^2+x_2^2-2)-4x_1x_2^2,\\ \dot{x}_2 &= x_2(x_1^2+x_2^2-2)+4x_1^2x_2.\end{aligned} \tag{2.332}$$

Let us choose $g_1(x_1,x_2)$ and $g_2(x_1,x_2)$ in such a way that one could eliminate the odd-order terms in system equations. Thus, let:

$$\begin{aligned}g_1(x_1,x_2) &= x_1,\\ g_2(x_1,x_2) &= x_2,\end{aligned} \tag{2.333}$$

then:

$$\begin{aligned}&g_1(x_1,x_2)dx_1+g_2(x_1,x_2)dx_2=0 \quad x_1dx_1+x_2dx_2=0\\ &\frac{1}{2}(x_1^2+x_2^2)=c=u(x_1,x_2),\end{aligned} \tag{2.334}$$

which is a closed contour and encloses $x_e=0$. Note that:

$$g_1(x_1,x_2)f_1(x_1,x_2)+g_2(x_1,x_2)f_2(x_1,x_2)=(x_1^2+x_2^2)(x_1^2+x_2^2-2). \tag{2.335}$$

Thus for $x_1^2+x_2^2<2$, the above equation would be negative definite, which implies asymptotic stability for $x_e=0$ of (2.332) inside the circle with radius 2 (largest attraction region), and unstable outside this circle. (Why?). ■

To verify the method, one may use the Lyapunov first stability theorem, which implies the stability of the linear part of this system, thus the $x_e=0$ of a nonlinear system would be stable in a small regain around the $x_e=0$.

Also, note that one may choose $V(x_1,x_2)=x_1^2+x_2^2$ as an LF candidate for the above nonlinear system and come up with the same stability condition as in Example 2.26.

**Example 2.30:**

Consider the following nonlinear system:

$$\begin{aligned}\dot{x}_1 &= x_2,\\ \dot{x}_2 &= -\sin(x_1)-0.5x_2.\end{aligned} \tag{2.336}$$

One may choose $g_i(x_1, x_2)$ for i = 1,2 in such a way that one could omit the $\sin(x_1)$ function in the $\sum_{i=1}^{2} f_i(x_1, x_2) g_i(x_1, x_2)$. Therefore let:

$$
\begin{aligned}
g_1(x_1, x_2) &= \sin(x_1), \\
g_2(x_1, x_2) &= x_2.
\end{aligned}
\tag{2.337}
$$

Thus:

$$
g_1(x_1, x_2) f_1(x_1, x_2) + g_2(x_1, x_2) f_2(x_1, x_2) = -0.5x_2^2 \leq 0 . \tag{2.338}
$$

Also:

$$
\begin{aligned}
& g_1(x_1, x_2)dx_1 + g_2(x_1, x_2)dx_2 = 0 \quad \sin(x_1)dx_1 + x_2 dx_2 = 0, \\
& -\cos(x_1) + 0.5x_2^2 = c = c_1 - 1, \\
& 1 - \cos(x_1) + 0.5x_2^2 = c_1,
\end{aligned}
\tag{2.339}
$$

which is a closed curve around $x_e = 0$. The relations (2.337) and (2.339) imply only the stability of (2.336).

For a double check, one may use the first Lyapunov method for the linear part of system (2.336). The characteristic equation of the linear system is $\lambda^2 + 0.5\lambda + 1 = 0$, which implies only the Lyapunov stability of the nonlinear system (2.336).

Application of the direct method of Lyapunov in this case is another approach to conclude the stability of this system. For the second-order system of the form $\ddot{x} + \frac{1}{2}\dot{x} + \sin x = 0$, which could be the Kirchhoff's Voltage Law (KVL) of an resistance, inductance, and capacitance (RLC) network, the following function:

$$
V(x_1, x_2) = \frac{1}{2} x_2^2 + \int_0^{x_1} \sin(y) dy, \tag{2.340}
$$

would be a suitable LF candidate to prove the stability of (2.336). In this case:

$$
\dot{V}(x_1, x_2) = -0.5x_2^2 \leq 0, \tag{2.341}
$$

which is a negative semidefinite function. Another $V(x_1, x_2)$ is:

$$
V(x_1, x_2) = 1 - \cos(x_1) + \frac{1}{2} x_2^2. \tag{2.342}
$$

Also, in this case:

$$
\dot{V}(x_1, x_2) = -\frac{1}{2} x_2^2 \leq 0, \tag{2.343}
$$

which is again a negative semidefinite function implying Lyapunov stability of (2.336). ■

**Example 2.31:**

Investigate the stability of the following system:

$$\dot{x}_1 = e^{x_1}\sin(x_2) - 4x_1^3,$$
$$\dot{x}_2 = e^{x_2}\sin(x_1) + x_2^7. \tag{2.344}$$

Let:

$$g_1(x_1, x_2) = x_2^7,$$
$$g_2(x_1, x_2) = 4x_1^3. \tag{2.345}$$

Then:

$$g_1(x_1, x_2)f_1(x_1, x_2) + g_2(x_1, x_2)f_2(x_1, x_2)$$
$$= x_2^7 e^{x_1}\sin(x_2) + 4x_1^3 e^{x_2}\sin(x_1), \tag{2.346}$$

which is positive definite in a small neighborhood of $x_e = 0$. Therefore, if the equipotential curve, $u(x_1, x_2) = c$ is closed, then (2.345) would be unstable. But:

$$g_1(x_1, x_2)dx_1 + g_2(x_1, x_2)dx_2 = x_2^7 dx_1 + 4x_1^3 dx_2 = 0,$$
$$x_2^7 x_1 + 4x_1^3 x_2 = x_1 x_2 (x_2^6 + 4x_1^2) = c. \tag{2.347}$$

Is this a closed curve? (Why?) ■

### 2.5.1.2 Linear Systems

Let us consider the application of the new approach to the linear system stability analysis.

Consider the following second-order LTI system:

$$\begin{pmatrix} \dot{x}_1 \\ \dot{x}_2 \end{pmatrix} = \begin{pmatrix} a_{11} & a_{12} \\ a_{21} & a_{22} \end{pmatrix} \begin{pmatrix} x_1 \\ x_2 \end{pmatrix}. \tag{2.348}$$

Let:

$$g_1(x_1, x_2) = 2p_{11}x_1 + 2p_{12}x_2,$$
$$g_2(x_1, x_2) = 2p_{12}x_1 + 2p_{22}x_2, \tag{2.349}$$

thus:

$$g_1(x_1, x_2)f_1(x_1, x_2) + g_2(x_1, x_2)f_2(x_1, x_2)$$
$$= 2(p_{11}a_{11} + p_{12}a_{21})x_1^2 + 2(a_{11}p_{12} + a_{21}p_{22} + p_{11}a_{12} + p_{12}a_{22})x_1x_2$$
$$+ 2(p_{12}a_{12} + p_{22}a_{22})x_2^2. \tag{2.350}$$

This equation could be written as follows:

$$(x_1,x_2)\left[\begin{pmatrix} a_{11} & a_{21} \\ a_{12} & a_{22} \end{pmatrix}\begin{pmatrix} p_{11} & p_{12} \\ p_{12} & p_{22} \end{pmatrix}+\begin{pmatrix} p_{11} & p_{12} \\ p_{12} & p_{22} \end{pmatrix}\begin{pmatrix} a_{11} & a_{12} \\ a_{21} & a_{22} \end{pmatrix}\right]\begin{pmatrix} x_1 \\ x_2 \end{pmatrix}. \quad (2.351)$$

It is obvious that (2.350) is negative definite if for a positive definite Q matrix, one has:

$$\begin{pmatrix} a_{11} & a_{21} \\ a_{12} & a_{22} \end{pmatrix}\begin{pmatrix} p_{11} & p_{12} \\ p_{12} & p_{22} \end{pmatrix}+\begin{pmatrix} p_{11} & p_{12} \\ p_{12} & p_{22} \end{pmatrix}\begin{pmatrix} a_{11} & a_{12} \\ a_{21} & a_{22} \end{pmatrix}=-\begin{pmatrix} q_{11} & q_{12} \\ q_{12} & q_{22} \end{pmatrix}=-Q, \quad (2.352)$$

This is exactly the same relation obtained for a linear system, using the Lyapunov method (see [2.12]). This will be completed if one can show the closed condition of the equipotential curve; therefore,

$$\begin{aligned} &g_1(x_1,x_2)dx_1 + g_2(x_1,x_2)dx_2 \\ &= 0 \quad (2p_{11}x_1 + 2p_{12}x_2)dx_1 + (2p_{12}x_1 + 2p_{22}x_2)dx_2 = 0, \end{aligned} \quad (2.353)$$

which is a complete integrable function. Thus:

$$u(x_1,x_2) = p_{11}x_1^2 + 2p_{12}x_1x_2 + p_{22}x_2^2 = c_i. \quad (2.354)$$

For $u(x_1,x_2)$ to be a closed curve one, should have:

$$p_{11} > 0 \text{ and } p_{11}p_{22} - p_{12}^2 > 0, \quad (2.355)$$

which is the condition of positive definiteness of $P$ matrix. This completes the proof of the case.

#### 2.5.1.3 Higher-Order Systems

Consider the following $n$th-order system:

$$\begin{aligned} \dot{x}_1 &= f_1(x_1,\ldots,x_n) \\ \dot{x}_2 &= f_2(x_1,\ldots,x_n) \\ &\vdots \\ \dot{x}_n &= f_n(x_1,\ldots,x_n) \end{aligned} \quad (2.356)$$

---

**Definition 2.6:**

The nth-dimensional closed surfaces with the algebraic relation of $u(x_1, x_2, \ldots, x_n) = c_i$ are called equipotential surfaces because they are the response of the following $n$th-order Pfaffian differential equation:

$$\sum_{i=1}^{n} F_i.dx_i = 0, \text{ where } F_i \triangleq \frac{\partial u(x_1, x_2)}{\partial x_i}.$$ ■

---

**Theorem 2.21:**

Assume there are clockwise closed curves $u(x_1, \ldots, x_n) = C$ for all $C > 0$, that enclose $x_e = 0$, then the equilibrium state of the system (2.356) is asymptotically stable iff the following inequality holds [f1]:

$$u.\dot{x} < 0 \qquad (2.357)$$ ■

**Proof:**

The proof of this theorem is quite similar to that of Theorem 2.20. ■

Similar to the low-order case, one may come up with the following algorithm:

---

**Algorithm 2.2:**

Choose the following auxiliary system:

$$\begin{aligned} \dot{x}_1 &= g_1(x_1, \ldots, x_n), \\ \dot{x}_2 &= g_2(x_1, \ldots, x_n), \\ &\vdots \\ \dot{x}_n &= g_n(x_1, \ldots, x_n), \end{aligned} \qquad (2.358)$$

such that the equilibrium state ($x_e = 0$) is enclosed by the response of the following Pfaffian differential equation:

$$g_1(x_1, \ldots, x_n)dx_1 + \cdots + g_n(x_1, \ldots, x_2)dx_n = 0. \qquad (2.359)$$

Then, the equilibrium state of the system (2.356) is asymptotically stable iff:

$$g_1(x_1,\dots,x_n)f_1(x_1,\dots,x_n)+\dots+g_n(x_1,\dots,x_2)f_n(x_1,\dots,x_n)<0. \qquad (2.360)$$ ■

**Example 2.32:**

Consider the following system:

$$\begin{cases} \dot{x}_1=\sigma(x_2-x_1), \\ \dot{x}_2=rx_1-x_2-x_1x_3, \\ \dot{x}_n=x_1x_3-bx_3. \qquad \sigma,r,b>0. \end{cases} \qquad (2.361)$$

Let:

$$\begin{aligned} g_1(x_1,x_2,x_3)&=rx_1, \\ g_2(x_1,x_2,x_3)&=\sigma x_2, \\ g_3(x_1,x_2,x_3)&=\sigma x_3. \end{aligned} \qquad (2.361b)$$

Thus:

$$\begin{aligned} &\int g_1(x_1,x_2,x_3)dx_1+\int g_2(x_1,x_2,x_3)dx_2+\int g_3(x_1,x_2,x_3)dx_3 \\ &=\frac{1}{2}(rx_1^2+\sigma x_2^2+\sigma x_3^2)=c, \end{aligned} \qquad (2.362)$$

which is a closed curve containing (0,0,0) as its zero equilibrium (ZE) state. Now:

$$\begin{aligned} &f_1(x_1,x_2,x_3)g_1(x_1,x_2,x_3)+f_2(x_1,x_2,x_3)g_2(x_1,x_2,x_3) \\ &+f_3(x_1,x_2,x_3)g_3(x_1,x_2,x_3)=2r\sigma x_1x_2-\sigma x_2^2-b\sigma x_3^2-r\sigma x_1^2<0 \end{aligned} \qquad (2.363)$$

If:

$$x^t\begin{pmatrix} -r\sigma & \sigma r & 0 \\ \sigma r & -\sigma & 0 \\ 0 & 0 & -br \end{pmatrix}x<0, \qquad (2.364)$$

which would be negative definite if r < 1.

For a double check, one may linearize the system (2.361), then:

$$A=\begin{pmatrix} -\sigma & \sigma & 0 \\ r & -1 & 0 \\ 0 & 0 & -b \end{pmatrix}, \qquad (2.365)$$

which implies the stable system.

The third approach is as follows:

$$V(x) = k_1x_1^2 + k_2x_2^2 + k_3x_3^3, \tag{2.366}$$

then, if $k_1 = 2$ and $k_2 = \sigma = k_3$:

$$\dot{V}(x) = 2\sigma r x_1 x_2 - \sigma x_2^2 - b\sigma x_3^2 - r\sigma x_1^2 < 0, \tag{2.367}$$

which implies the asymptotic stability of the zero equilibrium state of the system (2.362). ■

## Example 2.33:

Consider the following nonlinear system [k1]:

$$\begin{cases} \dot{x}_1 = x_2, \\ \dot{x}_2 = -a\sin x_1 - k\,x_1 - dx_2 - cx_3, \\ \dot{x}_3 = x_2 - x_3. \end{cases}$$

Let us define the following cosystem:

$$g_1(x) = (k+a)x_1,$$

$$g_2(x) = x_2,$$

$$g_3(x) = cx_3.$$

Then, using Theorem 2.21 (or Algorithm 2.2) in the small neighborhood of $x_1 = 0$ yields:

$$f_1g_1 + f_2g_2 + f_3g_3 = a\,x_1x_2 - ax_2\sin x_1 - dx_2^2 - cx_3^2 \approx -dx_2^2 - cx_3^2 \leq 0.$$

Also note that:

$$g_1dx_1 + g_2dx_2 + g_3dx_3 = 0.$$

Thus:

$$\frac{k+a}{2}x_1^2 + \frac{1}{2}x_2^2 + \frac{c}{2}x_3^2 = c_1,$$

which is a closed curve, therefore the zero equilibrium state of this system is asymptotically stable.

For verification, one may either linearize the system and use the Routh–Hurwitz method, or define $V(x) \triangleq 2a\int_0^{x_1}\sin x_1\,dx_1 + kx_1^2 + x_2^2 + cx_3^2 > 0$. ■

## 2.6 LYAPUNOV STABILITY ANALYSIS OF A TRANSFORMED NONLINEAR SYSTEM

In this section, two different theorems are presented to analyze the stability of a transformed nonlinear system using the Lyapunov direct method.

---

**Theorem 2.22: (Sangrody–Nikravesh Theorem 1)**

Let $\boldsymbol{x} = \boldsymbol{0}$ be a stable equilibrium state of the dynamical system $\dot{x} = f(x)$ and $V(x)$ be a Lyapunov function for this system, then the origin is a stable equilibrium state for the dynamical system $\dot{x} = H^{-1}f(x)$ if there exists a positive function $g(x)$ that satisfies the following conditions.

$$\int g(x)\left( \quad \left(V^2(x)\right)\right)^T H\,dx > 0 \tag{2.368}$$

Where $H$ is $n \times n$ positive definite matrix $\left(H = \begin{bmatrix} H_{c1} & H_{c2} & \cdots & H_{cn} \end{bmatrix}\right)$. ■

**Proof:**

Let $V(x)$ be an energy-like function for $\dot{x} = f(x)$ in an interval $(x_{li} \le x_i \le x_{ui})$ and suppose that there is an energy-like function $W(x)$ for $\dot{x} = H^{-1}f(x)$, therefore the following conditions are satisfied:

$$\left.\begin{array}{l} V(x) > 0 \\ \dfrac{d}{dt}V(x) < 0 \end{array}\right\} \quad V(x)\left( \quad V(x)\right)^T f(x)$$

$$= V(x)\left( \quad V(x)\right)^T HH^{-1}f(x) < 0. \tag{2.369}$$

If there exists a Lyapunov function $W(x)$ such that for the system $\dot{x} = H^{-1}f(x)$, one has the following inequalities:

$$\left.\begin{array}{l} W(x) > 0 \\ \dfrac{d}{dt}W(x) < 0 \end{array}\right\} \quad W(x)\left( \quad W(x)\right)^T H^{-1}f(x) < 0. \tag{2.370}$$

It can be concluded from (2.369) and (2.370) that there must be a positive function $g(x)$ such that:

$$\frac{W(x)\left( \quad W(x)\right)^T H^{-1}f(x)}{V(x)\left( \quad V(x)\right)^T HH^{-1}f(x)} = g(x).$$

Note that the scalar function $V(x)(\quad V(x))^T HH^{-1} f(x) \neq 0$. One solution for this equation is as follows:

$$\left(\quad W^2(x)\right)^T = g(x)\left(\quad V^2(x)\right)^T H \tag{2.371}$$

where:

$$W^2(x) = \left(\begin{matrix} \dfrac{\partial W^2(x)}{\partial x_1} & \dfrac{\partial W^2(x)}{\partial x_2} & \cdots & \dfrac{\partial W^2(x)}{\partial x_n} \end{matrix}\right)^T$$

$W^2(x)$, which is a positive function, can be obtained using (2.371).

$$W^2(x) = \int g(x)\left(\quad V^2(x)\right)^T H\, dx > 0 \tag{2.372}$$

This completes the proof. ■

---

**Theorem 2.23: (Sangrody–Nikravesh Theorem 2)**

In the Theorem 2.22, $g(x)$ exists, if the following condition is satisfied:

$$\left(\begin{matrix} \left(\dfrac{\partial \quad V^2(x)}{\partial x_1} + \dfrac{\partial \ln(g(x))}{\partial x_1} \quad V^2(x)\right)^T \\ \left(\dfrac{\partial \quad V^2(x)}{\partial x_2} + \dfrac{\partial \ln(g(x))}{\partial x_2} \quad V^2(x)\right)^T \\ \vdots \\ \left(\dfrac{\partial \quad V^2(x)}{\partial x_n} + \dfrac{\partial \ln(g(x))}{\partial x_n} \quad V^2(x)\right)^T \end{matrix}\right) \times H : be\,Symmetric \tag{2.373}$$

■

**Proof:**

$W^2(x)$ exists if the following condition is satisfied:

$$\frac{\partial\left(\dfrac{\partial W^2(x)}{\partial x_j}\right)}{\partial x_i} = \frac{\partial\left(\dfrac{\partial W^2(x)}{\partial x_i}\right)}{\partial x_j} \tag{2.374}$$

replacing (2.371) into (2.374), it can be concluded that:

$$\frac{\partial\left(g(x)\left(\nabla V^2(x)\right)^T H_{cj}\right)}{\partial x_i}=\frac{\partial\left(g(x)\left(\nabla V^2(x)\right)^T H_{ci}\right)}{\partial x_j} \tag{2.375}$$

expanding (2.375), one has the following:

$$\frac{\partial g(x)}{\partial x_i}\left(\nabla V^2(x)\right)^T H_{cj}+g(x)\frac{\partial\left(\left(\nabla V^2(x)\right)^T H_{cj}\right)}{\partial x_i}$$
$$=\frac{\partial g(x)}{\partial x_j}\left(\nabla V^2(x)\right)^T H_{ci}+g(x)\frac{\partial\left(\left(\nabla V^2(x)\right)^T H_{ci}\right)}{\partial x_j},$$

$$\frac{1}{g(x)}\frac{\partial g(x)}{\partial x_i}\left(\nabla V^2(x)\right)^T H_{cj}+\frac{\partial\left(\left(\nabla V^2(x)\right)^T H_{cj}\right)}{\partial x_i}$$
$$=\frac{1}{g(x)}\frac{\partial g(x)}{\partial x_j}\left(\nabla V^2(x)\right)^T H_{ci}+\frac{\partial\left(\left(\nabla V^2(x)\right)^T H_{ci}\right)}{\partial x_j},$$

therefore:

$$\left(\frac{\partial \nabla V^2(x)}{\partial x_j}+\frac{\partial \ln(g(x))}{\partial x_j}\nabla V^2(x)\right)^T H_{ci}$$
$$=\left(\frac{\partial \nabla V^2(x)}{\partial x_i}+\frac{\partial \ln(g(x))}{\partial x_i}\nabla V^2(x)\right)^T H_{cj} \tag{2.376}$$

In this equation $H_{ci}$ and $H_{cj}$ are the $i$th and the $j$th columns of the $H$ matrix. Relation (2.373) is obtained using (2.376) for all $i$ and $j$. This condition assures the existence of $W(x)$. It must be noted that this theorem can be used for dynamical systems for which $V(x)$ is known. ■

### Example 2.34:

Let the following system be given. The origin is a stable equilibrium state for this system and $V^2(x)=x_2\sin(x_2)+2x_1^2+2x_1x_2+4x_2^2$ is a Lyapunov function in the intervals $-\frac{\pi}{2}\le x_1\le\frac{\pi}{2}$ and $-\frac{\pi}{2}\le x_2\le\frac{\pi}{2}$. It can be shown that a dynamical system given by (2.378), which is the transformed system of the original one, that is, (2.377), has stable zero equilibrium state in the interval given above.

$$f(x)=\begin{pmatrix}-2x_1+6x_2+\frac{3}{2}\sin(x_2)+x_2\cos(x_2)\\ -\sin(x_2)-4x_1-2x_2\end{pmatrix} \tag{2.377}$$

The transformed system is as follows:

$$\begin{pmatrix} \dot{x}_1 \\ \dot{x}_2 \end{pmatrix} = H^{-1}f(x) = \begin{pmatrix} \frac{7}{4}x_2 + \frac{1}{2}\sin(x_2) + \frac{x_2}{4}\cos(x_2) \\ -\frac{1}{2}\sin(x_2) - 2x_1 - x_2 \end{pmatrix} \tag{2.378}$$

According to Theorem 2.22, and considering the two dynamical systems, $H^{-1}$ is obtained as follows:

$$H^{-1} = \begin{pmatrix} \frac{1}{4} & -\frac{1}{8} \\ 0 & \frac{1}{2} \end{pmatrix}$$

If $g(x)$ is considered as a positive constant, then the condition (2.373) is satisfied as is shown below:

$$\begin{pmatrix} 16 & 8 \\ 8 & -2x_2\sin(x_2) + 4\cos(x_2) + 18 \end{pmatrix} \text{ is symmetric}$$

Since this matrix is symmetric, therefore $W(x)$ exists and can be obtained according to Theorem 2.22 as follows.

$$W^2(x) = \int \left( V^2(x) \right)^T H\, dx = 8x_1^2 + 8x_1x_2 + 9x_2^2 + 2x_2\sin(x_2)$$

This function is positive definite, so the second condition is satisfied and the origin is stable equilibrium state for the second dynamical system. This function's time derivation is given below:

$$\dot{W}(x) = -4(2x_1 + x_2)^2 - (7 + \cos(x_2))x_2\sin(x_2) - \sin^2(x_2)$$

which is a negative definite function. ■

---

**Corollary 2.5:**

Suppose that $x_e = 0$ is a stable equilibrium state for a dynamical system $\dot{x} = f(x)$ and $V^2(x) = x^T Px$ is a Lyapunov function for it. Then $\dot{x} = H^{-1}f(x)$ has stable zero equilibrium state if $PH$ is positive definite and symmetric matrix. In this situation, the Lyapunov function of the new dynamical system will be as follows:

$$W^2(x) = \int x^T PH\, dx > 0 \tag{2.379}$$ ■

**Proof:**

$V^2(x)$ is a positive definite quadratic function, therefore $P$ must be positive definite and symmetric and also $\left( V^2(x)\right)^T = 2x^T P$. If one let $g(x)$ be 1, then from (2.373) one gets the following:

$$\begin{pmatrix} \left(\frac{\partial \ V^2(x)}{\partial x_1}\right)^T \\ \left(\frac{\partial \ V^2(x)}{\partial x_2}\right)^T \\ \vdots \\ \left(\frac{\partial \ V^2(x)}{\partial x_n}\right)^T \end{pmatrix} \times H = 2PH \quad \textit{is symmetric}$$

This equation shows that the $PH$ matrix must be symmetric. From the Theorem 2.22, it can be concluded that the $PH$ matrix must be positive definite, thus:

$$\int \left( V^2(x)\right)^T H\,dx = \int 2x^T PH\,dx = 2x^T PHx > 0$$

Therefore, the $PH$ must be positive definite and symmetric. ■

**Example 2.35:**

Let $x_e = 0$ be stable equilibrium state of the following dynamical system, and $V^2(x) = x_1^2 + x_2^2$ be its Lyapunov function.

$$\begin{aligned} \dot{x}_1 &= x_2 - x_1\left(x_1^2 + x_2^2\right) \\ \dot{x}_2 &= -x_1 - x_2\left(x_1^2 + x_2^2\right) \end{aligned} \tag{2.380}$$

It can be shown that dynamical system (2.381) has the stable zero equilibrium state and its Lyapunov function is given by: $W^2(x) = 5x_1^2 + 2x_1x_2 + x_2^2$.

$$\begin{aligned} \dot{x}_1 &= x_1 + 2x_2 + (2x_2 - x_1)\left(x_1^2 + x_2^2\right) \\ \dot{x}_2 &= -5x_1 - 2x_2 + (2x_1 - 5x_2)\left(x_1^2 + x_2^2\right) \end{aligned} \tag{2.381}$$

Considering the above two dynamical systems, the $PH$ matrix is positive definite and symmetric which is given below:

$$PH = \begin{pmatrix} 5 & 2 \\ 2 & 1 \end{pmatrix}$$

Therefore, the new system has stable zero equilibrium state and its Lyapunov function would be as follows:

$$5x_1^2 + 2x_1x_2 + x_2^2 .$$

Consider the following dynamical equation:

$$\dot{x} = H^{-1}\left( Ax + \begin{bmatrix} x^T N_1 x \\ x^T N_2 x \\ \vdots \\ x^T N_n x \end{bmatrix} \right) \tag{2.382}$$

where $H$, $A$ and $N_i$ are constant matrices. Using the Theorem 2.22 and Corollary 2.5, one can express the following corollary in order to achieve a useful Lyapunov function for (2.382). ■

---

**Corollary 2.6:**

Let $x_e = 0$ be a zero equilibrium state for the dynamical system (2.382). This zero equilibrium state is stable if for every symmetric and positive definite matrix $Q$, the following conditions have a positive definite and symmetric matrix $P$ as their solutions.

$$A^T P + PA = -Q : \textit{Positive definite matrix} \tag{2.383}$$

$$P \times H \quad \text{be Symmetric and positive definite matrix} \tag{2.384}$$ ■

**Proof:**

The equation (2.383) shows the existence of a Lyapunov function for the linear part of the model (i.e., $Ax$). The relation (2.384) is the condition of Corollary 2.5, and the relation (2.385) is needed to neutralize the quadratic parts of dynamical equation ($x^T N_i x$) in derivation of the Lyapunov function. ■

**PROBLEMS:**

**2.1:** Using Theorem 2.2, see if the ZES of the following system is stable:

$$\dot{x}_1 = -x_1 - x_2,$$

$$\dot{x}_2 = x_1 - x_2^3.$$ ■

**2.2:** Using Theorem 2.3, see if the ZES of the following system is stable:

$$\dot{x}_1 = -x_1 - 2x_2,$$

$$\dot{x}_2 = x_1 - 4x_2.$$ ■

**2.3:** Using the linearization method, see if the equilibrium states of the following systems are stable:

a.

$$\dot{x}_1 = -x_1 + x_1^2 x_2,$$

$$\dot{x}_2 = x_1 - x_2.$$

b.

$$\dot{x}_1 = x_2 + x_1(x_1^2 + x_2^4),$$

$$\dot{x}_2 = -x_1 + x_2(x_1^2 + x_2^4).$$

c.

$$\dot{x}_1 = x_1 - x_1 x_2,$$

$$\dot{x}_2 = -x_2 + x_1 x_2.$$ ■

**2.4:** Verify the stability of the ZES of the following systems using Krasovskii's method:

a.

$$\dot{x}_1 = -6x_1 + 2x_2,$$

$$\dot{x}_2 = 2x_1 - 6x_2 - 2x_2^3.$$

b.

$$\dot{x}_1 = \alpha \sin x_2 x_3,$$

$$\dot{x}_2 = \beta \sin x_1 x_3,$$

$$\dot{x}_3 = c \sin x_1 x_2.$$ ■

**2.5:** Using Theorem 2.4, see if the ZES of the following systems are stable (see Figure 2.3):

a.

$$G(s) = \frac{s+1}{(s+2)(s+3)(s+5)} \quad \text{and } f(e) = e^3.$$

b.

$$G(s)=\frac{(s+3)^2}{(s+1)(s+4)(s+5)} \text{ and } f(e)=1-exp(-e).$$ ■

**2.6:** Using Ingwerson's method, see if the ZES of the following system is stable:

$$\dot{x}_1 = x_2,$$

$$\dot{x}_2 = x_3,$$

$$\dot{x}_3 = x_1,$$

$$\dot{x}_4 = \sin x_1 + \sin x_2 + x_4.$$ ■

**2.7:** Using Szego's method, see if the ZES of the following systems are stable:

a.

$$\dot{x}_1 = x_2,$$

$$\dot{x}_2 = x_2(-x_2^4 - x_1^2) - x_1 x_1^6.$$

b.

$$\dot{x}_1 = x_2,$$

$$\dot{x}_2 = -x_1 - x_1^2 x_2 - x_2.$$ ■

**2.8:** Determine the stability of the ZES of the following systems using the variable gradient method:

a.

$$\dot{x}_1 = -x_1 + x_2^2,$$

$$\dot{x}_2 = -x_2.$$

b.

$$\dot{x}_1 = -2x_1,$$

$$\dot{x}_2 = -2x_2 + 2x_1 x_2^2.$$

c.

$$\dot{x}_1 = -x_1 - 2x_1^2 x_2,$$

$$\dot{x}_2 = -x_2.$$ ■

**2.9:** Verify if the ZES of the following systems are stable using the energy metric method of Wall and Moe:

a.

$$\dot{x}_1 = -x_1 - x_2,$$

$$\dot{x}_2 = x_1 - x_2^3.$$

b.

$$\dot{x}_1 = x_2,$$

$$\dot{x}_2 = a x_1^2 x_2.$$

■

**2.10:** Using Leighton's method (Section 2.3.11), see if the ZES of the following system is stable:

$$\ddot{x} + \dot{x}^5 + x^7 + \dot{x} x^2 = 0$$

■

**2.11:** Using Zubov's method, see if the ZES of the following system is stable:

$$\dot{x}_1 = -x_1 - x_2 + x_1 x_2,$$

$$\dot{x}_2 = -x + x_2 + x_1 x_2.$$

■

**2.12:** Using the LaSalle Invariance Principle, see if the following system is stable:

$$\dot{x}_1 = -x_1 + \frac{1}{3} x_1^3 + x_2,$$

$$\dot{x}_2 = -x_1.$$

■

**2.13:** Using Theorem 2.8, see if the ZES of the following system is stable:

$$\dot{x}_1 = x_2 - x_1^7 (x_1^4 + 2x_2^2 - 10),$$

$$\dot{x}_2 = -x_1^3 - 3x_2^5 (x_1^4 + 2x_2^2 - 10).$$

Hint: note that the set defined by $x_1^4 + 2x_2^2 = 10$ is invariant. ■

**2.14:** Use LaSalle's invariant set (Theorem 2.8) to find the region of attraction in the following nonlinear systems:

a.

$$\dot{x}_1 = x_1 (x_1^2 + x_2^2 - 2) - 4 x_1 x_2^2,$$

$$\dot{x}_2 = 4 x_1^2 x_2 + x_2 (x_1^2 + x_2^2 - 2).$$

b.

$$\dot{x}_1 = x_2 - x_1^7 (x_1^4 + 2x_2^2 - 10),$$

$$\dot{x}_2 = -x_1^3 - 3x_2^5 (x_1^4 + 2x_2^2 - 10).$$

■

**2.15:** Apply Theorem 2.9 to show the stability of the following systems:

a. $\dot{x} = -x$.

b. $\ddot{x} + 2\dot{x} + x = 0$.

Use $V(x) = x^2$ for both cases. ■

**2.16:** Use Theorem 2.11 to verify the stability of the ZES of the following systems:
a. $\dot{x}+x=0$ .
b. $\ddot{x}+2\dot{x}+2x=0$ . ■

**2.17:** Show that:

a. $g(u)=\begin{pmatrix} 1 & 0 & -1 \\ 0 & 2 & 1 \\ 0 & 0 & 3 \end{pmatrix}u,$ and $g(u)=\begin{pmatrix} 1 & & 0 \\ 0 & 2 & 0 \\ 0 & 0 & 3 \end{pmatrix}u$ ,

are not of class *W*. ■

**2.18:** Evaluate the stability of the following system using the VLF concept:

$$\dot{x}_1=-x_1+x_2,$$

$$\dot{x}_2=-x_2,$$

$$\dot{x}_3=-x_1^2+x_3.$$ ■

**2.19:** Using higher-order derivatives of the LF candidate (Theorem 2.13), see if the ZES of the following system is stable:

$$\dot{x}_1=x_2,$$

$$\dot{x}_2=-x_1-2x_2,$$

Use $V(x)=\frac{1}{2}(x_1^2+x_2^2)$ as the LF candidate. ■

**2.20:** Using Theorem 2.15, verify the stability of the ZES of the following nonlinear system:

$$\dot{x}=-x^3-x^5-x^7 \cdot$$ ■

**2.21:** Consider the following second-order linear system:

$$\begin{bmatrix} \dot{x}_1 \\ \dot{x}_2 \end{bmatrix}=\begin{bmatrix} -2 & -2 \\ 1 & -2 \end{bmatrix}\begin{bmatrix} x_1 \\ x_2 \end{bmatrix}.$$

Use Theorem 2.16 with $V(x)=x_1^2+x_2^2$, to determine the stability of the ZES of this system. ■

**2.22:** For the following nonlinear system, check the Δ-homogeneity of zero degree with respect to weight $r = (1,2)$,

$$\dot{x}_1=ax_1,$$

$$\dot{x}_2=bx_1^2+cx_2, \quad b\neq 0.$$

Find the parameters' ranges for the ZES stability. ■

**2.23:** For the zero equilibrium state of the following nonlinear systems, using conventional and Lyapunov's higher-order derivatives methods, specify the attraction region. Consider $V(x) = x_1^2 + x_2^2$ as the LF candidate.

a.

$$\dot{x}_1 = -x_1 + x_2^3,$$

$$\dot{x}_2 = x_1^3 - x_2.$$

b.

$$\dot{x}_1 = -x_1 + x_1^3 + x_1 x_2^2,$$

$$\dot{x}_2 = -x_2 + x_2^3 + x_2 x_1^2.$$

■

**2.24:** Consider the following nonlinear systems:

a.

$$\dot{x}_1 = x_1(x_1^2 - x_2^2 - 1) + x_1 x_2^2,$$

$$\dot{x}_2 = x_2(x_1^2 - x_2^2 - 1) - x_1^2 x_2.$$

b.

$$\dot{x}_1 = x_1(x_1^2 - 2ax_1 + a^2 + x_2^2 - 2bx^2 + b^2 - 1) + x_1 x_2^2,$$

$$\dot{x}_2 = x_2(x_1^2 - 2ax_1 + a^2 + x_2^2 - 2bx^2 + b^2 - 1) - x_1^2 x_2.$$

Find the necessary and sufficient conditions for the stability of the ZES using Theorem 2.20 (or Algorithm 2.1). ■

# ENDNOTES

1. All materials in this subsection are directly obtained from Nikravesh and Hoft [n1], but since this article may not be available to everyone, the original papers are also quoted.
2. Note that a quadratic function, whether its corresponding matrix is symmetric or not, is always equal to a quadratic function with a symmetric matrix (since the quadratic function with a skew symmetric matrix is always zero [s1]). Therefore, for the positive definiteness test of a function, the corresponding matrix should be written with a symmetric matrix. For example: $x_1^2 + 6x_1x_2 + 4x_2^2$, which is an indefinite function, for a test, must be written as $(x_1 \;\; x_2)\begin{pmatrix} 1 & 3 \\ 3 & 4 \end{pmatrix}\begin{pmatrix} x_1 \\ x_2 \end{pmatrix}$, with a symmetric matrix, otherwise if it is written as $(x_1 \;\; x_2)\begin{pmatrix} 1 & 0 \\ 6 & 4 \end{pmatrix}\begin{pmatrix} x_1 \\ x_2 \end{pmatrix}$, with a nonsymmetric $P$ matrix, it looks to be a positive definite function, which is incorrect, although all its minors determinants, are positive. The symmetric property of the matrix

is also necessary for having real positive eigenvalues for the positive definiteness of a matrix. A counter example would be $\begin{pmatrix} 0 & 1 \\ -2 & 3 \end{pmatrix}$.

3. All materials in this section are directly obtained from Nikravesh and Hoft [n1], but since this article may not be available to everyone, the original papers are also quoted.
4. Since the eigenvalues of $J$ both are negative then, the zero equilibrium state of this linear system is obviously asymptotically stable. If zero was the sole isolated equilibrium state, the stability would be global.
5. The global asymptotic stability for a linear system with an equilibrium state might be meaningful, but for the nonlinear system with more than an equilibrium state it is meaningless.
6. Equation (2.27) could be written as: $(x_1\ x_2)\begin{pmatrix} -4 & 2x_1 \\ 2x_2 & -4 \end{pmatrix}\begin{pmatrix} x_1 \\ x_2 \end{pmatrix}$, therefore $\dot{V}(x)$ is negative definite if $16-4x_1x_2<0$. Note that this region is not a closed region, so (2.26) which was suitable for the linearized system (2.24) is not a suitable Lyapunov function for stability analysis of a nonlinear system (2.22).
7. The Aizerman conjecture turns out not to be always true. A better conjecture would be a Kalman conjecture in which in addition to having a limit for the slope of nonlinear function, $f(e)$, the slope of it, that is, $f'(e) \triangleq \frac{df(e)}{de}$ should also be limited according to the following equations $k_1 \le \frac{f(e)}{e} \le k_2,\quad k_3 \le \frac{d}{de}f(e) \le k_4,\quad \text{and} \quad k_3 \le k_1 \le k_2 \le k_4.$ Thus, Kalman's conjecture implies Aizerman's conjecture but the converse is not true. Also, it was shown that this conjecture for some cases turns out not to be true [h1].
8. $Q(x)$ for up to and including forth-order systems of the form: $y^{(n)} + a_1(x)y^{(n-1)} + .... + a_{n-1}(x)\ y = 0$ are given by Ingwerson [i1].
9. Note that: $\dot{V}(x) = \frac{dV(x)}{dt} = \nabla V(x)^T f(x) \rightarrow dV = \nabla V^T f(x)dt = \nabla V^T dx.$
10. Locally Positive Definite Function.
11. For the definitions of $\mathcal{K}$, $\mathcal{K}_\infty$ and $\mathcal{KL}$ functions see Section 3.1.
12. Peuteman and Aeyels [p3] called this lemma the *Generalized Comparison Principle.*
13. Right-hand derivative of $D^+V(x)$ is given by $[v_1(x), v_2(x), \ldots, v_m(x)]^T \left[\frac{\partial}{\partial x_1}, \ldots, \frac{\partial}{\partial x_n}\right]$.
14. Therefore, each $\dot{v}_i(x)$ could be sign indefinite.
15. Note that this representation could be considered as a new VLF representation if $v_i(x) \triangleq V^{(i-1)}(x)$, $\forall i = 1,\ldots,m$. However, in this case, the first component of VLF is LPDF and other components might be even sign

indefinite, which in this sense, is different from the VLF previously introduced.

16. It is worth noting that choosing $Q=2I$ in (2.2-22) for this LTI system implies $P=\begin{pmatrix} 3/10 & 1/5 \\ 1/5 & 23/10 \end{pmatrix}>0$, which in turn implies asymptotic stability of this system.
17. Where $u_{-1}(t)$ is the unit step function. $u_{-n}(t)$ is the $n$th-order time integral of $u_{-1}(t)$and $u_{+n}(t)$ is its $n$th-order time derivative of $u_{-1}(t)$.
18. Preliminary work on stabilizing nonlinear systems using Ä-homogeneous approximation is given in Appendix A3.

# 3 Stability Analysis of Nonautonomous Systems

## 3.1 PRELIMINARIES

Consider the following nonlinear time-varying system with the equilibrium state at the origin:

$$\dot{x} = f(t,x), \quad x \in R^n, \qquad f(t,0) = 0, \quad \forall t \geq t_0. \tag{3.1}$$

A function $\phi : R_+ \to R_+$ is of class K and is denoted by $\phi \in$ K if $\phi$ is continuous, strictly increasing, and $\phi(0) = 0$. It is of class K infinity and is denoted by $\phi \in \mathbf{K}_\infty$ if $\phi(p) \to +\infty$ as $p \to +\infty$.

A function $\beta : R_+ \times R_+ \to R_+$ is of class KL and is denoted by $\beta \in$ L if $\beta(p,t)$ is continuous, strictly increasing with respect to $p$, strictly decreasing with respect to $t$, $\beta(0,t) = 0$, and $\beta(p,t) \to 0$ as $t \to +\infty$.

---

**Lemma 3.1 [k1]:**

The zero equilibrium state of $\dot{x} = f(t,x)$:

(i) Is uniformly stable (US) iff there exists an $\alpha \in$ K and a scalar $c > 0$ independent of $t_0$, such that:

$$\| x(t) \| \leq \alpha(\| x(t_0) \|), \quad \forall t \geq t_0, \quad \forall \| x(t_0) \| < c, \tag{3.2}$$

(ii) Is uniformly asymptotically stable (UAS) iff there exists a $\beta \in$ L and a scalar $c > 0$ independent of $t_0$, such that:

$$\| x(t) \| \leq \beta(\| x(t_0) \|, t - t_0), \quad \forall t \geq t_0, \quad \forall \| x(t_0) \| < c, \tag{3.3}$$

(iii) Is uniformly globally asymptotically stable iff (3.3) is satisfied for all $x(t_0)$. ■

---

**Corollary 3.1 [m3]:**

If Lemma 3.1 is considered with $\alpha_0 \in$ K and $\beta_{t0} \in \mathbf{KL}$ depending on initial time $t_0$, then only the stability of the zero equilibrium state (not uniform) is concluded. ■

**Definition 3.1 [m2]:**

Consider a time-varying function, $V(t,x)$ then: ■

(i) $V(t,x)$ is a (locally) positive definite function or briefly (L)PDF if $V(t,0) = 0$ and for some $\phi_1 \in K$ (and $r > 0$) one has:

$$V(t,x) \geq \phi_1(\|x\|), \quad \forall x \in R^n, \quad (\|x\| < r). \tag{3.4a}$$

(ii) $V(t,x)$ is radially unbounded (RU) if in (3.4a), $\phi_1 \in K_\infty$.
(iii) $V(t,x)$ is a negative definite function (NDF) if $-V(t,x)$ is PDF.
(iv) $V(t,x)$ is indefinite if its sign changes.
(v) $V(t,x)$ is decrescent if for some $\psi \in K$ and $0 < r \leq +\infty$ one has:

$$V(t,x) \leq \psi(\|x\|), \quad \forall \|x\| < r. \tag{3.4b}$$

■

If $V(t,x)$ is of class $C^1$, then its total time derivative $\dot{V}(t,x)$ along the solutions of (3.1) is computed by:

$$\dot{V}(t,x) \triangleq [\partial V(t,x)/\partial x]^T f(t,x) + \partial V(t,x)/\partial t \tag{3.5}$$

**Theorem 3.1 [m2]: (Lyapunov Theorem)**

(a) If there exists a class $C^1$ decrescent LPD $V(t,x)$ function such that $\dot{V}(t,x)$ is ND, then the zero equilibrium state of (3.1) is uniformly asymptotically stable.
(b) Moreover, if $V(t,x)$, is also RU then, the zero equilibrium state of (3.1) is uniformly globally asymptotically stable. ■

A $V(t,x)$ function with properties given in Theorem 3.1 is called a *Lyapunov function* (LF). Any PDF $V(t,x)$, which is a candidate to have the properties of this theorem is a Lyapunov function (LF) candidate.

The following lemma, which is known as Barbalat's lemma, is useful in obtaining asymptotic stability of the zero equilibrium state of nonlinear systems.

**Lemma 3.2 [k1]:**

Let $\phi : R \rightarrow R$ be a uniformly continuous function on $[0, +\infty]$. Assume that $\int_0^t \phi(\tau)d\tau$ exists and is finite, then:

$$\phi(t) \rightarrow 0 \text{ as } t \rightarrow +\infty \tag{3.6}$$

A function $\phi : R \to R$ is uniformly continuous if for each $\varepsilon > 0$, there exists a $\delta > 0$ such that:

$$\forall x, y \in R, \quad |x - y| < \delta \quad \Rightarrow \quad |\phi(x) - \phi(y)| < \varepsilon \tag{3.7}$$

It is known that if $|\partial\phi(x)/\partial x|$ is bounded for a class $C^1$ function $\phi : R \to R$ then, it is uniformly continuous. ■

---

**Lemma 3.3 [k1]: (Comparison Lemma)**

Consider the following scalar differential equation:

$$\dot{u} = g(t, u), \quad u(t_0) = u_0, \tag{3.8}$$

where $g(t,u)$ is continuous in $t$ and locally Lipschitz in $u$, for all $t \geq 0$ and all $u(t) \in J \subset R$. Let $[t_0, T)$ ($T$ could be infinity) be the maximal interval of existence of solution $u(t)$. Assume $u(t) \in J$ for all $t \in [t_0, T)$. Let $v(t)$ be a continuous function where its upper right-hand derivative, that is, $D^+ v(t)$* satisfies the following differential inequality:

$$D^+ v(t) \leq g(t, v(t)), \quad v(t_0) \leq u_0 \tag{3.9}$$

with $v(t) \in J$ for all $t \in [t_0, T)$, then $v(t) \leq u(t)$ for all $t \in [t_0, T)$. ■

The above lemma has a useful generalization to a vector form, that is, both $v(t), u(t) \in R^m$. Unfortunately, not all vector functions $g : R \times R^m \to R^m$ could be implemented in this case. The following definition introduces the useful vector functions for a vector-formed comparison lemma.

---

**Definition 3.2 [p3]:**

Let $a = (a_1, \ldots, a_m)^T \in R^m$ and $b = (b_1, \ldots, b_m)^T \in R^m$. A map $g : R \times R^m \to R^m$ is of the class (quasi-monotone nondecreasing) if:

$$\forall i = 1, 2, \ldots, m, \quad \begin{pmatrix} a_i = b_i \\ a_j \leq b_j, \text{ for } j \neq i \end{pmatrix} \Rightarrow g_i(t, a) \leq g_i(t, b). \tag{3.10}$$

■

It is clear that every scalar function $g(t,u)$ is of class $W$.

---

* See the footnote to Lemma 2.1.

**Lemma 3.4 [p3]: (Generalized Comparison Principle)**

Consider the following vector differential equation:

$$\dot{u} = g(t,u), \quad u(t_0) = u_0, \quad u \in R^m, \tag{3.11}$$

where $g(t,u)$ is continuous in $t$ and locally Lipschitz in $u$. Let $V(t)$ be a continuous vector function whose upper right-hand derivative, that is, $D^+v(t)$ satisfies the following differential inequality component-wise:

$$D^+v(t) \le g(t,v(t)), \quad v(t_0) \le u_0, \quad v \in R^m \tag{3.12}$$

If $g(t, u)$ is of the class, then $v(t) \le u(t)$ for $t \ge t_0$. ■

## 3.2 RELAXED LYAPUNOV STABILITY CONDITIONS

In the following, almost all the methods of Section 2.4 are generalized for the time-varying cases. The generalized methods use some time-varying decrescent functions $V_i(t,x)$ in the stability analysis of the zero equilibrium state, that is,

$$V_i(t,x) \le \psi_i(\|x\|), \qquad \psi_i \in \mathrm{K}, \qquad \forall i \tag{3.13}$$

Although the application of decrescent functions for a vector case may be somewhat confusing, this property is used to meet the following requirement for stability analysis:

"$V_i(t, x) \to 0$ when $\|x\| \to 0$."

The above requirement is used in the proof of the theorems in this section.

In Section 2.4, a simpler requirement was needed for autonomous systems because the continuity of $V_i(x)$ implies:

$$\text{if } V_i(0) = 0 \qquad V_i(x) \le \psi_i(\|x\|), \quad \psi_i \in \mathrm{K}, \tag{3.14}$$

$\psi_i : R_+ \to R_+$ could be defined as $\psi_i(r) \triangleq r + \sup_{\|x\| \le r} V_i(x)$ in (3.14).

The method in Section 2.4.1, "LaSalle Invariance Principle," may not be applied directly in time-varying systems. Nevertheless, it has been tried in several papers to prove asymptotic stability of the zero equilibrium state of a nonlinear system (3.1) providing $\dot{V}(t,x) \le 0$ for a given $V(t,x) > 0$ LF candidate [a3].

### 3.2.1 AVERAGE DECREMENT OF FUNCTION

The time-varying version of the method in Section 2.4.2 is primarily introduced in the literature for time-varying systems. All theorems in Section 2.4.2 are applicable to the nonlinear time-varying systems, using a time dependent LF candidate $V(t, x)$, which is decrescent, that is, (3.4) is satisfied. Let us recall these theorems for the time-varying case.

---

**Theorem 3.2 [m2]:**

Let $\dot{x} = f(t, x)$ be a nonlinear system with a zero equilibrium state. If there exists a decrescent LPDF $V(t, x)$, a $T > 0$ and a $\gamma \in$ K such that:

$$V(t + T, x(t + T)) - V(t, x(t)) \leq -\gamma(\|x(t)\|) < 0, \quad \forall t, \tag{3.15}$$

then, the zero equilibrium state of $\dot{x} = f(t, x)$ is uniformly asymptotically stable. ■

This theorem requires an average decrement of $V$ function on each time interval $[t, t + T]$. But the following theorem reduces the set of such intervals to an infinitely countable set.

---

**Theorem 3.3 [m3, p3]:**

Let $\dot{x} = f(t, x)$ be a nonlinear system with a zero equilibrium state. If there exists a decrescent LPDF $V(t, x)$, a $T > 0$, $\gamma \in$ K, and a strictly increasing sequence of times $\{t_k^*\}_{-\infty}^{+\infty}$ with the properties: $0 < t_{k+1}^* - t_k^* < T$, $t_k^* \to \pm\infty$ as $k \to \pm\infty$, such that:

$$V(t_{k+1}^*, x(t_{k+1}^*)) - V(t_k^*, x(t_k^*)) \leq -\gamma(\| x(t_k^*) \|) < 0 \tag{3.16}$$

then the zero equilibrium state of $\dot{x} = f(t, x)$ is uniformly asymptotically stable. ■

The following theorem obtains exponential stability of the zero equilibrium state using the average decrement method.

---

**Theorem 3.4 [m3,a2]:**

If there exist positive numbers in addition to conditions of Theorem 2.10 such that:

$$\begin{cases} V(t_{k+1}^*, x(t_{k+1}^*)) - V(t_k^*, x(t_k^*)) \leq -r \| x(t_k^*) \|^2 < 0, \\ \lambda_{\min} \| x \|^2 \leq V(t, x) \leq \lambda_{\max} \| x \|^2, \end{cases} \tag{3.17}$$

then the zero equilibrium state of $\dot{x} = f(t, x)$ is exponentially stable. ■

The application of this theorem is not trivial, since one needs the solution of the system's trajectory and an upper bound for it.

### 3.2.2 Vector Lyapunov Function

In this section, the method from Section 2.4.3, "Vector Lyapunov Function [VLF]," is modified here. In order to be applicable to a time-varying case, the modifications are as follows: Let the state vector of a time-varying nonlinear system $\dot{x} = f(t,x)$ be decomposed into $m$ subvectors, that is, $x = (x_1^T, \ldots, x_m^T)^T$, then a VLF for $\dot{x} = f(t,x)$ is defined as the following vector function.

$$V(t,x) = [V_1(t,x), V_2(t,x), \ldots, V_m(t,x)]^T \tag{3.18}$$

where:

$$\phi_i(\| x_i \|) \leq V_i(t,x) \leq \psi_i(\| x_i \|) \quad , \quad \phi_i, \psi_i \in \mathbf{K} \tag{3.19}$$

Note that the left-hand side inequality in (3.19) represents the decrescent property of $V_i(t, x)$ with respect to $x_i$.

Using the above definition for a VLF, Theorem 2.12 in Section 2.4.3 is converted to the following theorem for the case of time-varying systems.

---

**Theorem 3.5 [m2]: (Time-Varying Version of Theorem 2.12)**

Consider the nonlinear system $\dot{x} = f(t,x)$ and the VLF (3.18) with properties given by (3.19). Also, let the inequality $\dot{V}(t,x) \leq g[V(t,x)]$ be satisfied component-wise and the mapping $g: R^m \rightarrow R^m$ be of the class. If the zero equilibrium state of $\dot{u}(t) = g[u(t)]$ is:

(i) Stable in the sense of Lyapunov, then the zero equilibrium state of $\dot{x} = f(t,x)$ is uniformly stable in the sense of Lyapunov.
(ii) Asymptotically stable, then the zero equilibrium state of $\dot{x} = f(t,x)$ is uniformly asymptotically stable. ■

**Example 3.1 [k5]:**

Consider the following scalar system $\dot{x}(t) = -(e^{-t} + 1)^{1/2}\gamma(x)$, in which:
$\gamma(.) = R^+ \rightarrow R$ is any given nondecreasing bounded function in $W$ satisfying $x\gamma(x)\rangle 0, \gamma(0) = 0$ where $W = \{x \in R : |x| \leq a,\ a > 0\}$.
Let:

$$V(t,x) = |x(t)|^3.$$

Then:

$$V(x+h\dot{x}+hy)=\left|x+h\dot{x}+hy\right|^{3}.$$

Thus:

$${}_{h,y}\sqrt{V(t,x)}=\frac{\left|x+h\dot{x}+hy\right|^{3}-|x|^{3}}{h}.$$

But, it is obvious that $|a|-|b|\leq|a+b|\leq|a|+|b|$, then:

$$\frac{\left(|x|-h|\dot{x}+y|\right)^{3}-|x|^{3}}{h}\leq\Delta_{h,y}V(t,x)\leq\frac{\left(|x|+h|\dot{x}+y|\right)^{3}-|x|^{3}}{h}.$$

But:

$$\frac{(|x|-h|\dot{x}+y|)^{3}-|x|^{3}}{h}$$
$$=\frac{(|x|^{3}-3h|x|^{2}|\dot{x}+y|+3h^{2}|x||\dot{x}+y|^{2}-h^{3}|\dot{x}+y|^{3}-|x|^{3}}{h},$$

or:

$$\frac{(|x|+h|\dot{x}+y|)^{3}-|x|^{3}}{h}$$
$$=\frac{(|x|^{3}+3h|\dot{x}+y||x|^{2}+3h^{2}|x||\dot{x}+y|^{2}+h^{3}|\dot{x}+y|^{3}-|x|^{3}}{h}.$$

Therefore:

$$\inf\Delta_{h,y}V(t,x)=\frac{-3h|x|^{2}|\dot{x}+y|+3h^{2}|x||x+y|^{2}-h^{3}|\dot{x}+y|^{3}}{h},$$

or:

$$\inf_{h\to 0}\Delta_{h,y}V(t,x)=-3|x|^{2}|\dot{x}+y|.$$

Therefore:

$$D\_V(t,x)=\lim_{y\to 0}\ \inf_{h\to 0}\Delta_{h,y}V(t,x)=\lim_{y\to 0}-3|\dot{x}+y||x|^{2}=-3|\dot{x}|x^{2}$$

$D\_V(t,x) = -3(e^{-t}+1)^{1/2}\,\gamma(x) \le -3x^2|\gamma(x)|$, which implies uniform asymptotic stability of the system. ■

### 3.2.3 Higher-Order Derivatives of a Lyapunov Function Candidate

The first $m$ higher-order derivatives of $V(t, x)$ along the solutions of $\dot{x} = f(t,x)$ are computed iteratively using:

$$V^{(i)}(t,x) \triangleq [\partial V^{(i-1)}/\partial x]^T f(t,x) + \partial V^{(i-1)}/\partial t, \quad i = 1,2,\ldots,m, \tag{3.20}$$

All time-varying versions of the theorems in Section 2.4.4, "Higher-Order Derivatives of an LF Candidate," are also applicable here. The generalized versions of these theorems are given as follows:

---

**Theorem 3.6 [m2]: (Time-Varying Version of Theorems 2.14 and 2.15)**

Let $\dot{x} = f(t,x)$ be a nonlinear system with a zero equilibrium state, and $V(t, x)$ be a function satisfying the following inequality:

$$V^{(m)}(t,x) \le g_m(t,V,\dot{V},\ldots,V^{(m-1)}) \tag{3.21}$$

Also assume:

(i) $\alpha_1(\| x \|) \le V(t,x) \le \alpha_2(\| x \|), \quad \| x \| < r, \quad \forall t, \quad \alpha_1,\alpha_2 \in \mathbf{K}$
(ii) The following nonlinear system, that is, the *Comparison Equation* (CE) is of the class W:

$$u^{(m)}(t) = g_m(t,u,\dot{u},\ldots,u^{(m-1)}) \tag{3.22}$$

Now the following hold:

(a) If for identically equal initial conditions:

$$u^{(i)}(t_0) = V^{(i)}(t_0,x_0), \quad i = 0,1,\ldots,m-1, \quad \forall \| x_0 \| < r \tag{3.23}$$

one has:

$$u(t) < \alpha_3\left(u(t_0)\right), \quad t \ge t_0, \quad \alpha_3 \in \mathbf{K} \tag{3.24}$$

then, the zero equilibrium state of $\dot{x} = f(t,x)$ is uniformly stable (US) in the sense of Lyapunov.

(b) If for identically equal initial conditions (3.23) one has:

$$u(t) < \alpha_3\left(u(t_0), t - t_0\right), \quad t \geq t_0, \quad \alpha_3 \in \mathbf{KL}, \tag{3.25}$$

then the zero equilibrium state of $\dot{x} = f(t,x)$ is uniformly asymptotically stable.

(c) If the nonlinear system (3.22) is uniformly asymptotically stable and when defining $\mathrm{V}(t,x) \triangleq [V(t,x), \dot{V}(t,x), ..., V^{(m-1)}(t,x)]^T$ one has:

$$\|\mathrm{V}(t,x)\| \leq \psi(\|x\|),^{*} \quad \psi \in \mathrm{K}, \tag{3.26}$$

then, the zero equilibrium state of $\dot{x} = f(t,x)$ is uniformly asymptotically stable. ■

**Example 3.2:**

Consider the following system:

$$\dot{x}(t) = -t\,x(t)$$

Let $V(x) = x^2(t)$.
Obviously:

$$\frac{1}{2}x^2 \leq x^2(t) \leq 2x^2(t).$$

Also, the following derivatives are obtainable:

$$\begin{aligned} \dot{V}(x) &= 2x\,\dot{x} = -2t\,x^2, \\ \ddot{V} &= -2x^2 + 4t^2x^2, \\ \dddot{V} &= 12tx^2 - 8t^3x^2. \end{aligned}$$

Therefore:

$$\dddot{V} + 6\dot{V} = -8t^3x^2 \leq 0,$$

* The vector Lyapunov function V(t,x) in this case would be decrescent if (3.26) holds $\forall t$ and $\forall x \in B_r$ for $0 < r < \infty$.

which implies:

$$\dddot{V} \le -6\dot{V}.$$

Now consider the following linear system:

$$\dddot{u} + 6\dot{u} = 0 \text{ with } u(0) = x^2(0), \quad \dot{u}(0) = 0 \text{ and } \ddot{u}(0) = -2x^2(0).$$

The eigenvalues of the linear system are, $\lambda = 0$ and $\pm j\sqrt{6}$, then the solution is:

$$u(t) = (u(0) + \frac{1}{6}\ddot{u}(0) + \frac{1}{\sqrt{6}}\dot{u}(0)\sin\sqrt{6}\,t - \frac{1}{6}\ddot{u}(0)\cos\sqrt{6}\,t,$$

thus:

$$u(t) = \frac{2}{3}x^2(0) + \frac{1}{3}x^2(0)\cos\sqrt{6}\,t < 10x^2(0) = |10|\,u(0).$$

Since:

$$\alpha_3 = |.| \in K \text{ (see Equation [3.24])},$$

then the zero equilibrium state of the system is uniformly stable. ■

**Exercise 3.1:**

Prove Theorem 3.6. (Hint: See proofs of Theorems 2.14 and 2.15.) ■

---

**Theorem 3.7 [m2, m3]: (Time-Varying Version of Theorem 2.16)**

Consider the nonlinear system $\dot{x} = f(t,x)$ and a $C^1$ m-vector function V($t$, $x$) of the following form:

$$\mathrm{V}(t,x) = [\mathrm{V}_1(t,x), \mathrm{V}_2(t,x), \ldots, \mathrm{V}_m(t,x)]^T, \tag{3.27}$$

whose time derivative $\dot{\mathrm{V}}(t,x)$ along the solutions of $\dot{x} = f(t,x)$ satisfies the following differential controllable canonical form inequality:

$$\begin{bmatrix} \dot{\mathrm{V}}_1(t,x) \\ \dot{\mathrm{V}}_2(t,x) \\ \vdots \\ \dot{\mathrm{V}}_m(t,x) \end{bmatrix} \le \begin{bmatrix} 0 & 1 & \cdots & 0 \\ 0 & 0 & \ddots & \vdots \\ 0 & 0 & \cdots & 1 \\ -a_0 & -a_1 & \cdots & -a_{m-1} \end{bmatrix} \begin{bmatrix} \mathrm{V}_1(t,x) \\ \mathrm{V}_2(t,x) \\ \vdots \\ \mathrm{V}_m(t,x) \end{bmatrix} \tag{3.28}$$

In addition, assume V($t$, 0) = 0 and all the roots of the following characteristic equation are negative real numbers:

$$s^m + a_{m-1}s^{m-1} + \cdots + a_1 s + a_0 = 0, \tag{3.29}$$

therefore:

(i) If $V_1(t,x)$ is LPDF, then the zero equilibrium state of $\dot{x} = f(t,x)$ is asymptotically stable.
(ii) If $V_1(t,x)$ is PDF and RU, then the zero equilibrium state of $\dot{x} = f(t,x)$ is globally asymptotically stable. ■

**Exercise 3.2:**

Prove Theorem 3.7. (Hint: See proof of Theorem 2.16.) ■

---

**Corollary 3.2 [m2,m3]:**

Consider the nonlinear system $\dot{x} = f(t,x)$ and let $V(t, x)$ be smooth enough such that the higher-order derivatives $\dot{V}(t,x), \ddot{V}(t,x), \ldots, V^{(m)}(t,x)$, are well defined and satisfy the following differential inequality:

$$V^{(m)}(t,x) + a_{m-1}V^{(m-1)}(t,x) +; \cdots ,+ a_1\dot{V}(t,x) + a_0 V(t,x) \le 0. \tag{3.30}$$

Also, the characteristic equation (3.29) has only negative real roots.

(i) If $V(t, x)$ is LPDF, then the zero equilibrium state of $\dot{x} = f(t,x)$ is asymptotically stable.
(ii) If $V(t, x)$ is PDF and RU, then the zero equilibrium state of $\dot{x} = f(t,x)$ is globally asymptotically stable.

**Proof:**

The proof of this corollary is similar to the proof of Corollary 2.1. ■

**Example 3.3 [m2,m3]:**

A nonlinear time-varying system is given below:

$$\begin{cases} \dot{x} = -g_1(x) - 2xy + ke^{-pt}x, \\ \dot{y} = -g_2(y) + x^2 + ke^{-pt}y, \qquad k > 1, p > 0. \end{cases} \tag{3.31}$$

Assume:

$$\sigma^2 \le g_i(\sigma)\sigma \le 2\sigma^2, \forall i = 1,2. \tag{3.32}$$

Let $g_i(\sigma)$ be a nonsmooth function. Find the conditions of the globally asymptotic stability of the zero equilibrium state of this system.

**Solution:**

Using $V_1 = x^2 + 2y^2$ as a PDF, computing $\dot{V}_1$ and eliminating the $g_i(\sigma)$ terms yields:

$$\begin{aligned} \dot{V}_1 &= 2x\dot{x} + 4y\dot{y} = -2xg_1(x) - 4yg_2(y) + (x^2 + 2y^2)2ke^{-pt} \\ &\le -2(x^2 + 2y^2) + (x^2 + 2y^2)2ke^{-pt} = 2V_1(ke^{-pt} - 1) \triangleq V_2. \end{aligned} \tag{3.33}$$

Clearly $\dot{V}_1$ is not negative definite because at $t = 0$, $\dot{V}_1$ is positive for $k > 1$. Thus, the Lyapunov direct method fails to prove the stability of the ZES of this system using this function. However, $\dot{V}_1$ approaches a negative definite function. The $\dot{V}_1$ function is not smooth, but the $V_2$ function in (3.33) is smooth. Computing $\dot{V}_2$ and rearranging terms yields:

$$\begin{aligned} \dot{V}_2 &= -2pV_1ke^{-pt} + 2\dot{V}_1(ke^{-pt} - 1) = -2pV_1ke^{-pt} \\ &\quad + 2[-2xg_1(x) - 4yg_2(y) + 2V_1ke^{-pt}](ke^{-pt} - 1) \\ &= -(p+2)2V_1ke^{-pt} + 4V_1(ke^{-pt})^2 \\ &\quad + 4[xg_1(x) + 2yg_2(y)] - 4[xg_1(x) + 2yg_2(y)]ke^{-pt}. \end{aligned} \tag{3.34}$$

Then, using the bounds of (3.32) and eliminating the $g_i(\sigma)$ functions from $\dot{V}_2$, yields:

$$\begin{aligned} \dot{V}_2 &\le -(p+2)2V_1ke^{-pt} + 4V_1(ke^{-pt})^2 + 8[x^2 + 2y^2] - 4[x^2 + 2y^2]ke^{-pt} \\ &= -(p+4)2V_1ke^{-pt} + 4V_1(ke^{-pt})^2 + 8V_1. \end{aligned} \tag{3.35}$$

Now by use of (3.33) and (3.35), arrange the following linear combination:

$$\begin{aligned} \dot{V}_2 + a_1 V_2 + a_0 V_1 &\le -(p+4)2\, V_1ke^{-pt} + 4\, V_1(ke^{-pt})^2 + 8\, V_1 + a_1 2\, V_1(ke^{-pt} - 1) + a_0 V_1 \\ &= 2\, V_1ke^{-pt}[2ke^{-pt} - (p+4) + a_1] + V_1[8 + a_0 - 2a_1] \\ &\le 2\, V_1ke^{-pt}\underbrace{[2k - p - 4 + a_1]}_{=0} + V_1\underbrace{[8 + a_0 - 2a_1]}_{=0}. \end{aligned} \tag{3.36}$$

Next, set the right-hand side brackets of (3.36) equal to zero, as follows:

$$\begin{cases} 2k - p - 4 + a_1 = 0, \\ 8 + a_0 - 2a_1 = 0, \end{cases} \Rightarrow \begin{cases} a_1 = p - 2k + 4, \\ a_0 = 2p - 4k. \end{cases} \tag{3.37}$$

Now, using (3.33) and (3.36) along with the above values for and yields:

$$\begin{bmatrix} \dot{V}_1 \\ \dot{V}_2 \end{bmatrix} \le \begin{bmatrix} 0 & 1 \\ -a_0 & -a_1 \end{bmatrix} \begin{bmatrix} V_1 \\ V_2 \end{bmatrix}. \tag{3.38}$$

It is desirable for $\mathrm{V}(t,x) = [V_1(t,x), V_2(t,x)]^T$ to satisfy the conditions of part (ii) in Theorem 3.7 by the following approach: The $V_1(t,x)$ function is PDF and RU and $V_1(t,0) = V_2(t,0) = 0$. Also, the inequality (3.38) is in the form of (3.28), and its characteristic equation is of the form $\Delta(s) = s^2 + a_1 s + a_0 = 0$. The polynomial $\Delta(s)$ has negative real roots, if:

$$\begin{cases} a_1 = p - 2k + 4 > 0, \\ a_0 = 2p - 4k > 0, \\ \Delta_1 = a_1^2 - 4a_0 = (p - 2k)^2 + 16 \ge 0. \end{cases} \tag{3.39}$$

Solving the above inequalities yields:

$$p > 2k. \tag{3.40}$$

Thus, (3.40) is a sufficient condition for global asymptotical stability of the zero equilibrium state of (3.31). ■

---

**Theorem 3.8 [m3,m5]: (Meigoli–Nikravesh Theorem)**

Consider an $m$-vector of class $C^1$ function $\mathrm{V}(t,x) = [V_1(t,x), V_2(t,x), \ldots, V_m(t,x)]^T$ with the following properties:

(i) The first component $V_1(t,x)$ of $\mathrm{V}(t, x)$ is PDF and RU, that is, $V_1(t,0) = 0$, $\forall t \ge 0$ and there is a function $\phi_1 \in \mathbf{K}_\infty$ such that:

$$V_1(t,x) \ge \phi_1(\| x \|), \quad \forall x \in R^n, \forall t \ge 0. \tag{3.41}$$

(ii) The $V_i(t,x)$ components are all decrescent, that is, there are functions $\psi_i \in \mathbf{K}$ for $i = 2,\ldots,m$ such that:

$$V_i(t,x) \le \psi_i(\| x \|), \quad \forall x \in R^n, \quad \forall t \ge 0, \tag{3.42}$$

then:

(1) If the following differential inequality satisfies the time derivative $\dot{V}(t,x)$ along the trajectories of $\dot{x} = f(t,x)$:

$$\begin{bmatrix} a_{11} & 0 & 0 & \cdots & 0 \\ a_{21} & a_{22} & 0 & 0 & \vdots \\ \vdots & a_{ij} & \ddots & 0 & 0 \\ a_{m-1,1} & \cdots & & a_{m-1,m-1} & 0 \\ a_{m1} & \cdots & & a_{m,m-1} & a_{mm} \end{bmatrix} \begin{bmatrix} \dot{V}_1 \\ \dot{V}_2 \\ \vdots \\ \dot{V}_{m-1} \\ \dot{V}_m \end{bmatrix} \le \begin{bmatrix} V_2 \\ V_3 \\ \vdots \\ V_m \\ -\phi_2(\| x \|) \end{bmatrix} \tag{3.43}$$

where $\phi_2 \in \mathbf{K}$ and $A = [a_{ij}]_{m \times m}$ is a lower triangular matrix with the following properties:

$$a_{ij} \begin{cases} = 0 \quad , & \text{if } i < j, \\ > 0 \quad , & \text{if } i = j, \\ \ge 0 \quad , & \text{if } i > j, \end{cases} \tag{3.44}$$

then, the zero equilibrium state of $\dot{x} = f(t,x)$ is uniformly globally asymptotically stable.

(2) If the above conditions hold only locally, that is, for $\|x\| < r$ and a given $r > 0$, then the zero equilibrium state of $\dot{x} = f(t,x)$ is uniformly asymptotically stable.

For the proof of Theorem 3.8, see Appendix A2. ■

---

**Corollary 3.3 [m3, m5]:**

Consider the nonlinear system $\dot{x} = f(t,x)$ and let $V(t,x)$ be smooth enough such that the higher-order derivatives $\dot{V}(t,x), \ddot{V}(t,x), \ldots, V^{(m)}(t,x)$ are well defined.

(i) If $V(t,x)$ is PDF and RU, and the $V^{(i)}(t,x)$ functions for $i = 0,1,\ldots,m-1$ are all decrescent, and:

(ii) $$\sum_{i=1}^{m} a_i V^{(i)}(t,x) \leq -\phi_2(\| x \|), \quad \phi_2 \in \mathbf{K}, \quad \forall x \in R^n, \tag{3.45}$$

for:

$$\begin{cases} a_i \geq 0 & ,i = 1,2,\ldots,m-1, \\ a_m > 0, \end{cases} \tag{3.46}$$

then, the zero equilibrium state of $\dot{x} = f(t,x)$ is globally uniformly asymptotically stable.

(iii) If the above conditions hold only locally, that is, for $\|x\| < r$ for some $r > 0$, then, the zero equilibrium state of $\dot{x} = f(t,x)$ is uniformly asymptotically stable. ■

**Proof:**

For the proof, refer to the proof of Corollary 2.2. ■

**Example 3.4:**

Consider the following LTI system:

$$\begin{bmatrix} \dot{x}_1 \\ \dot{x}_2 \end{bmatrix} = \begin{bmatrix} -1 & 1 \\ -4 & -1 \end{bmatrix} \begin{bmatrix} x_1 \\ x_2 \end{bmatrix}.$$

The eigenvalues of this system are $s_{1,2} = -1 \pm 2j$. Thus, the system is asymptotical stable using the first method of Lyapunov. Now, consider the following PDF function:

$$V(x) = x_1^2 + x_2^2,$$
$$\dot{V}(x) = 2x_1\dot{x}_1 + 2x_2\dot{x}_2 = -2x_1^2 - 6x_1x_2 - 2x_2^2,$$

$\dot{V}(x)$ is not negative definite, since $\Delta = (-6)^2 - 4(-2)(-2) = 20 > 0$. So the Lyapunov direct method fails using this Lyapunov function candidate. The $V(x)$ function and any of the higher-order Lyapunov function derivatives are also in quadratic forms. So, one might find a linear dependency between them. The computation yields:

$$\ddot{V}(x) = 28x_1^2 + 24x_1x_2 - 2x_2^2,$$
$$\dddot{V}(x) = -152x_1^2 + 24x_1x_2 + 28x_2^2.$$

Using the bases $\{x_1^2, x_1x_2, x_2^2\}$, one has:

$$\begin{bmatrix} V & \dot{V} & \ddot{V} & \dddot{V} \end{bmatrix} = \begin{bmatrix} x_1^2, x_1x_2, x_2^2 \end{bmatrix} \begin{bmatrix} 1 & -2 & 28 & -152 \\ 0 & -6 & 24 & 24 \\ 1 & -2 & -2 & 28 \end{bmatrix}.$$

Let us find a linear combination of $V^{(i)}$ as follows:

$$r_0V + r_1\dot{V} + r_2\ddot{V} + \dddot{V} = \begin{bmatrix} V\dot{V}\ddot{V}\dddot{V} \end{bmatrix}\begin{bmatrix} r_0r_1r_2 1 \end{bmatrix}^T = 0.$$

Substituting into this equation from its above equation implies:

$$\begin{bmatrix} 1 & -2 & 28 & -152 \\ 0 & -6 & 24 & 24 \\ 1 & -2 & -2 & 28 \end{bmatrix} \begin{bmatrix} r_0 \\ r_1 \\ r_2 \\ 1 \end{bmatrix} = 0 \quad \text{or} \quad \begin{bmatrix} 1 & -2 & 28 \\ 0 & -6 & 24 \\ 1 & -2 & -2 \end{bmatrix} \begin{bmatrix} r_0 \\ r_1 \\ r_2 \end{bmatrix} = \begin{bmatrix} 152 \\ -24 \\ -28 \end{bmatrix}.$$

Thus, solving for $r_0$, $r_1$ and $r_2$ yields:

$$40V + 28\dot{V} + 6\ddot{V} + \dddot{V} = 0.$$

The characteristic equation of this system, that is,

$$\Delta(s) = s^3 + 6s^2 + 28s + 40 = (s+2)[(s+2)^2 + 16] = 0,$$

is a Hurwitz polynomial. Using Theorem 3.7 implies that the zero equilibrium state of this system is globally asymptotically stable, which coincides with the eigenvalue analysis given above.

The roots of $\Delta(s)$ are the eigenvalues of the given LTI system, namely $s_{1,2} = -1 \pm 2j$. This property is preserved for all LTI systems. ■

The next example introduces a generalized homogeneous nonlinear system with zero degree of homogeneity (see also Meigoli and Nikravesh [m2]).

### Example 3.5:

Consider the following nonlinear differential equation with unknown real parameters *a*, *b*, and *c*.

$$\begin{cases} \dot{x}_1 = ax_1, \\ \dot{x}_2 = bx_1^2 + cx_2. \end{cases}$$

The linearization of this system in the neighborhood of the origin, that is, $\{\dot{x}_1 = ax_1,\ \dot{x}_2 = cx_2\}$ represents an asymptotical stable system iff $a$ and $c$ are both negative. The zero equilibrium state of this nonlinear system is therefore asymptotically stable if $a$ and $c$ are both negative. It is proven that the requirements that $a$ and $c$ both be negative are necessary and sufficient conditions for the global asymptotic stability of zero equilibrium state as well.

The independence of the stability condition from the parameter $b$, must not be confusing, since this nonlinear system is in a lower triangular form and it can be considered as a cascade combination of the two linear systems $\{\dot{x}_1 = ax_1\}$ and $\{\dot{x}_2 = cx_2 + u\}$ with $u = bx_1^2$.

This nonlinear system is in a form that the $x_1^2$ and $x_2$ can be viewed as interchanged terms. Let $V(x) = (x_1^2)^2 + (x_2)^2$ be an LF candidate; computation yields:

$$\dot{V}(x) = 4x_1^3\dot{x}_1 + 2x_2\dot{x}_2 = 4ax_1^4 + 2bx_1^2x_2 + 2cx_2^2,$$

$$\ddot{V}(x) = (16a^2 + 2b^2)x_1^4 + (4ab + 6bc)x_1^2x_2 + (4c^2)x_2^2,$$

$$\dddot{V}(x) = (64a^3 + 12ab^2 + 6cb^2)x_1^4 + (8a^2b + 16abc + 14bc^2)x_1^2x_2 + (8c^3)x_2^2.$$

The $V(x)$, $\dot{V}(x)$, $\ddot{V}(x)$ and $\dddot{V}(x)$ functions (and all other derivatives) are a linear combination of $\{x_1^4,\ x_1^2x_2,\ x_2^2\}$. Thus, one has:

$$\left[V\ \dot{V}\ \ddot{V}\ \dddot{V}\right] =$$

$$\{x_1^4,\ x_1^2x_2,\ x_2^2\}\begin{bmatrix} 1 & 4a & 16a^2 + 2b^2 & 64a^3 + 12ab^2 + 6cb^2 \\ 0 & 2b & 4ab + 6bc & 8a^2b + 16abc + 14bc^2 \\ 1 & 2c & 4c^2 & 8c^3 \end{bmatrix}.$$

A linear dependence of these functions can be found by solving for $r_0$, $r_1$, $r_2$ in the following equation:

$$\begin{bmatrix} 1 & 4a & 16a^2 + 2b^2 & 64a^3 + 12ab^2 + 6cb^2 \\ 0 & 2b & 4ab + 6bc & 8a^2b + 16abc + 14bc^2 \\ 1 & 2c & 4c^2 & 8c^3 \end{bmatrix}\begin{bmatrix} r_0 \\ r_1 \\ r_2 \\ 1 \end{bmatrix} = 0.$$

Solving this equation by the Gauss method yields:

$$\begin{cases} r_0 = -(16a^2c + 8ac^2), \\ r_1 = 8a^2 + 16ac + 2c^2, \\ r_2 = -(6a + 3c). \end{cases}$$

So, the following linear dependence is obtained:

$$-(16a^2c + 8ac^2)V + (8a^2 + 16ac + 2c^2)\dot{V} - (6a + 3c)\ddot{V} + \dddot{V} = 0.$$

The new characteristic polynomial is given as:

$$\Delta(s)=s^3-(6a+3c)s^2+(8a^2+16ac+2c^2)s-(16a^2c+8ac^2)$$
$$=(s-2c)(s-4a)[s-(2a+c)].$$

Thus, the sufficient conditions for global asymptotic stability of the zero equilibrium state are given:

$$c<0,\ \ a<0.$$ ■

**Example 3.6 [m3, m5]:**

Consider the following time-varying nonlinear system (Example 3.3 system):

$$\begin{cases}\dot{x}_1=-g_1(x_1)-2x_1x_2+ke^{-pt}x_1,\\ \dot{x}_2=-g_2(x_2)+x_1^2+ke^{-pt}x_2,\end{cases}\quad t\geq 0,\quad (k,p>0). \tag{3.47}$$

The functions $g_i(\sigma)$ for $i=1,2$ are continuous (and possibly nonsmooth) functions and $g_i(0)=0,\ \ g_i(\sigma)\sigma>0, \forall\sigma\neq 0$. Consider the following functions:

$$V_i(t,x)\triangleq(2ke^{-pt})^{(i-1)}(x_1^2+2x_2^2)\ ,i=1,2,\ldots,m \tag{3.48}$$

Clearly, $V_1(t,x)=x_1^2+2x_2^2$ is PDF and RU and all $V_i(t,x)$'s are decrescent. Computation of $\dot{V}_i(t,x)$ along the solutions of (3.47) yields:

$$\begin{aligned}\dot{V}_i(t,x)&=-p(i-1)(2ke^{-pt})^{(i-1)}(x_1^2+2x_2^2)+(2ke^{-pt})^{(i-1)}(2x_1\dot{x}_1+4x_2\dot{x}_2)\\&=-(2ke^{-pt})^{(i-1)}[2x_1g_1(x_1)+4x_2g_2(x_2)+(p(i-1)-2ke^{-pt})(x_1^2+2x_2^2)],\end{aligned} \tag{3.49}$$

$\dot{V}_1(t,x)$ might be sign indefinite, since $k>0$ (e.g., consider the case of $g_1(\sigma)=g_2(\sigma)=\sigma^3$). Use of (3.49) yields:

$$\dot{V}_i(t,x)\leq(2ke^{-pt})^{(i)}(x_1^2+2x_2^2)=V_{i+1}(t,x), \tag{3.50}$$

Setting up a linear combination $\Sigma_{i=1}^m a_i\dot{V}_i(t,x)$ using (3.49) with the positive $a_i$ weights yields:

$$\begin{aligned}\sum_{i=1}^m a_i\dot{V}_i(t,x)&=-[2x_1g_1(x_1)+4x_2g_2(x_2)]\sum_{i=1}^m a_i(2ke^{-pt})^{(i-1)}\\&\quad+(x_1^2+2x_2^2)\sum_{i=1}^m a_i(2ke^{-pt})^{(i)}\\&\quad-(x_1^2+2x_2^2)\sum_{i=2}^m a_i(2ke^{-pt})^{(i-1)}p(i-1)\ \Rightarrow\\ \sum_{i=1}^m a_i\dot{V}_i(t,x)&=-[2x_1g_1(x_1)+4x_2g_2(x_2)]\sum_{i=1}^m a_i(2ke^{-pt})^{(i-1)}\\&\quad+2ke^{-pt}(x_1^2+2x_2^2)[a_m(2ke^{-pt})^{m-1}\\&\quad+\sum_{i=1}^{m-1}(a_i-pa_{i+1}i)(2ke^{-pt})^{i-1}].\end{aligned} \tag{3.51}$$

In what follows, (3.51) is simplified by selecting appropriate $a_i$ weights. Substituting:

$$a_i - pa_{i+1}\ i = 0 \qquad \text{for } i = 2,3,\ldots,m-1, \tag{3.52}$$

into (3.51), yields:

$$\sum_{i=1}^{m} a_i\dot{V}_i(t,x) = -[2x_1g_1(x_1) + 4x_2g_2(x_2)]\sum_{i=1}^{m} a_i(2ke^{-pt})^{(i-1)}$$
$$+ 2ke^{-pt}(x_1^2 + 2x_2^2)[a_m(2ke^{-pt})^{m-1} + (a_1 - pa_2)]. \tag{3.53}$$

It is desirable for $\Sigma_{i=1}^{m} a_i\dot{V}_i(t,x)$ to be negative definite. Therefore, the following is assumed:

$$a_m(2k)^{m-1} + a_1 - pa_2 = \sup_{t\geq 0}[a_m(2ke^{-pt})^{m-1} + (a_1 - pa_2)] = 0. \tag{3.54}$$

Substituting this relation into (3.53) and performing some manipulations yields:

$$\sum_{i=1}^{m} a_i\dot{V}_i(t,x) \leq -a_1(2x_1g_1(x_1) + 4x_2g_2(x_2)) \tag{3.55}$$

Setting and solving (3.52) and (3.54) for $a_i = 2,\ldots,m$ yields:

$$a_1 = 1,\ a_2 = (m-1)!\,p^{m-2}\big/[p^{m-1}(m-1)! - (2k)^{m-1}],$$
$$a_i = a_2\big/(i-1)!\,p^{i-2},\ i = 3,\ldots,m. \tag{3.56}$$

The relations (3.50) and (3.55) both are of the form (3.43). Using Theorem 3.19, if $a_1 > 0$ and $a_i \geq 0, i = 2,\ldots,m$ in (3.55), then it would render the zero equilibrium state of (3.47) to be uniformly globally asymptotically stable.

Using (3.56), a sufficient condition for positiveness of coefficients is given as $p^{m-1}(m-1)! - (2k)^{m-1} > 0$, or

$$p > 2k\big/\sqrt[m-1]{(m-1)!} \tag{3.57}$$

The above procedure can be repeated for all integers $m \geq 2$, and it is observed that $2k\big/\sqrt[m-1]{(m-1)!} \to 0$ as $m \to +\infty$. Therefore, for every $p$, $k > 0$ a positive integer $m \geq 2$ can be found to satisfy (3.57). Hence, the uniform global asymptotic stability of the zero equilibrium state of (3.47) is proved for each $k$, $p > 0$.

The above results can be verified using the Lyapunov direct method. In the above procedure, concentrating on the case of $m \to +\infty$ and manipulating (3.56) yields:

$$a_2p\left[1 - (2k/p)^{m-1}\tfrac{1}{(m-1)!}\right] = 1.$$

Letting $m \to +\infty$ in this relation, would imply $a_2 \to 1/p$ and,

$$\begin{cases} a_1 = 1, \\ a_i \to 1/(i-1)!\,p^{i-1},\ i = 2,3,\ldots, \end{cases} \quad \text{as } m \to +\infty. \tag{3.58}$$

Define the following summation using the weights in (3.58):

$$s(t,x) \triangleq \sum_{i=1}^{+\infty} a_i V_i(t,x) = (x_1^2 + 2x_2^2) \sum_{i=1}^{+\infty} \left( \frac{2ke^{-pt}}{p} \right)^{(i-1)} \frac{1}{(i-1)!}. \tag{3.59}$$

This is an expansion of $s(t,x) = (x_1^2 + 2x_2^2)\exp[2ke^{-pt}/p]$, which is clearly a decrescent, PDF and RU function. Using $s(t,x)$ as an LF candidate for (3.47) would imply:

$$\dot{s}(t,x) = -(2x_1 g_1(x_1) + 4x_2 g_2(x_2))\exp[2ke^{-pt}/p], \tag{3.60}$$

which is negative definite. Note that the finding of the appropriate $s(t,x)$ as an LF candidate is a consequence of higher-order derivatives of a given inappropriate LF candidate $V(t,x)$. ■

## 3.3 NEW STABILITY THEOREMS (FATHABADI–NIKRAVESH TIME-VARYING METHOD)

This section is the time-varying version of Section 2.5. Thus, the theorems and their proofs will closely follow similar ones of that section.

---

**Theorem 3.9: (Expanded Fathabadi–Nikravesh Theorem)**

Consider the following $n$th order nonautonomous system:

$$\begin{cases} \dot{x}_1 = f_1(x_1, x_2, \ldots, x_n, t), \\ \dot{x}_2 = f_2(x_1, x_2, \ldots, x_n, t), \\ \cdot \\ \cdot \\ \dot{x}_n = f_n(x_1, x_2, \ldots, x_n, t), \end{cases} \tag{3.61}$$

The equilibrium state $x_e = 0$ of the system is asymptotically stable iff, there are closed surfaces with the algebraic relation of the form $u(x_1, x_2, \ldots, x_n, t) = c$ such that:

$$\nabla u.\dot{x} \langle 0, \quad \forall t \geq t_0, \quad \forall x \in \Omega, \tag{3.62}$$

where $\Omega$ is a closed neighborhood of $x_e = 0$ $[f1, f2]$. ■

---

**Theorem 3.10: (Expanded Fathabadi–Nikravesh Theorem)**

Consider the $n$th order nonautonomous system (3.61). The equilibrium state of this system is unstable iff, there are closed surfaces with the algebraic relation of the form $u(x_1, x_2, \ldots, x_n, t) = c$ such that:

$$\nabla u.\dot{x} \rangle 0, \ \forall t \geq t_0, \ \forall x \in \Omega, \tag{3.63}$$

where $\Omega$ is an open neighborhood of $x_e = 0$ $[f1, f2]$. ■

**Expanded Algorithm 3.1:**

Choose the following auxiliary system:

$$\begin{cases} \dot{x}_1 = g_1(x_1, x_2, \dots, x_n, t), \\ \dot{x}_2 = g_2(x_1, x_2, \dots, x_n, t), \\ \cdot \\ \cdot \\ \dot{x}_n = g_n(x_1, x_2, \dots, x_n, t), \end{cases} \tag{3.64}$$

such that:

(a)

$$g_1(x_1, x_2, \dots, x_n, t) f_1(x_1, x_2, \dots, x_n, t) + \dots \\ + g_n(x_1, x_2, \dots, x_n, t) f_n(x_1, x_2, \dots, x_n, t) \langle\ 0, \tag{3.65}$$

and:

(b) The equilibrium state ($x_e = 0$) is enclosed by the response of the following Pfaffian differential equation:

$$g_1(x_1, x_2, \dots, x_n, t) dx_1 + \dots + g_n(x_1, x_2, \dots, x_n, t) dx_n = 0, \tag{3.66}$$

then, the equilibrium state of the system (3.61) is asymptotically stable [$f1, f2$]. ■

**Expanded Algorithm 3.2:**

Choose the following auxiliary system:

$$\begin{cases} \dot{x}_1 = g_1(x_1, x_2, \dots, x_n, t), \\ \dot{x}_2 - g_2(x_1, x_2, \dots, x_n, t), \\ \cdot \\ \cdot \\ \dot{x}_n = g_n(x_1, x_2, \dots, x_n, t), \end{cases} \tag{3.67}$$

such that:

(a)

$$g_1(x_1, x_2, \dots, x_n, t) f_1(x_1, x_2, \dots, x_n, t) + \dots \\ + g_n(x_1, x_2, \dots, x_n, t) f_n(x_1, x_2, \dots, x_n, t)\ \rangle\ 0, \tag{3.68}$$

and,

(b) The equilibrium state ($x_e = 0$) is enclosed by the response of the following Pfaffian differential equation:

$$g_1(x_1, x_2, \ldots, x_n, t)dx_1 + \ldots + g_n(x_1, x_2, \ldots, x_n, t)dx_n = 0, \tag{3.69}$$

then, the equilibrium state of the system (3.61) is unstable [f1, f2]. ■

**Example 3.7:**

Consider the following nonlinear nonautonomous system:

$$\begin{cases} \dot{x}_1 = -2x_1^3 + e^{-3t}x_1x_2^2, \\ \dot{x}_2 = -2x_2^3. \end{cases} \tag{3.70}$$

Choosing the following auxiliary system:

$$\begin{cases} \dot{x}_1 = g_1(x_1, x_2) = 2x_1, \\ \dot{x}_2 = g_2(x_1, x_2) = 2x_2, \end{cases} \tag{3.71}$$

yields:

$$\begin{aligned} & g_1(x_1, x_2)f_1(x_1, x_2) + g_2(x_1, x_2)f_2(x_1, x_2) \\ & = -4x_1^4 + 2e^{-3t}x_1^2x_2^2 - 4x_2^4 \langle 0, \quad \forall t \geq 0, \end{aligned} \tag{3.72}$$

On the other hand, (3.71) yields:

$$u_1(x_1, x_2, t) = (x_1^2 + x_2^2) = c_1$$

It is clear that the above set of closed curves encloses $x_e = 0$, and therefore the equilibrium state of the system (3.70) is asymptotically stable.

For verification, let:

$$V(x) = \int g_1 dx_1 + \int g_2 dx_2 = x_1^2 + x_2^2.$$

Thus:

$$\dot{V}(x) = 2x_1\dot{x}_1 + 2x_2\dot{x}_2 = -4x_1^4 + 2x_1^2x_2^2e^{-3t} - 4x_2^4.$$

It is clear that $e^{-3t} \leq 1$ for $t \geq 0$, thus:

$$\dot{V}(x) \leq -4x_1^4 + 8x_1^2x_2^2 - 4x_2^4 = -\left(2x_1^2 - 2x_2^2\right)^2 \leq 0,$$

which implies stability of the zero equilibrium state of the system using the Lyapunov method. ■

Note that even though for either positive or negative $c_i$, the curves $u_i = c_i$ are closed, but for the stable system, $c_i$ should be only positive.

### Example 3.8:

Consider the following nonlinear nonautonomous system:

$$\begin{cases} \dot{x}_1 = -2x_1^3 + e^{-3t}x_1x_2^2, \\ \dot{x}_2 = -2x_2^3, \\ \dot{x}_3 = -(3-e^{-t})x_3^5, \\ \dot{x}_4 = (2-e^{-2t})x_3^4x_4 - x_4^3. \end{cases} \tag{3.73}$$

Choosing the following auxiliary system:

$$\begin{cases} \dot{x}_1 = g_1(x_1,x_2,x_3,x_4,t) = 2x_1 \\ \dot{x}_2 = g_2(x_1,x_2,x_3,x_4,t) = 2x_2 \\ \dot{x}_3 = g_3(x_1,x_2,x_3,x_4,t) = 4x_3^3 \\ \dot{x}_4 = g_4(x_1,x_2,x_3,x_4,t) = x_4 \end{cases} \tag{3.74}$$

yields:

$$\begin{aligned} & g_1(x_1,x_2,\dots,x_4,t)f_1(x_1,x_2,\dots,x_4,t) + \dots \\ & + g_4(x_1,x_2,\dots,x_4,t)f_4(x_1,x_2,\dots,x_4,t) = \\ & -4x_1^4 + 2e^{-3t}x_1^2x_2^2 - 4x_2^4 - 4.(3-e^{-t})x_3^8 \\ & +(2-e^{-2t})x_3^4x_4^2 - x_4^4 < 0, \qquad \forall t \geq 0, \end{aligned} \tag{3.75}$$

On the other hand, (3.74) yields:

$$u_1(x_1,x_2,x_3,x_4,t) = x_1^2 + x_2^2 + x_3^4 + 0.5x_4^2 = c_1.$$

It is clear that the above set of closed surfaces encloses $x_e = 0$, and therefore, the equilibrium state of the system (3.73) is asymptotically stable.

For verification, let:

$$V(x) = \frac{1}{2}\left(x_1^2 + x_2^2 + x_3^4 + x_4^2\right).$$

Thus:

(i)

$$\frac{1}{4}(x_1^2 + x_2^2 + x_3^4 + x_4^2) \leq V(x) \leq x_1^2 + x_2^2 + x_3^4 + x_4^2$$

Now:

(ii)

$$
\begin{aligned}
\dot{V}(x) &= x_1(-2x_1^3 + e^{-3t}x_1x_2^2) - 2x_2^4 - 2(3-e^{-t})x_3^8 + (2-e^{-2t})x_3^4x_2^2 - x_4^4 \\
&= -2x_1^4 + e^{-3t}x_1^2x_2^2 - 2x_2^4 - 2(3-e^{-t})x_3^8 + (2-e^{-2t})x_3^4x_2^2 - x_4^4 \\
&\le -2x_1^4 + x_1^2x_2^2 - 2x_2^4 - x_3^8 + 2x_3^4x_2^2 - x_4^4 \\
&= -(2x_1^4 - x_1^2x_2^2 + 2x_2^4 + x_3^8 - 2x_3^4x_2^2 + x_4^4) \\
&= \begin{bmatrix} x_1^2 & x_2^2 & x_3^4 & x_4^2 \end{bmatrix} \begin{bmatrix} 2 & -1/2 & 0 & 0 \\ -1/2 & 2 & -1 & 0 \\ 0 & -1 & 1 & 0 \\ 0 & 0 & 0 & 1 \end{bmatrix} \begin{bmatrix} x_1^2 \\ x_2^2 \\ x_3^4 \\ x_4^2 \end{bmatrix}
\end{aligned}
$$

which is a continuous negative definite function, which in turn implies a uniformly asymptotically stable zero equilibrium state of the system. ■

## Example 3.9:

Consider the following nonlinear nonautonomous system:

$$
\begin{cases}
\dot{x}_1 = -4x_1^5 + 2\sin(t)e^{-3t}x_1^2x_2^4, \\
\dot{x}_2 = -x_2^5, \\
\dot{x}_3 = -(2+te^{-t})x_3^5.
\end{cases}
\tag{3.76}
$$

Choosing the following auxiliary system:

$$
\begin{cases}
\dot{x}_1 = g_1(x_1,x_2,x_3,t) = x_1, \\
\dot{x}_2 = g_2(x_1,x_2,x_3,t) = 4x_2^3, \\
\dot{x}_3 = g_3(x_1,x_2,x_3,t) = x_3,
\end{cases}
\tag{3.77}
$$

yields:

$$
\begin{aligned}
&g_1(x_1,x_2,x_3,t)f_1(x_1,x_2,x_3,t) + \ldots + g_3(x_1,x_2,x_3,t)f_3(x_1,x_2,x_3,t) \\
&= -4x_1^6 + 2\sin(t)e^{-3t}x_1^3x_2^4 - 4x_2^8 - (2+te^{-t})x_3^6 \le 0, \quad \forall t \ge 0,
\end{aligned}
\tag{3.78}
$$

On the other hand, (3.77) yields:

$$
u_1(x_1,x_2,x_3,t) = x_1^2 + x_2^4 + x_3^2 = c_1.
$$

It is clear that the above set of closed surfaces encloses $x_e = 0$, and therefore, the equilibrium state of the system (3.77) is asymptotically stable. ■

## 3.4 APPLICATION OF PARTIAL STABILITY THEORY IN NONLINEAR NONAUTONOMOUS SYSTEM STABILITY ANALYSIS [c1]

In this section, the definitions, notations, and notion of partial stability are reviewed.

Consider the following nonlinear autonomous dynamical system:

$$\begin{aligned} \dot{x}_1(t) &= f_1(x_1(t), x_2(t)), \qquad x_1(0) = x_{10}, \\ \dot{x}_2(t) &= f_2(x_1(t), x_2(t)), \qquad x_2(0) = x_{20}, \end{aligned} \tag{3.79}$$

where $x_1 \in D$, $D \subseteq R^{n_1}$, $x_2 \in R^{n_2}$, $f_1 : D \times R^{n_2} \to R^{n_1}$ is such that for every $x_2 \in R^{n_2}$, $f_1(0, x_2) = 0$ and $f_1(\cdot, x_2)$ is locally Lipschitz in $x_1$, $f_2 : D \times R^{n_2} \to R^{n_2}$ is such that for every $x_1 \in D$, $f_2(x_1, \cdot)$ is locally Lipschitz in $x_2$.

Note that under the above assumptions, the solution $(x_1(t), x_2(t))$ of (3.79) exists and is unique. The following definition introduces partial stability; that is, the stability with respect to $x_1$, for the nonlinear dynamical system (3.79).

---

**Definition 3.3:**

(i) The nonlinear dynamical system (3.79) is Lyapunov stable with respect to $x_1$ if for every $\varepsilon > 0$ and $x_{20} \in R^{n_2}$, there exists $\delta = \delta(\varepsilon, x_{20}) > 0$ such that $\|x_{10}\| < \delta$ implies that $\|x_1(t)\| < \varepsilon$ for all $t \geq 0$.

(ii) The nonlinear dynamical system (3.79) is Lyapunov stable with respect to $x_1$ uniformly in $x_{20}$ if for every $\varepsilon > 0$ there exists $\delta = \delta(\varepsilon) > 0$ such that $\|x_{10}\| < \delta$ implies that $\|x_1(t)\| < \varepsilon$ for all $t \geq 0$ and for all $x_{20} \in R^{n_2}$.

(iii) The nonlinear dynamical system (3.79) is asymptotically stable with respect to $x_1$ if it is Lyapunov stable with respect to $x_1$ and for every $x_{20} \in R^{n_2}$ there exists $\delta = \delta(x_{20}) > 0$ such that $\|x_{10}\| < \delta$ implies $\lim x_1(t) \to 0$, as $t \to \infty$.

(iv) The nonlinear dynamical system (3.79) is asymptotically stable with respect to $x_1$ uniformly in $x_{20}$ if it is Lyapunov stable with respect to $x_1$ uniformly in $x_{20}$ and there exists $\delta > 0$ such that $\|x_{10}\| < \delta$ implies that $\lim x_1(t) = 0$, as $t \to \infty$. for all $x_{20} \in R^{n_2}$.

(v) The nonlinear dynamical system (3.79) is globally asymptotically stable with respect to $x_1$ uniformly in $x_{20}$ if it is Lyapunov stable with respect to $x_1$ uniformly in $x_{20}$ and $\lim x_1(t) \to 0$, as $t \to \infty$ for all $x_{10} \in R^{n_1}$ and $x_{20} \in R^{n_2}$.

(vi) The nonlinear dynamical system (3.79) is exponentially stable with respect to $x_1$ uniformly in $x_{20}$ if there exist scalars $\alpha$, $\beta$, $\delta > 0$ such that $\|x_{01}\| < \delta$ implies that $\|x_1(t)\| \leq \alpha \|x_{10}\| e^{-\beta t}, t \geq 0$, for all $x_{20} \in R^{n_2}$.

(vii) The nonlinear dynamical system (3.79) is globally exponentially stable with respect to $x_1$ uniformly in $x_{20}$ if there exist scalars $\alpha$, $\beta > 0$ such that $\|x_1(t)\| \leq \alpha \|x_{10}\| e^{-\beta t}, t \geq 0$, for all $x_{10} \in R^{n_1}$ and $x_{20} \in R^{n_2}$.

Next, the sufficient conditions for partial stability of the nonlinear dynamical system (3.79) are presented. For the following result, recall the definitions of class and class $\mathbf{K}_\infty$ function and define $\dot{V}(x_1,x_2) \triangleq V'(x_1,x_2)f(x_1,x_2)$, where $f(x_1,x_2) \triangleq [f_1^T(x_1,x_2)\ f_2^T(x_1,x_2)]^T$, for a given continuously differentiable function $V : D \times R^{n_2} \to R$. Furthermore, it is assumed that the solution $x(t)$ of (3.79) exists and is unique for all $t \geq 0$. It is important to note that unlike standard theory, the existence of a Lyapunov function $V(x_1, x_2)$ satisfying the conditions in Theorem 3.11 is not sufficient to ensure that all solutions of (3.79) starting in $D \times R^{n_2}$ can be extended to infinity since none of the states of (3.79) serve as an independent variable. Note, however, continuous differentiability of $f_1(\cdot,\cdot)$ and $f_2(\cdot,\cdot)$ provides a sufficient condition for the existence and uniqueness of solutions of (3.79) for all $t \geq 0$.

---

**Theorem 3.11:**

Consider the nonlinear dynamical system given by (3.79), then the following statements hold:

(i) If there exists a continuously differentiable function $V : D \times R^{n_2} \to R$ and a class function $\alpha(\cdot)$ such that:

$$V(0,x_2) = 0, \quad x_2 \in R^{n_2}, \tag{3.80}$$

$$\alpha(\|x_1\|) \leq V(x_1,x_2), \quad (x_1,x_2) \in D \times R^{n_2}, \tag{3.81}$$

$$\dot{V}(x_1,x_2) \leq 0, \quad (x_1,x_2) \in D \times R^{n_2}, \tag{3.82}$$

then the nonlinear dynamical system given by (3.79) is stable in the sense of Lyapunov with respect to $x_1$.

(ii) If there exists a continuously differentiable function $V : D \times R^{n_2} \to R$ and class functions $\alpha(\cdot)$ and $\beta(\cdot)$ such that (3.81) and (3.82) hold and;

$$V(x_1,x_2) \leq \beta(\|x_1\|), \quad (x_1,x_2) \in D \times R^{n_2}, \tag{3.83}$$

then the nonlinear dynamical system given by (3.79) is stable in the sense of Lyapunov with respect to $x_1$ uniformly in $x_{20}$.

(iii) If there exists a continuously differentiable function $V : D \times R^{n_2} \to R$ and class K functions $\alpha(\cdot)$, $\beta(\cdot)$ and $\gamma(.)$ such that (3.81) and (3.83) hold and;

$$\dot{V}(x_1,x_2) \leq -\gamma(\|x_1\|), \quad (x_1,x_2) \in D \times R^{n_2}, \tag{3.84}$$

then the nonlinear dynamical system given by (3.79) is asymptotically stable with respect to $x_1$ uniformly in $x_{20}$.

(iv) If $D \times R^{n_1}$ and there exists a continuously differentiable function $V : R^{n_1} \times R^{n_2} \to R$, a class K function $\gamma(\cdot)$, and class $K_\infty$ functions $\alpha(\cdot)$ and

$\beta(.)$ such that (3.81), (3.83), and (3.84) hold, then the nonlinear dynamical system given by (3.79) is globally asymptotically stable with respect to $x_1$ uniformly in $x_{20}$.

(v) If there exists a continuously differentiable function $V: D \times R^{n_2} \rightarrow R$ and positive constants $\alpha$, $\beta$, $\gamma$, and $p$ such that $p \geq 1$ and for all

$$(x_1, x_2) \in D \times R^{n_2},$$

$$\alpha \|x_1\|^p \leq V(x_1, x_2) \leq \beta \|x_1\|^p, \tag{3.85}$$

$$\dot{V}(x_1, x_2) \leq -\gamma \|x_1\|^p, \tag{3.86}$$

then the nonlinear dynamical system given by (3.79) is exponentially stable with respect to $x_1$ uniformly in $x_{20}$.

(vi) If $D \times R^{n_1}$ and there exists a continuously differentiable function $V: R^{n_1} \times R^{n_2} \rightarrow R$, and positive constants $\alpha$, $\beta$, $\gamma$, and $p$ such that $p \geq 1$ and (3.85) and (3.86) hold, then the nonlinear dynamical system given by (3.79) is globally exponentially stable with respect to $x_1$ uniformly in $x_{20}$.

**Proof:**

(i) Let $\varepsilon > 0$ be such that $B_\varepsilon \triangleq \{x_1 \in R^{n_1} : \|x_1\| < \varepsilon\} \subset D$, define $\eta \triangleq \alpha(\varepsilon)$, and define $D_\eta \triangleq \{x_1 \in B_\varepsilon$ there exists $x_2 \in R^{n_2}$ such that $V(x_1, x_2) < \eta\}$. Now, since $\dot{V}(x_1, x_2) \leq 0$ it follows that $V(x_1(t), x_2(t))$ is a nonincreasing function of time and hence, $D_\eta \times R^{n_2}$ is a positive invariant set with respect to (3.79). Next, since $V(\cdot,\cdot)$ is continuous and $V(0, x_{20}) = 0$, $x_{20} \in R^{n_2}$ there exists $\delta = \delta(\varepsilon, x_{20}) > 0$ such that $V(x_1, x_{20}) < \eta$, $x_1 \in B_\delta$, or equivalently $B_\delta \subset D_\eta$. Hence, for all $(x_{10}, x_{20}) \in B_\delta \times R^{n_2}$ it follows that:

$$\alpha(\|x_1(t)\|) \leq V(x_1(t), x_2(t)) \leq V(x_{10}, x_{20}) < \eta = \alpha(\varepsilon),$$

and thus $x_1(t) \in B_\varepsilon$, $t \geq 0$, establishing stability in the sense of Lyapunov with respect to $x_1$.

(ii) Let $\varepsilon > 0$ and let $B_e$ and $\eta$ be given as in the proof of (i). Now, let $\delta = \delta(\varepsilon) > 0$ be such that $\beta(\delta) = \alpha(\varepsilon)$. Hence, it follows from (3.83) that for all, $(x_1, x_{20}) \in B_\delta \times R^{n_2}$, $\alpha(\|x_1(t)\|) \leq V(x_1(t), x_2(t)) \leq V(x_{10}, x_{20}) < \beta(\delta) = \alpha(\varepsilon)$,

and thus:

$$x_1(t) \in B_e, \quad t \geq 0.$$

(iii) Lyapunov stability uniformly in $x_{20}$ follows from (ii). Next, let $\varepsilon > 0$ and $\delta = \delta(\varepsilon) > 0$ be such that for every $x_{10} \in B_\delta$, $x_1(t) \in B_e, t \geq 0$, (the existence of such a $(\delta,\varepsilon)$ pair follows from Lyapunov uniform stability) and assume that (3.84) holds. Since (3.84) implies (3.82), it follows that for every $x_{10} \in B_\delta$, $V(x_1(t), x_2(t))$ is a nonincreasing function of time and since $V(\cdot,\cdot)$ is bounded from below, it follows from the Bolzano–Weierstass theorem [r5] that there exists $L \geq 0$ such that $\lim_{t\to\infty} V(x_1(t), x_2(t)) = L$. Now, suppose for some $x_{10} \in B_\delta$, *ad absurdum** $L > 0$ so that $D_L \triangleq \{x_1 \in B_e$: there exists $x_2 \times R^{n2}$ such that $V(x_1,x_2) \leq L\}$ is nonempty and $x_1(t) \notin D_L, t \geq 0$. Thus, as in the proof of (i), there exists $\delta > 0$ such that $B_\delta \subset D_L$. Hence, it follows from (3.84) that for every $x_{10} \in B_\delta \backslash D_L$ and $t \geq 0$,

$$
\begin{aligned}
V(x_1(t),x_2(t)) &= V(x_{10},x_{20}) + \int_0^t \dot{V}(x_1(s),x_2(s))ds \\
&\leq V(x_{10},x_{20}) - \int_0^t \gamma(\|x_1(s)\|)ds \\
&\leq V(x_{10},x_{20}) - \gamma(\delta)t.
\end{aligned}
$$

Letting $t \geq \dfrac{V(x_{10},x_{20}) - L}{\gamma(\delta)}$ it follows that $V(x_1(t),x_2(t)) \leq L$, which is a contradiction. Hence, L = 0, which implies that $V(x_1(t), x_2(t)) \to 0$, as $t\to\infty$. for all $x_{10} \in B_\delta$. Now, since $V(x_1(t), x_2(t)) \geq \alpha(\|x_1(t)\|) \geq 0$ it follows that $\alpha(\|x_1(t)\|) \to 0$ or, equivalently, $x_1(t) \to 0$ as $t\to\infty$, establishing asymptotic stability with respect to $x_1$.

(iv) Let $v > 0$ be such that $\|x_{10}\| < V$. Now, it follows from (3.84) that $V(x_1(t), x_2(t))$ is a nonincreasing function of time and hence $x_1(t) \in D_V \triangleq \{x_1 \in R^{n_1}$ there exists $x_2 \in R^{n_2}$ such that $V(x_1,x_2) \leq V\}, t \geq 0$. Next, since $\alpha(\cdot)$ is a class $K_\infty$ function it follows that there exists $\delta > 0$ such that $\beta(V) < \alpha(\delta)$ and it follows from (3.83) that $\alpha(\| x_1(t)\|) \leq V(x_1(t), x_2(t)) \leq \beta(V) < \alpha(\delta), \ \forall\ t \geq 0$.

Hence $x_1(t) \in B_\delta, \forall\ t \geq 0$.
Now, with $\varepsilon = \delta$, the proof follows as in the proof of (iii).

(v) Let $\varepsilon < 0$ and $\eta \triangleq \alpha(\varepsilon)$, be given as in the proof of (i). Now, (3.86) implies that $\dot{V}(x_1, x_2) \leq 0$ and hence it follows that $V(x_1(t), x_2(t))$ is a nonincreasing function of time and $D_\eta \times R^{n_2} \subset D \times R^{n_2}$ is a positive invariant set with respect to (3.79). Thus, it follows from (3.85) and (3.86) that for all $t \geq 0$ and $(x_{10},x_{20}) \in D_\eta \times R^{n2}$, then:

$$
\dot{V}(x_1(t), x_2(t)) \leq -\gamma\|x_1(t)\|^p \leq -\frac{\gamma}{\beta} V(x_1(t), x_2(t)),
$$

* *ad absurdum*: disproof of the proposition by claming an absurdity to which it leads when carried to its logical conclusion.

which implies that:

$$V(x_1(t),x_2(t)) \le V(x_{10},x_{20})e^{-\frac{\gamma}{\beta}t}$$

Now, it follows from (3.85) that:

$$\alpha(\| x_1(t)\|^p \le V(x_1(t),x_2(t)) \le V(x_{10},x_{20})e^{-\frac{\gamma}{\beta}t} \le \beta\| x_{10}\|^p e^{-\frac{\gamma}{\beta}t},$$

and hence:

$$\| x_1(t)\|^p \le \left(\frac{\beta}{\alpha}\right)^{1/p} \| x_{10}\| e^{-\frac{\gamma}{\beta p}t},$$

establishing exponential stability with respect to $x_1$ uniformly in $x_{20}$.

(vi) The proof follows as in (iv) and (v). ■

**Remark 3.1:**

By setting $n_1 = n$ and $n_2 = 0$, Theorem 2.7 is specialized for the case of nonlinear autonomous systems form $\dot{x}_1(t) = f_1(x_1(t))$. In this case, Lyapunov (asymptotic) stability with respect to $x_1$ and Lyapunov (asymptotic) stability with respect to $x_1$ uniformly in $x_{20}$ are equivalent to the classical Lyapunov (asymptotic) stability of nonlinear autonomous systems. Furthermore, note that in this case there exists a continuously differentiable function $V{:}D{\to}R$ such that (3.81), (3.83), and (3.84) hold iff $V(\cdot)$ is such that $V(0) = 0$, $V(x_1)>0, \forall x_1 \ne 0$, and $\frac{\partial V(x_1)}{\partial x_1} V'(x_1)f_1(x_1)<0, \forall x_1 \ne 0$ [k1]. In addition, if $D = R^{n_1}$ and there exist class $\mathbf{K}_\infty$ functions $\alpha(\cdot)$, $\beta(\cdot)$ and a continuously differentiable function $V(\cdot)$ such that (3.81), (3.83), and (3.84) hold iff $V(\cdot)$ is such that $V(0) = 0$, $V(x_1)>0, \forall\ x_1 \ne 0$, $\frac{\partial V(x)}{\partial x} f_1(x_1)<0, \quad \forall x_1 \ne 0$, and $V(x_1)\to\infty$ as $\| x_1\| \to\infty$. Hence, in this case, Theorem 3.11 collapses to the classical Lyapunov stability theorem for autonomous systems [k1].

Note that in the ordinary partial stability theory $x_{10}$ and $x_{20}$ lie in a neighborhood of the origin, whereas in Definition 3.3, $x_{20}$ can be chosen arbitrarily.

In the case of time-invariant systems the Barbashin–Krasovskii–LaSalle invariance theorem [k1] shows that bounded system trajectories of a nonlinear dynamical system approach the largest invariant set $M$, characterized by the set of all points in a compact set $D$ of the state space where the Lyapunov derivative identically vanishes. In the case of partially stable systems, however, it is not generally clear on how to define a set $M$ since $\dot{V}(x_1,x_2)$ is a function of both $x_1$ and $x_2$. However, if $\dot{V}(x_1,x_2) \le -W(x_1) \le 0$, where $W : D \to R$ is continuous and nonnegative definite mapping, then a set $R$ can be defined as the set of all points where $W(x_1)$ identically vanishes there; that is, $R=\{x_1 \in D: W(x_1)=0\}$. In this case, as shown in the next theorem, the partial system trajectories $x_1(t)$ approach $R$ as $t$ tends to infinity.

---

**Theorem 3.12:**

Consider the nonlinear dynamical system given by (3.79) and assume $D \times R^{n_2}$ is a positive invariant set with respect to (3.79). Furthermore, assume that there exist functions $V: D \times R^{n_2} \to R$, $W$, $W_1$, $W_2: D \to R$ such that $V(\cdot,\cdot)$ is continuously differentiable, $W_1(\cdot)$ and $W_2(\cdot)$ are continuous and positive definite, $W(\cdot)$ is continuous and nonnegative definite, and for all $(x_1, x_2) \in D \times R^{n_2}$,

$$W_1(x_1) \le V(x_1, x_2) \le W_2(x_1), \tag{3.87}$$

$$\dot{V}(x_1, x_2) \le -W(x_1). \tag{3.88}$$

Then there exists $D_0 \subseteq$ D such that for all $(x_{10}, x_{20}) \in D_0 \times R^{n_2}$, $x_1(t) \to R \triangleq \{x_1 \in D: W(x_1)=0\}$ as $t \to \infty$. If, in addition, $D \times R^{n_1}$ and $W_1(\cdot)$ is radially unbounded, then for all $(x_{10}, x_{20}) \in R^{n_1} \times R^{n_2}$, $x_1(t) \to R \triangleq \{x_1 \in R^{n_1}: W(x_1)=0\}$ as $t \to \infty$. ■

**Proof:**

Assume (3.87) and (3.88) hold. Then it follows from Theorem 3.11 that the nonlinear dynamical system given by (3.79) is Lyapunov stable with respect to $x_1$ uniformly in $x_{20}$. Let $\varepsilon > 0$ be such that $B_\varepsilon \subset D$ and Let $\delta = \delta(\varepsilon) > 0$ be such that if $x_{10} \in B_\delta$, then $x_1(t) \in B_\varepsilon$, $t \ge 0$. Now, since $V(x_1(t), x_2(t))$ is monotonically nonincreasing and bounded from below by zero, it follows from the Bolzano–Weierstrass theorem [r5] that $\lim_{t\to\infty} V(x_1(t), x_2(t))$ exists and is finite. Hence, since for every $t \ge 0$,

$$\int_0^t W(x_1(\tau))d\tau \le -\int_0^t \dot{V}(x_1(\tau), x_2(\tau))d\tau = V(x_{10}, x_{20}) - V(x_1(t), x_2(t)),$$

It follows that $\lim_{t\to\infty} \int_0^t W(x_1(\tau))d\tau$ exists and is finite. Next, since $x_1(t)$ is uniformly continuous and $W(\cdot)$ is continuous on a compact set $B_\varepsilon$, it follows that $W(x_1(t))$ is uniformly continuous at every $t \ge 0$. Now, it follows from the Barbalat's lemma [k1] that $W(x_1(t)) \to 0$ as $t \to \infty$. Finally, if in addition $D = R^{n_1}$ and $W_1(\cdot)$ is radially unbounded, then as in the proof of (iv) of Theorem 3.11, for every $x_{10} \in R^{n_1}$ there exist $\varepsilon$ and $\delta$ both positive such that $x_{10} \in B_\delta$ implies $x_1(t) \in B_\varepsilon, t \ge 0$. Now, the proof follows by repeating the above arguments. ■

**Remark 3.2:**

Theorem 3.12 shows that the partial system trajectories $x_1(t)$ approach $R$ as t tends to infinity. However, since the positive limit set of the partial trajectory $x_1(t)$ is a subset of $R$, Theorem 3.12 is a much weaker result than the standard invariance principle wherein one would conclude that the partial trajectory $x_1(t)$ approaches the *largest invariant set M* contained in $R$. This is not true in general for partially stable systems since the positive limit set of a partial trajectory $x_1(t)$, $t \geq 0$, is not an invariant set. However, in the case where $f_1(.,x_2)$ is periodic, almost-periodic, or asymptotically independent of $x_2$, then an invariance principle for partial stable systems can be derived.

### 3.4.1 Unified Stability Theory for Nonlinear Time-Varying Systems

In this subsection, the results of Section 3.4 are used to prove the classical results on Lyapunov's direct method for nonlinear time-varying systems, thereby providing a unification between partial stability theory for autonomous systems and stability theory for time-varying systems. Specifically, consider the nonlinear time-varying dynamical system

$$\dot{x}(t) = f(t, x(t)), \quad x(t_0), \; t \geq t_0, \tag{3.89}$$

where $x(t) \in D, D \subseteq R^n$, $t \geq t_0$ such that $0 \in D, f:[t_0, t_1) \times D \rightarrow R^n$ is such that $f(\cdot,\cdot)$ is jointly continuous in $t$ and $x(t)$ and for every $t \in (t_0, t_1)$, $f(t,0) = 0$ and $f(t,\cdot)$ is locally Lipschitz in $x(t)$, uniformly in t for all t in compact subsets of $[0, \infty)$. Note, that under the above assumptions the solution, $x(t)$, $t \geq t_0$ to (3.89) exists and is unique over the interval $(t_0, t_1)$ [k1]. The following standard definition provides seven types of stability for the nonlinear time-varying dynamical system (3.89).

---

**Definition 3.4:**

(i) The nonlinear time varying dynamical system (3.89) is Lyapunov stable if for every $\varepsilon > 0$ and every $t_0 \in [0, \infty)$, there exists a $\delta = \delta(\varepsilon, t_0) > 0$ such that $\|x_0\| < \delta$ implies that $\|x(t)\| < \varepsilon$ for all $t \geq t_0$.

(ii) The nonlinear time-varying dynamical system (3.89) is uniformly Lyapunov stable if for every $\varepsilon > 0$ there exists a $\delta = \delta(\varepsilon) > 0$ such that $\|x_0\| < \delta$ implies that $\|x(t)\| < \varepsilon$ for all $t \geq t_0$ and for all $t_0 \in [0, \infty)$.

(iii) The nonlinear time-varying dynamical system (3.89) is asymptotically stable if it is Lyapunov stable and for every $t_0 \in [0, \infty)$, there exists a $\delta = \delta(\varepsilon, t_0) > 0$ such that $\|x_0\| < \delta$ implies that $\lim_{t \to \infty} x(t) \to 0$.

(iv) The nonlinear time-varying dynamical system (3.89) is uniformly asymptotically stable if it is uniformly Lyapunov stable and there exists a $\delta > 0$ such that $\|x_0\| < \delta$ implies that $\lim_{t \to \infty} x(t) \to 0$ for all $t_0 \in [0, \infty)$.

(v) The nonlinear time-varying dynamical system (3.89) is globally uniformly asymptotically stable if it is uniformly Lyapunov stable and $\lim_{t\to\infty} x(t) \to 0$ for all $x_0 \in R^n$ and $t_0 \in [0, \infty)$.

(vi) The nonlinear time-varying dynamical system (3.89) is (uniformly) exponentially stable if there exists scalars $\alpha$, $\beta$, and $\delta$ (not depending on $t_0$) all positive such that $\|x_0\| < \delta$ implies that $\|x(t)\| \leq \alpha \|x_0\| e^{-\beta t}$, $t \geq t_0$ and $t_0 \in [0, \infty)$.

(vii) The nonlinear time-varying dynamical system (3.89) is globally (uniformly) exponentially stable if there exists scalars $\alpha$ and $\beta$ both positive such that $\|x(t)\| \leq \alpha \|x_0\| e^{-\beta t}, t \geq t_0$ for all $x_0 \in R^n$ and $t_0 \in [0, \infty)$. ■

Next, using Theorem (3.11) the sufficient conditions for stability of the nonlinear time-varying dynamical system (3.89) is presented. For the following result, define:

$$\dot{V}(t,x) \triangleq \frac{\partial V(t,x)}{\partial x} f(t,x) + \frac{\partial V(t,x)}{\partial t}$$

for a given continuously differentiable function $V:[0,\infty)\times D \to R$.

---

**Theorem 3.13:**

Consider the nonlinear time-varying dynamical system given by (3.89). Then the following statements hold (based on Definition 3.4):

(i) If there exists a continuously differentiable function $V:[0,\infty)\times D \to R$ and a class function $\alpha(\cdot)$ such that

$$V(t,0) = 0, \qquad t \in [0,\infty), \tag{3.90}$$

$$\alpha(\|x\|) \leq V(t,x), \qquad (t,x) \in [0,\infty)\times D, \tag{3.91}$$

$$\dot{V}(t,x) \leq 0, \qquad (t,x) \in [0,\infty)\times D, \tag{3.92}$$

then the nonlinear time-varying dynamical system given by (3.89) is Lyapunov stable.

(ii) If there exists a continuously differentiable function $V:[0,\infty)\times D \to R$ and class functions $\alpha(\cdot)$ and $\beta(\cdot)$, such that (3.91) and (3.92) hold, and

$$V(t,x) \leq \beta(\|x\|), \qquad \forall (t,x) \in [0,\infty)\times D, \tag{3.93}$$

then the nonlinear time-varying dynamical system given by (3.89) is uniformly Lyapunov stable.

(iii) If there exists a continuously differentiable function $V:[0,\infty)\times D\to R$ and class functions $\alpha(\cdot)$, $\beta(\cdot)$, and $\gamma(\cdot)$ such that (3.91) and (3.93) hold, and:

$$\dot{V}(t,x)\le -\gamma(\|x\|),\qquad \forall (t,x)\in[0,\infty)\times D, \tag{3.94}$$

then, the nonlinear time-varying dynamical system given by (3.89) is uniformly asymptotically stable.

(iv) If $D=R^n$ and there exists a continuously differentiable function $V:[0,\infty)\times R^n\to R$ a class K function $\gamma(\cdot)$, class $K_\infty$ functions $\alpha(\cdot)$ and $\beta(\cdot)$ such that (3.91), (3.93) and (3.94) hold, then the nonlinear time-varying dynamical system given by (3.89) is globally uniformly asymptotically stable.

(v) If there exists a continuously differentiable function $V:[0,\infty)\times D\to R$ and positive constants $\alpha$, $\beta,\gamma$ and $p$ such that $p\ge 1$ and for all $(t,x)\in[0,\infty)\times D$,

$$\alpha\|x\|^p\le V(t,x)\le\beta\|x\|^p, \tag{3.95}$$

$$\dot{V}(t,x)\le -\gamma\|x\|^p, \tag{3.96}$$

then, the nonlinear time-varying dynamical system given by (3.89) is uniformly exponentially stable.

(vi) If $D=R^n$ and there exists a continuously differentiable function $V$: $[0,\infty)\times R^n\to R$, and positive constants $\alpha,\beta,\gamma$ and $p$ such that $p\ge 1$ and (3.95) and (3.96) hold, then the nonlinear time-varying dynamical system given by (3.89) is globally uniformly exponentially stable.

**Proof:**

First note that in requiring the existence of a Lyapunov function $V$: $[0,\infty)\times D\to R$ satisfying the conditions above, it follows that there exists a unique solution to (3.89) for all $t\ge 0$ [k1]. Next, let $n_1=n$, $n_2=1$, $x_1(t-t_0)=x(t)$, $x_2(t-t_0)=t$, $f_1(x_1,x_2)=f(t,x)$, and $f_2(x_1,x_2)=1$. Now, note that with $\tau=t-t_0$, the solution $x(t)$, $t\ge t_0$, to the nonlinear time-varying dynamical system (3.89) is equivalently characterized by the solution $x_1(\tau)$, $\tau>0$, to the nonlinear autonomous dynamical system:

$$\dot{x}_1(\tau)=f_1(x_1(\tau),x_2(\tau)),\quad x_1(0)=x_0,\quad \tau\ge 0,$$
$$\dot{x}_2(\tau)=1,\quad x_2(0)=t_0,$$

where $\dot{x}_1(\cdot)$ and $\dot{x}_2(\cdot)$ denote differentiation with respect to $\tau$. Now the result is a direct consequence of Theorem 3.11.

In light of Theorem 3.13, it follows that Theorem 3.12 can be trivially extended to address partial stability for a time-varying dynamical system. Specifically, consider the nonlinear time-varying dynamical system:

$$\dot{x}_1(t)=f_1(t,x_1(t),x_2(t)),\quad x_1(t_0)=x_{10},\quad t\ge t_0 \tag{3.97}$$

$$\dot{x}_2(t)=f_2(t,x_1(t),x_2(t)),\quad x_2(t_0)=x_{20}, \tag{3.98}$$

where $x_1 \in D$, $x_2 \in R^{n_2}$, $f_1 : (t_0,t_1) \times D \times R^{n_2} \to R^{n_1}$ is such that for every $t \in [t_0,t_1)$ and $x_2 \in R^{n_2}$, $f_1(t,0,x_2) = 0$ and $f_1(t,\cdot,x_2) = 0$ is locally Lipschitz in $x_1$, and $f_2 : (t_0,t_1) \times D \times R^{n_2} \to R^{n_2}$ is such that for every $x_1 \in D$, $f_2(\cdot,x_1,\cdot)$ is locally Lipschitz in $x_2$. Next, let:

$$\hat{x}_1(t-t_0) = x_1(t),\ \hat{x}_2(t-t_0) = \left[x_2^T(t)\ t\right]^T,\ \hat{f}_1(\hat{x}_1,\hat{x}_2) = f_1(t,x_1,x_2),$$

and $\hat{f}_2(\hat{x}_1,\hat{x}_2) = [f_2^T(t,x_1,x_2)1]^T$. Now, note that with $\tau = t - t_0$, the solution $(x_1(t),\ x_2(t)), t \geq t_0$, to the nonlinear time-varying dynamical system (3.97), (3.98) is equivalently characterized by the solution $(\hat{x}_1(\tau),\ \hat{x}_2(\tau)), \tau \geq 0$ to the following nonlinear autonomous dynamical system:

$$\dot{\hat{x}}_1(\tau) = \hat{f}_1(\hat{x}_1(\tau),\hat{x}_2(\tau)),\quad \hat{x}_1(0) = x_0,\quad \tau \geq 0 \tag{3.99}$$

$$\dot{\hat{x}}_2(\tau) = \hat{f}_2(\hat{x}_1(\tau),\hat{x}_2(\tau)),\quad \hat{x}_2(0) = \left[x_{20}^T\quad t_0\right]^T, \tag{3.100}$$

where $\dot{\hat{x}}_1(\cdot)$ and $\dot{\hat{x}}_2(\cdot)$ denote differentiation with respect to $\tau$. Hence, Theorem (3.11) can be used to drive sufficient conditions for partial stability results for the nonlinear time-varying dynamical system of the forms (3.97) and (3.98). Of course, in this case it is important to note that partial stability may be uniform with respect to either or both of $x_{20}$ and $t_0$.

Finally, using Theorems 3.11 and 3.13, some insight on the complexity of linear, time-varying systems is presented. Specifically, let $f(t,x) = A(t)x$, where $A : [0,\infty) \to R^{n\times n}$ is continuous, so that (3.93) becomes:

$$\dot{x}(t) = A(t)x(t),\qquad x(t_0) = x_0,\qquad t \geq t_0. \tag{3.101}$$

Now, in order to analyze the stability of (3.101), let $n_1 = n$, $n_2 = 1$, $x_1(t - t_0) = x(t)$, $x_2(t - t_0) = t$, $f_1(x_1, x_2) = A(t)x$, and $f_2(x_1, x_2) = 1$. Hence, the solution $x(t)$, $t \geq t_0$, to (3.101) can be equivalently characterized by the solution $x_1(\tau)$, $\tau \geq 0$, to the nonlinear autonomous dynamical system:

$$\dot{x}_1(\tau) = A(x_2(\tau))\ x_1(\tau),\quad x_1(0) = x_0,\quad \tau \geq 0 \tag{3.102}$$

$$\dot{x}_2(\tau) = 1,\quad x_2(0) = t_0. \tag{3.103}$$

It is clear from (3.102) and (3.103) that in spite of the fact that (3.101) is a linear system, its solution is inherently characterized by a nonlinear system of the forms (3.102) and (3.103). In the case where $A(\cdot)$ is periodic, (3.102) and (3.103) retain the same structure which generally makes stability analysis of such systems simpler. ■

**PROBLEMS:**

**3.1:** As was proved in Chapter 2, the linear system $\dot{x}(t) = Ax(t),\ t \geq 0$ is stable iff all eigenvalues of $A$ have negative real parts. But, in a time-varying case, that is, $\dot{x} = A(t)x(t),\ t \geq 0$ this is not true. A counter example would be as follows:

$$\dot{x} = \begin{bmatrix} -\alpha & e^{\gamma t} \\ 0 & -\beta \end{bmatrix} x, \quad \text{for} \quad \alpha,\beta,\gamma \text{ and } t > 0.$$

Show that the linear time-varying system is asymptotically stable if the eigenvalues of $A(t) + A^T(t)$ have negative real parts for all $t \geq 0$. Is this stability asymptotic? Is it global? Hint: Show that $V(x) = x^T x$ is a Lyapunov function in this case.

**3.2:** Another counter example for Problem 3.1 is given below:

$$\dot{x}(t) = \begin{pmatrix} -1+1.5\cos^2 t & 1-1.5\sin t\cos t \\ -1-1.5\sin t\cos t & -1+1.5\sin^2 t \end{pmatrix} x,$$

Why?

**3.3:** Determine if the following system has a stable equilibrium state. Indicate whether the stability is asymptotic and/or whether it is global.

$$\dot{x} = \begin{pmatrix} -1 & 2\sin t \\ 0 & -t-1 \end{pmatrix} x.$$

**3.4:** For each of the following linear systems, use a quadratic Lyapunov function to show that the origin is exponentially stable:

(a) $\dot{x} = \begin{pmatrix} -1 & \alpha(t) \\ \alpha(t) & -2 \end{pmatrix} x, \quad |\alpha(t)| \leq 1.$

(b) $\dot{x} = \begin{pmatrix} -1 & \alpha(t) \\ -\alpha(t) & -2 \end{pmatrix} x.$

(c) $\dot{x} = \begin{pmatrix} 0 & 1 \\ -1 & -\alpha(t) \end{pmatrix} x, \quad \alpha(t) \geq 2$

(d) $\dot{x} = \begin{pmatrix} -1 & 0 \\ \alpha(t) & -2 \end{pmatrix} x.$

In all cases, $\alpha(t)$ is continuous and bounded for all $t \geq 0$.

**3.5:** Verify if the following $g(t, x)$ is of class $W$.

$$g(t,x) = \begin{pmatrix} x_1^3 \sin t + 4x_2 \cos t \\ x_2^2 + \dfrac{1}{t+1} x_1^3 \end{pmatrix}.$$

**3.6:** Using Theorem 3.2 see if the ZES of the following system is stable. Is it uniformly asymptotically stable?

$$\dot{x} = A(t/\varepsilon)x,$$

$$A(t) = \begin{pmatrix} 0 & 1 \\ -1 & f(t) \end{pmatrix},$$

$$f(t) = a_0 + \sum_{i=1}^{n} a_i \sin(\omega_i t + \varphi_i) \text{ for } \varepsilon, a_i, \omega_i > 0.$$

■

# 4 Stability Analysis of Time-Delayed Systems

The dynamic models of many practical phenomena in physics, biology, economics, and so forth, involve significant time delays. For example, time delays occur in engineering systems that include transportation or propagation of materials and data, such as chemical reactors, combustion engines, and communication networks. The presence of time delay leads to serious deterioration of the stability of the system. Therefore, considerable attention has been devoted recently to the stability analysis and control of time-delayed systems. This chapter briefly introduces the preliminary ideas of time-delayed systems, especially the concept of stability and the well-known methods of stability analysis of time-delayed systems. In addition, some new results are presented in stability analysis and stabilization (see Appendix A4) of time-delayed systems.

## 4.1 PRELIMINARIES [h4]

In delay (retarded) differential equations which are also known as *hereditary systems*, *systems with aftereffects*, or *systems with time lags*, the state evolution rate $\dot{x}(t)$ depends not only on the present state but also on the history of the state. The delay differential equation can generally be written as follows:

$$\dot{x}(t) = f(x_t, u_t, t), \tag{4.1}$$

where $x(t) \in R^n$ and $u(t) \in R^m$, in addition $x_t$ and $u_t$ are functions defined by:

$$\begin{aligned} x_t(\theta) &= x(t+\theta), \\ u_t(\theta) &= u(t+\theta). \end{aligned} \qquad t \geq 0, \tag{4.2}$$

where $\theta \in J$ and $J \subset [-\infty, 0]$ is a given interval. $x_t$ can be considered as the argument of the function $x$ at the left-hand side of point $t$, observed from this point. The transition from $x$ to $x_t$ with $J = [-\tau, 0]$ is shown in Figure 4.1. The initial condition to solve the delay differential equation must be a function defined on $J$, which is known as *initial function*, so time-delayed systems are also called *infinite-dimensional systems*.

In the following, first the existence and uniqueness theorem for the initial value problem of the delay differential equation is given. Then, the basic stability definition is presented and finally Lyapunov-like theorems to prove the stability of time-delayed systems are described.

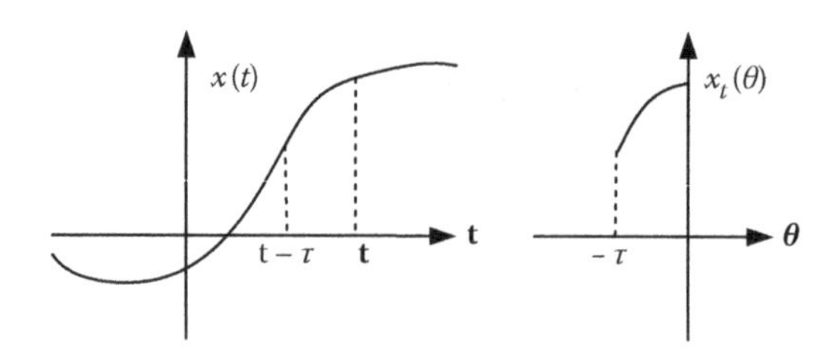


**FIGURE 4.1** Graphical interpretation of $x_t$.

---

**Theorem 4.1:**

Consider the following initial value problem:

$$\dot{x}(t) = f(x_t, t) \ , \ t \geq t_0, \qquad x_{t_0} = \phi. \tag{4.3}$$

Assume $\phi \in C[-\tau, t_0]$ and let $f : C[-\tau, t_0] \times [t_0, \infty) \to R^n$ be a continuous functional satisfying the Lipschitz condition in the first argument, then there is a $t_\phi \in (t_0, \infty]$ such that:

1. A solution $x$ of (4.3) exists on the interval $[t_0, t_\phi)$.
2. On any subinterval $[t_0, t_1]$ $[t_0, t_\phi)$, the solution is unique.
3. The solution $x$ depends continuously on $f$ and $\phi$. ■

Now, let $x_e = 0 \in C[-\tau, t_0]$ be an equilibrium state for the system described by (4.3). Assume that the solution of (4.3) is shown by $x(t; t_0, \phi)$ and $Q_\delta = \{v \in C[-\tau, 0] : \|v\| < \delta\}$. ■

---

**Definition 4.1:**

The equilibrium state $x_e = 0 \in C[-\tau, 0]$ is stable for a given $t_0 \in R$ if for any $\varepsilon > 0$ there is a $\delta = \delta(\varepsilon, t_0)$ such that for any initial function $\phi \in Q_\delta$ and $t \in [t_0, \infty)$ then, $\| x(t; t_0, \phi) \| \leq \varepsilon$. If, in the above definition, $\delta$ is independent of $t_0$, that is, $\delta = \delta(\varepsilon)$, such that $\| x(t; t_0, \phi) \leq \varepsilon$ for any initial function $\phi \in Q_\delta$ and $t_0 \in R$, $t \in [t_0, \infty)$, then the origin is uniformly stable. ■

---

**Definition 4.2:**

The equilibrium state is asymptotically stable for a given $t_0 \in R$ if it is stable and there is a $= (t_0) > 0$ such that $\lim_{t \to \infty} x(t; t_0, \phi) = 0$ for any initial function $\phi \in Q$ . If the equilibrium state is uniformly stable and there exists an $H > 0$ such that for any $\gamma > 0$ there is a $T(\gamma) > 0$ such that $\|x(t;t_0,\phi)\| \leq \gamma$ for any $t_0 \in R$, $t \geq t_0 + T(\gamma)$ and $\phi \in Q_H$, then the origin is uniformly asymptotically stable. ■

There are two different kinds of (asymptotic) stability for time-delayed systems: If the stability holds for different values of delays, the stability property is *delay-independent*; but if the stability is preserved for some values of delays and the system is unstable for other values, then the stability is *delay-dependent.*

The Lyapunov stability theorem is the underlying idea for stability analysis of ordinary differential equations. The following example demonstrates that using Lyapunov function $V(x,t)$, which depends only on the current state, is not sufficient to prove a time-delayed system's stability. Assume that the equilibrium state of the system $\dot{x}(t) = f(x_t)$ is uniformly asymptotically stable, then there exists a positive definite function $V$ such that:

$$\frac{\partial V}{\partial x} f < 0.$$

For every positive scalar $k$, the following holds:

$$\frac{\partial V}{\partial x}(k\,f) < 0,$$

so, the equilibrium state must be asymptotically stable for the system $\dot{x}(t) = k\,f(x_t)$. On the other hand, the simple delay differential equation $\dot{x}(t) = -k\,x(t-1)$ has the following characteristic equation:

$$s = -k\,e^{-s}$$

For $k < \pi/2$, all roots of the characteristic equation have negative real parts and for $k > \pi/2$ some roots have positive real parts, which means that the system is asymptotically stable only for some values of $k$ and unstable for other $k$.

**Example 4.1:**

Consider the following scalar nonlinear time-delayed system:

$$\dot{x}(t) = -x^3(t) - 2x^3(t-h)\sin(x(t-h)). \tag{4.4}$$

with the initial condition $\varphi(t) = 0.5\,sin(t)$, for $-h \leq t \leq 0$. The purpose of analysis is to investigate the effect of the delay value $h$ on the stability of the system (4.4). The simulation results demonstrate that the equilibrium state of the system (4.4) is stable for $h$ smaller than 15.

The simulation results show that the equilibrium state of the system (4.4) is locally stable with the derived conditions (see Figure. 4.2).

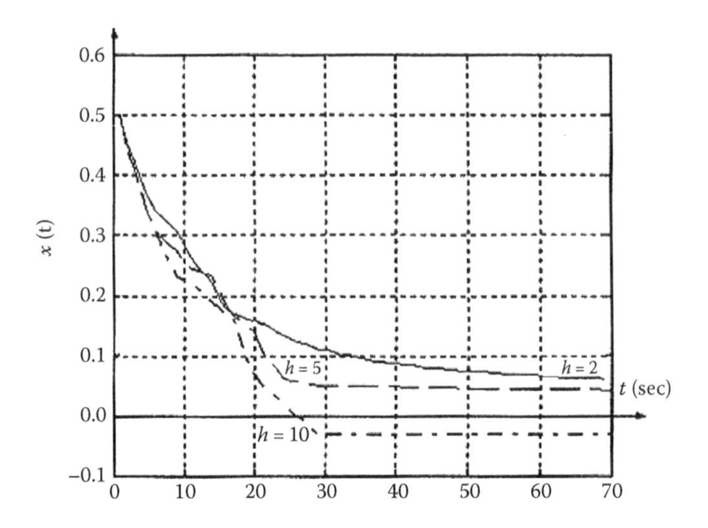


**FIGURE 4.2** The trajectories of the system (4.4) for $h \leq 10$.

Krasovskii was the first to generalize Lyapunov's method for time-delayed systems using $x_t$ as the state notion for delayed differential equations. He used functional instead of Lyapunov functions which are often called a Lyapunov–Krasovskii functional.

---

**Theorem 4.2 [h4]: (Lyapunov–Krasovskii Theorem)**

Assume $u, v, w : R \rightarrow R^+$ are nondecreasing continuous functions, such that $u(s)$, $v(s) > 0$ $\forall s$ and $u(0) = v(0) = 0$. If there exists the following continuous functional $V{:}R \times C[-\tau,0] \rightarrow R$ such that:

$$u(\|\phi(0)\|) \leq V(t,\phi) \leq v(\|\phi\|),$$

$$\dot{V}(t,\phi) \leq -w(\|\phi(0)\|),$$

then the equilibrium state of the system (4.3) is uniformly stable. If $w(s) > 0 \forall s$, then the equilibrium state is uniformly asymptotically stable.
Razumikhin introduces another method for stability analysis of time-delayed systems, which uses a function instead of a functional with additional condition. ■

---

**Theorem 4.3 [h4]: (Lyapunov–Razumikhin Theorem)**

Assume $u$,$v$ and $w{:}R \rightarrow R^+$ are nondecreasing continuous functions and $v : R \rightarrow R^+$ is strictly increasing, such that $u(s)$, and $v(s)$ both positive for $\forall s$ and $u(0) = v(0) = 0$.

If there is a continuous function $V: R\times R^n \rightarrow R$ such that $u(\|x\|) \leq V(t,x) \leq v(\|x\|)$, and:

$$\dot{V}(t,\phi(0)) \leq -w(\|\phi(0)\|), \text{ where: } \quad V(t+\theta,\phi(\theta)) \leq V(t,\phi(0)),$$

for $-\tau \leq \theta \leq 0$ then, the equilibrium state of the system (4.3) is uniformly stable. If $w(s)>0\ \forall s$, then the equilibrium state is uniformly asymptotically stable. ■

Note that the function $V$ in Razumikhin's theorem does not need to be nonincreasing along the system trajectories, but may indeed increase within a delay interval. The proof of Razumikhin's theorem is based on the fact that if $V$ is Razumikhin function, then $V^*(\psi) = \max_{-\tau\leq s\leq 0} V(\psi(s))$ is a Lyapunov–Krasovskii functional that is nonincreasing along the system trajectories.

## 4.2 STABILITY ANALYSIS OF LINEAR TIME-DELAYED SYSTEMS [p4]

In this section, the stability analysis of a linear time-delayed system is studied. A linear time-invariant system with time delay is defined as follows:

$$\dot{x}(t) = A_0 x(t) + A_1 x(t-h), \tag{4.5a}$$

$$\begin{aligned} y(t) &= Cx(t), \\ x(t) &= \quad (t), \qquad h \geq 0 \quad -h \leq t \leq 0, \end{aligned} \tag{4.5b}$$

where $A_i \in P^{n\times n}, i = 0,1,\ C \in P^{m\times n}$, and is a continuous function on the interval $[-h, 0]$. For stability analysis of the equilibrium state of such a system, the following Lyapunov–Krasovskii functional is considered:

$$V(t) = x^T(t)Px(t) + \int_{t-h}^{t} x^T(s)Qx(s)\,ds,$$

where $R$ and $Q$ are positive definite matrices. Taking the time derivative of $V(t)$ along the trajectory of (4.5a) yields:

$$\begin{aligned}
\dot{V}(t) &= (A_0x(t) + A_1x(t-h))^T Rx(t) + x^T(t)R(A_0x(t) + A_1x(t-h)) \\
&\quad + x^T(t)Qx(t) - x^T(t-h)Qx(t-h) \\
&= x^T(t)(A_0^T R + RA_0 + Q)x(t) + x^T(t-h)A_1^T Rx(t) \\
&\quad + x^T(t)RA_1x(t-h) - x^T(t-h)Qx(t-h) \\
&= \begin{pmatrix} x(t) & x(t-h) \end{pmatrix}^T \begin{pmatrix} A_0^T R + RA_0 + Q & RA_1 \\ A_1^T R & -Q \end{pmatrix} \begin{pmatrix} x(t) \\ x(t-h) \end{pmatrix},
\end{aligned}$$

The equilibrium state of system (4.5) is asymptotically stable if $\dot{V}(t) < 0$. Consequently, if $\dot{V}(t) < 0$, the matrix $\begin{pmatrix} A_0^T R + RA_0 + Q & RA_1 \\ A_1^T R & -Q \end{pmatrix}$ should be negative definite. ■

---

**Lemma 4.1 [b3]: (Schur Complement Lemma)**

Let $P, Q, S \in R^n$ be matrices such that $S > 0, \quad S = S^T$, then:

$$\begin{pmatrix} P & Q \\ Q^T & -S \end{pmatrix} < 0 \qquad P + QS^{-1}Q^T < 0.$$ ■

According to Lemma 4.1, the matrix inequality $\begin{pmatrix} A_0^T R + RA_0 + Q & RA_1 \\ A_1^T R & -Q \end{pmatrix} < 0$ can be represented by $A_0^T R + RA_0 + Q + RA_1 Q^{-1} A_1^T R < 0$.

By solving any of the above matrix inequalities, sufficient conditions for stability analysis of the system (4.5) are obtained.

### 4.2.1 Stability Analysis of Linear Time-Varying Time-Delayed Systems

Consider the following class of systems:

$$\dot{x}(t) = A_0(t)x(t) + A_1(t)x(t-h), \tag{4.6a}$$

$$y(t) = C(t)x(t), \qquad x(t) = \quad (t), \ h \geq 0, \quad -h \leq t \leq 0, \tag{4.6b}$$

where $A_i(t) : [t_0, \infty) \to R^{n \times n}, \ i = 0,1$ and $C(t) : [t_0, \infty) \to R^{m \times n}$ are continuous functions.is a continuous function on the interval $[t_0 - h, t_0]$.

---

**Definition 4.3 [p4]:**

Let $\alpha$ be a positive scalar, then the equilibrium state of the system (4.6a) is said to be $\alpha$-stable, if there is a function $\xi(.) : R^+ \to R^+$ such that for every $\phi(t)C([-h,0], R^n)$ the solution $x(t, \ )$ of the system satisfies:

$$\|x(t,\phi)\| \leq \xi(\|\phi\|)e^{-\alpha t}, \ \forall t \in R^+.$$ ■

---

**Theorem 4.4 [p4]:**

The equilibrium state of the linear time-varying time-delayed system (4.6a) is $\alpha$ stable if there is a symmetric matrix $P(t)>0$ such that any one of the following conditions holds:

Condition 1:

$$\dot{P}(t)+A_{0,\alpha}^T(t)P(t)+P(t)A_{0,\alpha}(t)+P(t)A_{1,\alpha}(t)A_{1,\alpha}^T(t)P(t)+2I=0.$$

Condition 2:

$$\begin{pmatrix} \dot{P}(t)+A_{0,\alpha}^T(t)P(t)+P(t)A_{0,\alpha}(t)+I & P(t)A_{1,\alpha}(t) \\ A_{1,\alpha}^T(t)P(t) & -I \end{pmatrix}<0,$$

where $A_{0,\alpha}(t)=A_0(t)+\alpha I$, and $A_{1,\alpha}(t)=e^{\alpha h}A_1(t)$.

The proof of this theorem and the proofs of the other theorems of this section are omitted here and the interested reader is referred to Phat Vu and Niamsup [p4]. Also, the proof of Theorem 4.8, which is the most complicated one among the theorems of this section, is given later in the section. ■

**Example 4.2:**

Consider the following scalar system:

$$\dot{x}(t)=-3e^t x(t)+0.5x(t-1),$$

$$y(t)=2x(t),$$

$$x(t)=0, \quad -1\le t\le 0,$$

for $\alpha=1$ and $P(t)=e^{-t}$, Condition 2 of Theorem 4.4 is satisfied, thus the equilibrium state of the above system is stable although some parameter of the system is unbounded.

The general case for a linear time-varying system with time delay is as follows:

$$\begin{aligned} \dot{x}(t)&=A_0(t)x(t)+A_1(t)x(t-\tau(t)), \\ y(t)&=C(t)x(t), \end{aligned} \tag{4.7a}$$

$$x(t)= \ (t), \quad \tau(t)\ge 0, \quad -\sup(\tau(t))\le t\le 0, \tag{4.7b}$$

where $\tau(t)\le h$ is the time-varying delay of the system and $h$ is constant.

For stability analysis of the equilibrium state of the system (4.7), the following theorems are proposed.

---

**Theorem 4.5 [p4]:**

Assume there exist constant values $h \geq 0, \overline{\tau} \geq 0$ such that for time-variable $\tau(t)$ one has: $\dot{\tau}(t) \leq h < 1,\ \tau(t) \leq \overline{\tau}$, then the equilibrium state of the system (4.7) is globally stable if the following matrix inequality holds:

Condition 1:

$$\begin{bmatrix} \dot{P}(t) + A_0^T(t)P(t) + P(t)A_0(t) + \dfrac{1}{1-h}I & P(t)A_1(t) \\ A_1^T(t)P(t) & -I \end{bmatrix} \leq 0.$$

Also, the global stability of the equilibrium state of the system (4.7) is obtained if for every positive semidefinite matrix $K(t)$, there is a symmetric positive definite matrix $P(t)$ satisfying the following Ricatti differential equation:

Condition 2:

$$\dot{P}(t) + A_0^T(t)P(t) + P(t)A_0(t) + P(t)A_1(t)A_1^T(t)P(t) + \frac{2}{1-h}I + K(t) = 0.$$ ■

---

**Theorem 4.6 [p4]:**

With the conditions of Theorem 4.5, the equilibrium state of the system (4.7) is globally asymptotically stable if there are symmetric positive definite matrix $P(t)$ and a positive definite matrix $K(t)$ such that one of the following conditions holds:

Condition 1:

$$\dot{P}(t) + A_0^T(t)P(t) + P(t)A_0(t) + P(t)A_1(t)A_1^T(t)P(t) + \frac{2}{1-h}I + K(t) = 0.$$

Condition 2:

$$\begin{bmatrix} \dot{P}(t) + A_0^T(t)P(t) + P(t)A_0(t) + \dfrac{1}{1-h}I & P(t)A_1(t) \\ A_1^T(t)P(t) & -I \end{bmatrix} < 0.$$ ■

Now the results are extended to multi-delay system. The multi-delay case for the system (4.7) is as follows:

$$\dot{x}(t) = A_0(t)x(t) + \sum_{i=1}^{r} A_i(t)x(t - h_i(t)), \tag{4.8a}$$

$$\begin{aligned} &y(t) = C(t)x(t), \\ &x(t) = \phi(t), \qquad \dot{h}_i(t) \leq \mu,\ \ 0 \leq h_i(t) \leq \tau,\ i = 1,\ldots,r, \end{aligned} \tag{4.8b}$$

where $\mu$ and $\tau$ are constants. For stability analysis of the system (4.8), the following theorems are presented and/or derived. ■

---

**Theorem 4.7 [p4]:**

With the conditions of Theorem 4.5, the equilibrium state of the system (4.8) is globally stable if there are symmetric positive definite and positive semidefinite matrices $P(t)$ and $K(t)$, respectively such that one of the following conditions holds:

Condition 1:

$$\dot{P}(t)+A_0^T(t)P(t)+P(t)A_0(t)+\sum_{i=1}^{r}P(t)A_i(t)A_i^T(t)P(t)+\frac{r}{1-\mu}I+K(t)=0.$$

Condition 2:

$$\begin{bmatrix} \chi & \alpha_1 & \alpha_2 & \cdots & \alpha_r \\ \alpha_1^T & -I & 0 & \cdots & 0 \\ \alpha_2^T & 0 & -I & \cdots & 0 \\ \vdots & \vdots & \vdots & \ddots & \vdots \\ \alpha_r^T & 0 & 0 & \cdots & -I \end{bmatrix} \leq 0.$$

where:

$$\chi = A_0^T(t)P(t)+\dot{P}(t)+P(t)A_0(t)+\frac{r}{1-\mu}I,$$

$$\alpha_i = P(t)A_i(t), \qquad 1 \leq i \leq r.$$

■

**Example 4.3:**

Consider the following scalar system:

$$\dot{x}(t) = -1.5x(t)+\sin(t)x(t-$$

$$x(t)=1, \quad for \quad -2 \leq t \leq 0.$$

If P = 1, then the Riccati differential equation condition of Theorem 4.7 holds. Therefore the equilibrium state of the system is asymptotically stable. This system is simulated and the following figure presents the resulted trajectory (Figure 4.3).

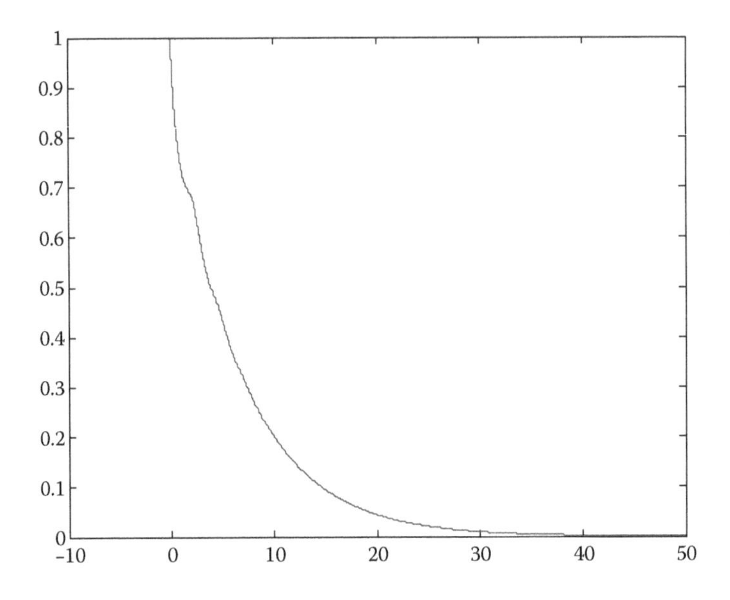


**FIGURE 4.3** The stable output trajectory of Example 4.3.

---

**Theorem 4.8 [p4]:**

With the conditions of Theorem 4.5, the equilibrium state of the system (4.8) is globally asymptotically stable if there are symmetric positive definite matrices $P(t)$ and $K(t)$ such that one of the following conditions holds:

Condition 1:

$$\dot{P}(t) + A_0^T(t)P(t) + P(t)A_0(t) + \sum_{i=1}^{r} P(t)A_i(t)A_i^T(t)P(t) + \frac{r}{1-\mu}I + K(t) = 0.$$

Condition 2:

$$\begin{bmatrix} \chi & \alpha_1 & \alpha_2 & \cdots & \alpha_r \\ \alpha_1^T & -I & 0 & \cdots & 0 \\ \alpha_2^T & 0 & -I & \cdots & 0 \\ \vdots & \vdots & \vdots & \ddots & \vdots \\ \alpha_r^T & 0 & 0 & \cdots & -I \end{bmatrix} < 0,$$

where:

$$\chi = A_0^T(t)P(t) + \dot{P}(t) + P(t)A_0(t) + \frac{r}{1-\mu}I,$$

$$\alpha_i = P(t)A_i(t), \qquad 1 \le i \le r.$$

■

**Proof:**

Consider the following Lyapunov–Krasovskii functional:

$$V(t) = x^T(t)P(t)x(t) + \frac{1}{1-h}\sum_{i=1}^{r}\int_{t-\tau i(t)}^{t}\|x(s)\|^2\,ds$$

Taking the derivative of $V(t)$ with respect to $t$ along the trajectory (4.8) and replacing (4.8) for $\dot{x}$, then one has:

$$\dot{V} \le \dot{x}^T Px + x^T P\dot{x} + \frac{r}{1-h}\|x\|^2 - \sum_{i=1}^{r}\|x(t-\tau_i(t))\|^2$$

$$\dot{V} \le x^T A_0^T Px + \sum_{i=1}^{r}\underline{x^T(t-\tau_i(t))A_{i1}^T Px} + x^T PA_0 x + \sum_{i=1}^{r}\underline{x^T PA_i x(t-\tau_i(t))}$$

$$+\frac{r}{1-h}\|x\|^2 - \sum_{i=1}^{r}\|x(t-\tau_i(t))\|^2 + x^T P\cdot x.$$

This inequality can be written as follows:

$$\dot{V}(t) = F^T(t)\begin{bmatrix} \chi & \alpha_1 & \alpha_2 & \cdots & \alpha_r \\ \alpha_1^T & -I & 0 & \cdots & 0 \\ \alpha_2^T & 0 & -I & \cdots & 0 \\ \vdots & \vdots & \vdots & \ddots & \vdots \\ \alpha_r^T & 0 & 0 & \cdots & -I \end{bmatrix}F(t),$$

where:

$$\chi = A_0^T(t)P(t) + \dot{P}(t) + P(t)A_0(t) + \frac{r}{1-h}I,$$

$$\alpha_i = P(t)A_i(t), \quad 1 \le i \le r,$$

$$F(t) = \begin{bmatrix} x(t) \\ x(t-\tau_1(t)) \\ \vdots \\ x(t-\tau_r(t)) \end{bmatrix}.$$

Thus, if the Condition 2 of Theorem 4.8 holds, then $\dot{V}(t) < 0$, thus the equilibrium state of the system is globally asymptotically stable. ■

To continue the proof and obtain the first condition, one needs the following lemma which is from Boyd et al. [b3].

---

**Lemma 4.2 [h5]:**

For any vector $V_1, V_2 \in R^n$ and positive definite matrix $M \in R^{n\times n}$, the following inequality holds:

$$2V_1^T V_2 \le V_1^T M V_1 V_2^T M^{-1} V_2$$

Using the above lemma with $M = I$ yields:

$$\dot{V}(t) \le x^T(t)(\dot{P}(t) + A_0^T(t)P(t) + P(t)A_0(t)$$

$$+\sum_{i=1}^{r} P(t)A_i(t)A_i^T(t)P(t) + \frac{r}{1-h}I)x(t) \triangleq -x^T(t)K(t)x(t).$$

thus, the zero equilibrium state of the system is globally asymptotically stable if Condition 1 holds. This competes the proof. ■

**Example 4.4:**

Consider the following scalar system:

$$\dot{x}(t) = (-3 + \sin(t))x(t) + e^{-t}x(t-2),$$

$$x(t) = 1, \text{for } -2 \le t \le 0.$$

If P = 1, then the Riccati differential equation condition of Theorem 4.8 holds. Therefore, the equilibrium state of the system is asymptotically stable. This system is simulated and the resulted trajectory is shown in Figure 4.4. ■

## 4.3 DELAY-DEPENDENT STABILITY ANALYSIS OF NONLINEAR TIME-DELAYED SYSTEMS [v2,v3]*

This section proposes a new procedure for constructing a Lyapunov–Krasovskii functional. These functionals can analyze locally the delay-dependent stability analysis of the equilibrium state for some nonlinear time-delay systems. The final results have been obtained using the linear matrix inequalities (LMI) method. This new method can also be applied to a class of nonlinear time-delay systems, which contain a stable linear term in the scalar case. So, this algorithm can be used for a special

* For stabilizing the special class of nonlinear delay systems, the interested reader is referred to Appendix A4.

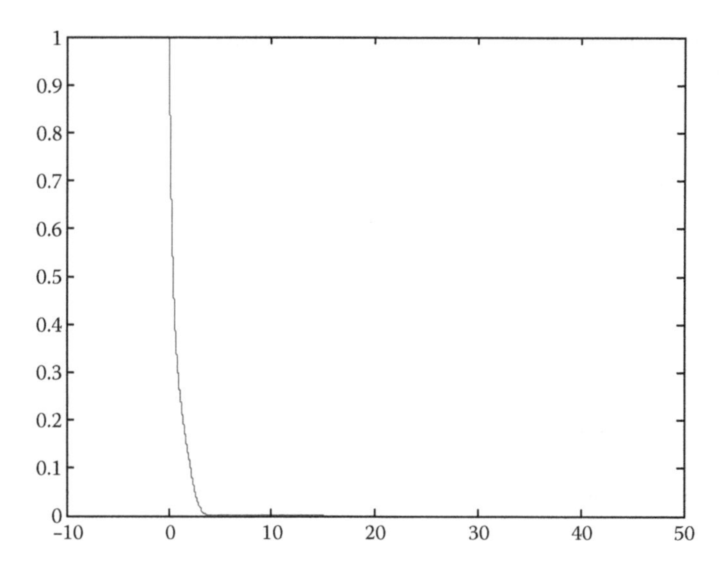


**FIGURE 4.4** The state trajectory of Example 4.4.

type of nonlinear time-delay systems having the same functions for its retarded and nonretarded parts. In this special case, the delay-independent stability results have been derived. It is possible to expand the procedure for other types of nonlinear time-delay systems. In this part, this method is applied to a more general class of nonlinear time-delay systems in which different functions could be considered for retarded and nonretarded parts. The final part of this section is dedicated to the delay-dependent stability analysis of scalar time-varying delay systems.

### 4.3.1 Vali–Nikravesh Method of Generating the Lyapunov–Krasovskii Functional for Delay-Dependent System Stability Analysis

Generally, the time-delay systems are represented as follows:

$$\begin{cases} \dot{x}(t) = f(t, x_t(\theta)), & t \geq 0, \\ x_t(\theta) = x(t+\theta), & -h \leq \theta \leq 0, \end{cases} \tag{4.9}$$

where $x \in R^n$ is the state vector of the system, $h \in R^+$ is the delay, $f$ is a function of class $C^1$, $x(t) = \varphi(t)$ is the initial condition of the system for the time interval $-h \leq \theta \leq 0$, and $x_t$ is of class $C^1$ for $t \geq 0$. Having an appropriate Lyapunov functional $V : R^+ \times C^1 \to R$, the sufficient conditions for stability of (4.9) are:

$$i)\ u\,(|\varphi(0)|) \leq V\,(t,\varphi), \tag{4.10}$$

$$ii)\ \dot{V}(t,\phi) \leq 0, \tag{4.11}$$

where $|\phi(0)| = \max_{-h\le\tau\le 0} |\phi(\tau)|$ and $u: R^+ \to R^+$ is a function for which $u(0) = 0$ and $u(s) > 0,\ \forall s>0$.

In order to establish the stability condition, the offered Lyapunov functional should meet the conditions (4.10) and (4.11) for $t \ge 0$.

In this section, the time-delay nonlinear dynamic equation is defined as follows:

$$\dot{x}(t) = BF(x(t)) + CG(x(t-h)), \qquad \forall\ t \ge 0, \tag{4.12}$$

where $x \in R^n$ is the state vector, $h \in R^+$ is the delay, and $B$ and $C$ are the diagonal matrices with real negative elements. It is obvious that the procedure is easily extendable to the case where $B$ and $C$ are constant matrices, not necessary diagonals. $x = 0$ is the equilibrium state of the system (4.12). $F = [f_1 \quad f_2 \quad \dots \quad f_n]^T$ and $G = [g_1 \quad g_2 \quad \dots \quad g_n]^T$ are vector functions with $n$ components, which are continuous functions of the state vector. It is assumed that the following relations hold for this system and for each component of a vector $y \in R^n - \{0\}$:

$$0 < \frac{f_i(y)}{y_i} \le m_{1i} \quad , \quad 0 < \frac{g_i(y)}{y_i} \le m_{2i}, \quad \forall y_i \ne 0 \quad , \quad i = 1,2,.\ n \tag{4.13}$$

where each $m_{1i}, m_{2i} (i = 1,2,\dots,n)$ are positive constants. It should be noted that due to the existence of the conditions (4.13), the system stability will be treated locally. Note that the conditions (4.13) are similar to the Lipschitz conditions. From relations (4.13), one has:

$$\begin{cases} |f_i(y)| \le m_{1i}|y_i|, \\ |g_i(y)| \le m_{2i}|y_i|, \end{cases} \qquad \forall \quad y_i \ne 0, \quad i = 1,2,\dots,n.$$

Regarding this fact that $F$, $G$, and $y$ are $n$ components vectors, it can be proven that the following relations will hold for any given negative definite matrix $K \in R^{n\times n}$:

$$\begin{cases} y^T KF \le F^T M_1^{-1} KF, & or & F^T Ky \le F^T K M_1^{-1} F, \\ y^T KG \le G^T M_2^{-1} KG, & or & G^T Ky \le G^T K M_2^{-1} G, \end{cases} \tag{4.14}$$

where $M_1 = diag(m_{1i})$ and $M_2 = diag(m_{2i})$ are positive diagonal matrices.

For the system (4.12), the Lyapunov functional $V: C_1 \to R$ is assumed to be as follows:

$$V(x_t) = V_1(x_t) + V_2(x_t),$$

where

$$V_1 = \frac{1}{2}\left[x(t) + \int_{t-h}^{t} R(t,u)G(x(u))\,du\right]^T \left[x(t) + \int_{t-h}^{t} R(t,u)G(x(u)\right.$$

$R(t,u)$ is generally an unknown nonconstant $n\times n$ matrix with time-dependent coefficients, which is determined as the algorithm is proceeded. One should first calculate the derivative of $V_1$ before reaching the final conclusion. Hence:

$$\dot{V}_1 = \frac{1}{2}\left[x(t) + \int_{t-h}^{t} R(t,u)G(x(u))\,du\right]^T \left[\int_{t-h}^{t} \frac{\partial R(t,u)}{\partial t} G(x(u))\,du + BF(x(t))\right.$$

$$\left. + R(t,t)G(x(t)) + (C - R(t,t-h))G(x(t-h))\right]$$

$$+ \frac{1}{2}\left[\int_{t-h}^{t} \frac{\partial R(t,u)}{\partial t} G(x(u))\,du + BF(x(t)) + R(t,t)G(x(t))\right.$$

$$\left. + (C - R(t,t-h))G(x(t-h))\right]^T \left[x(t) + \int_{t-h}^{t} R(t,u)G(x(u))\,du\right].$$

For simplicity, let:

$$R(t,t-h) = C. \tag{4.15}$$

In order to put $\dot{V}_1$ in quadratic form, let $\frac{\partial R}{\partial t}$ be in terms of $R$. Therefore, $\frac{\partial R}{\partial t}$ is considered as $\frac{\partial R(t,u)}{\partial t} = \Psi(t)R(t,u)$ where $\Psi(t)$ is an unknown matrix and for simplicity could be considered as a constant matrix $Q$. Applying condition (4.15), the matrix $R$ would be obtained as follows:

$$R(t,u) = Ce^{Q(u-t+h)}.$$

Let $Q$ be an unknown real matrix; its structure will be determined later on in the procedure for obtaining the stability conditions of (4.12). Two following relations are defined to simplify the results:

$$G_1(x(t)) \triangleq \int_{t-h}^{t} R(t,u)G(x(u))\,du = C\int_{t-h}^{t} e^{Q(u-t+h)}G(x(u))\,du,\ C_1 \triangleq Ce^{Qh}.$$

Then, $\dot{V}_1$ would be as follows:

$$\dot{V}_1 = x^T(t)C_1G(x(t)) + x^T(t)BF(x(t)) - G_1^T(x(t))Q^Tx(t)$$

$$+ G_1^T(x(t))C_1G(x(t)) + G_1^T(x(t))BF(x(t)) - G_1^T(x(t))QG_1(x(t)).$$

---

**Remark 4.1:**

$Q$ is the only constant matrix in $V_1$ and $\dot{V}_1$, which can be selected arbitrarily. Since, the sign of $\dot{V}_1$ is not important, the structure of $Q$ is chosen in such a way that $V_1$ meets the required constraint of (4.10). Therefore, if this constraint is satisfied by $V$ in addition to $V_1$, it must also be satisfied by $V_2$. So, the optimum choice for $V_2$ can be made according to the following Lemma 4.3: ■

---

**Lemma 4.3:**

For the continuous function $k(t)$, one has the following equation:

$$\frac{d}{dt}\left[\int_{-h}^{0}\int_{t+s}^{t} k(u)\,du\,ds\right] = h\,k(t) - \int_{-h}^{0} k(t+s)\,ds = h\,k(t) - \int_{t-h}^{t} k(\tau)\,d\tau,$$

in which $h$ is some positive constant. ■

The following relation for $V_2$ could be considered according to the above lemma:

$$V_2 = \int_{-h}^{0}\int_{t+s}^{t} G^T(x(u))Pe^{Qs}G(x(u))\,du\,ds,$$

where $P$ is a constant unknown $n \times n$ matrix, which is assumed to be positive definite. Thus, the derivative of $V_2$ would be as follows:

$$\dot{V}_2 = G^T(x(t))PQ^{-1}\left[I - e^{-Qh}\right]G(x(t)) - \int_{t-h}^{t} G^T(x(u))Pe^{Q(u-t)}G(x(u))\,du. \quad (4.16)$$

Now, the last term of the above relation should be converted to the quadratic form of $G_1$.

---

**Lemma 4.4 [b3]:**

Linear Matrix Inequalities (LMI): Schur complement; let $T{:}V \to W$ be partitioned as follows:

$$T(x) = \begin{bmatrix} T_{11}(x) & T_{12}(x) \\ T_{21}(x) & T_{22}(x) \end{bmatrix}$$

$V$ is the vector space and $S$ is the set of matrices which:

$$W = \{M \mid \exists\, n > 0, M = M^T \in R^{n \times n}\}.$$

It is also assumed that $T_{11}(x)$ is an $r \times r$ nonsingular matrix. So, the matrix $N := T_{22} - T_{21}T_{11}^{-1}T_{12}$ is called the *Schur complement of* $T_{11}$ *in* $T$. Then, $T(x) > 0$ if and only if:

$$\begin{cases} T_{11}(x) > 0, \\ T_{22}(x) - T_{12}(x)\left[T_{11}(x)\right]^{-1} T_{21}(x) > 0. \end{cases} \tag{4.17}$$

Note that the second inequality in (4.17) is a nonlinear matrix inequality in $x$. ■

Applying Lemma 4.4 into the second term of equation (4.16), one may set up the following matrix:

$$A = \begin{bmatrix} A_1 & A_2 \\ A_3 & A_4 \end{bmatrix}, \tag{4.18}$$

in which:

$$A_1 \equiv \left(\int_{t-h}^{t} e^{Qu} du\right) P^{-1} e^{Q^T t}, \quad A_2 \equiv \int_{t-h}^{t} G^T(x(u)) e^{Q^T u} du, \quad A_3 \equiv \int_{t-h}^{t} e^{Qu} G(x(u)) du,$$

and

$$A_4 \equiv \int_{t-h}^{t} G^T(x(u)) P e^{Q(u-t)} G(x(u)) du.$$

If $Q$ and $P$ are diagonal matrices with nonzero elements, then the $A$ matrix will be a symmetrical matrix and thus the conditions of Lemma 4.4 will be satisfied. It is possible to show that the $A$ matrix is positive definite. Thus, the relations (4.17) are satisfied for the $A$ matrix. In other words, applying Lemma 4.4 for the $A$ matrix, then:

$$\int_{t-h}^{t} G^T(x(u)) P e^{Q(u-t)} G(x(u)) du \geq \left(\int_{t-h}^{t} G^T(x(u)) e^{Qu} du\right) \times \left(e^{-Qt} P\left[I - e^{-Qh}\right]^{-1} Q e^{-Qt}\right)\left(\int_{t-h}^{t} e^{Qu} G(x(u)) du\right).$$

Thus, one has:

$$\int_{t-h}^{t} G^T(x(u)) P e^{Q(u-t)} G(x(u)) du \geq G_1^T(x(t)) \left[C^{-1} e^{-Qh} P \times \left[I - e^{-Qh}\right]^{-1} Q e^{-Qh} C^{-1}\right] G_1(x(t)).$$

Therefore:

$$\dot{V}_2 \le G^T(x(t))PQ^{-1}\left[I-e^{-Qh}\right]G(x(t))-G_1^T(x(t))\times$$
$$\left[C^{-1}e^{-Qh}P\left[I-e^{-Qh}\right]^{-1}Qe^{-Qh}C^{-1}\right]G_1(x(t).$$

Finally, the derivative of the Lyapunov functional $V$ will be obtained as follows:

$$\dot{V} \le x^T(t)C_1G(x(t))+x^T(t)BF(x(t))-G_1^T(x(t))Q\,x(t)$$
$$+G_1^T(x(t))C_1G(x(t))+G_1^T(x(t))BF(x(t))+G^T(x(t))PQ^{-1}\times$$
$$\left[I-e^{-Qh}\right]G(x(t))-G_1^T(x(t))\left[Q+C^{-1}e^{-Qh}P\left[I-e^{-Qh}\right]^{-1}Qe^{-Qh}C^{-1}\right]G_1(x(t)).$$

If the diagonal elements of the $Q$ matrix are negative, then the following relations are held using (4.14):

$$x^T(t)C_1G(x(t)) \le G^T(t)M_2^{-1}C_1G(x(t)),$$
$$x^T(t)BF(x(t)) \le F^T(t)M_1^{-1}BF(x(t)),$$
$$-G_1^T(x(t))Qx(t) \le -G_1^T(x(t))QM_2^{-1}G(x(t)) \quad (4.20)$$

By substituting (4.20) into (4.19), one has:

$$\dot{V} \le G_1^T(x(t))\left[C_1-QM_2^{-1}\right]G(x(t))+G^T(x(t))\left[M_2^{-1}C_1+PQ^{-1}\left[I-e^{-Qh}\right]\right]$$
$$\times G(x(t))-G_1^T(x(t))\left[Q+C^{-1}e^{-Qh}P\left[I-e^{-Qh}\right]^{-1}Qe^{-Qh}C^{-1}+\frac{1}{4}M_1B\right]G_1(x(t).$$

In order for this inequality to become a quadratic form in terms of $G$ or $G_1$, one should have the following inequalities:

$$i)\ -\left[Q+C^{-1}e^{-Qh}P\left[I-e^{-Qh}\right]^{-1}Qe^{-Qh}C^{-1}+\frac{1}{4}M_1B\right] \le 0,$$
$$ii)\ M_2^{-1}C_1+PQ^{-1}\left[I-e^{-Qh}\right] \le 0 \quad (4.21)$$

Finally, $\dot{V}$ can be expressed as follows:

$$\dot{V} \le G^T(x(t))\Bigg[ M_2^{-1}C_1 + PQ^{-1}\left[I - e^{-Qh}\right] + \frac{1}{4}\left(C_1 - M_2^{-1}Q\right)\times \left(Q + C^{-1}e^{-Qh}P\left[I - e^{-Qh}\right]^{-1}Qe^{-Qh}C^{-1} + \frac{1}{4}M_1B\right)^{-1}\left(C_1 - QM_2^{-1}\right)\Bigg]G(x(t)), \tag{4.22}$$

or:

$$\dot{V} \le \lambda_M\Bigg[ M_2^{-1}C_1 + PQ^{-1}\left[I - e^{-Qh}\right] + \frac{1}{4}\left(C_1 - M_2^{-1}Q\right)\times \left(Q + C^{-1}e^{-Qh}P\left[I - e^{-Qh}\right]^{-1}Qe^{-Qh}C^{-1} + \frac{1}{4}M_1B\right)^{-1}\left(C_1 - QM_2^{-1}\right)\Bigg]\|G(x(t))\|_2^2, \tag{4.23}$$

where $\lambda_M[D]$ is the maximum eigenvalue of the square matrix $D$.

---

**Remark 4.2:**

In order to reach (4.22) (or [4.23]) and to obtain the least necessary conditions for $\dot{V}$ being negative definite, it is necessary that (4.21) is satisfied. The limits on matrix $P$ can be determined using (4.21). Undoubtedly, in order to find the exact limits of $P$ and $Q$, one needs to analyze (4.21) more deeply. This investigation is done for the scalar case in the next section. ■

Choosing $P$ and $Q$ from relations (4.21) properly, the system (4.12) is stable in the sense of Lyapunov if:

$$\lambda_M\Bigg[ M_2^{-1}C_1 + PQ^{-1}\left[I - e^{-Qh}\right] + \frac{1}{4}\left(C_1^T - M_2^{-1}Q\right)\times \left(Q + C^{-T}e^{-Qh}P\left[I - e^{-Qh}\right]^{-1}Qe^{-Qh}C^{-1} + \frac{1}{4}M_1B^T\right)^{-1}\Bigg]\left(C_1 - QM_2^{-1}\right) \le 0. \tag{4.24}$$

Using the above inequality, one can obtain the acceptable bound for the $M$ matrix (region of attraction) for the local stability properly of the system (4.12). The theoretical investigation and the simulation results showed that there are at least proper $P$ and $Q$ matrices for the stable system for which (4.24) is satisfied.

**Example 4.5:**

As an illustration, consider the following system:

$$\begin{bmatrix} \dot{x}_1(t) \\ \dot{x}_2(t) \end{bmatrix} = \begin{bmatrix} -4 & -2 \\ -2 & -3 \end{bmatrix} \begin{bmatrix} x_1^3(t) \\ \sin(x_2(t)) \end{bmatrix} + \begin{bmatrix} -3 & 2 \\ -4 & 1 \end{bmatrix} \begin{bmatrix} x_1(t-h) \\ x_2^3(t-h) \end{bmatrix}, \quad t \geq 0 \tag{4.25}$$

in which the initial conditions are assumed to be as follows:

$$\phi(t) = \begin{bmatrix} 0.01 \\ -0.02 \end{bmatrix} \sin(t), \quad -h \leq t \leq 0. \tag{4.26}$$

The objective here is to investigate the effect of the delay $(h)$ on the stability of the system given in Equation (4.25) and compare the simulation results with the one proposed by the stability theorem. By carrying out the simulation, one can observe that the system given in (4.25) would be stable for the $h$ up to 2.32 second. For example, if $h = 0.2$, we would choose $q = -0.1$ with $m_1 = 0.5$ and $m_2 = 0.2$ and obtain the bounds for $p$.

Choosing $p = 2$, the time derivative of the Lyapunov functional is as follows:

$$\dot{V} \leq [-2.543] \| F(x(t)) \|^2 ,$$

which is a good indication for stability of the zero equilibrium state of the system. $F$ is according to (4.12). The simulation results show that the system given in (4.25) is locally stable with the above conditions (Figure 4.5).

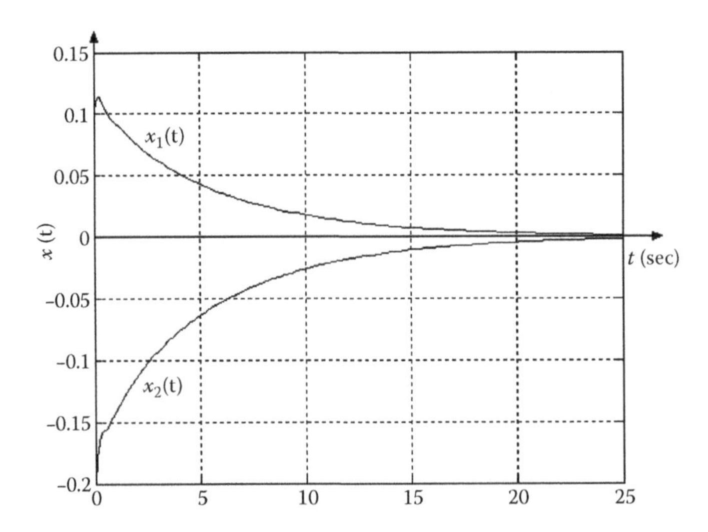


**FIGURE 4.5** The trajectories of the system (4.25) with $h = 0.2$. (The simulation results for Example 4.5.)

**Example 4.6:**

The scalar time-delay nonlinear dynamic equation is considered as follows:

$$\dot{x}(t) = bf(x(t)) + cf(x(t-h)), \quad t \geq 0, \tag{4.27}$$

where $x \in R$ is the state variable, $h \in R^+$ is the delay and $b,c$ are the real parameters of the system (4.27) with $b + c < 0$. $x = 0$ is the equilibrium point of (4.27), $f$ is a continuous function and the relation $0 \leq \frac{f(y)}{y}$ (for each $y \in R - \{0\}$) is satisfied for (4.27) (Lipschitz condition). $V_1$ and $V_2$ are considered as follows:

$$V_1 = \frac{1}{2}\left[x(t) + \int_{t-h}^{t} r(t,u)f(x(u))du\right]^2,$$

$$V_2 = \int_{-h}^{0}\int_{t+s}^{t} pe^{qs}f^2(x(u))du\,ds.$$

Applying the procedure from the preceding section, one has:

$$\dot{V}_1 = b_1x(t)f(x(t)) - qx(t)f_1(x(t)) + b_1f(x(t))f_1(x(t)) - qf_1^2(x(t)),$$

$$\dot{V}_2 = pf^2(x(t))\int_{t-h}^{t} e^{qs}\,ds - pe^{-qt}\int_{t-h}^{t} e^{qu}f^2(x(u))du.$$

Using the Schwarz integral inequality (instead of the LMI algorithm), one has:

$$\left[\int_{t-h}^{t} e^{qu}f(x(u))du\right]^2 \leq \int_{t-h}^{t} e^{qu}\,du \times \int_{t-h}^{t} e^{qu}f^2(x(u))du,$$

thus:

$$\int_{t-h}^{t} \geq \frac{\left[\int_{t-h}^{t} e^{qu}f(x(u))du\right]^2}{\int_{t-h}^{t} e^{qu}\,du} = \frac{\int_{t-h}^{t} e^{qt-2qh}}{c^2\int_{-h}^{0} e^{qs}\,ds}f_1^2(x(t)).$$

Therefore, the derivative of $V_2$ can be obtained as follows:

$$\dot{V}_2 \leq \frac{p}{q}(1-e^{-qh})f^2(x(t)) - \frac{pqe^{-2qh}}{c^2(1-e^{-qh})}f_1^2(x(t)).$$

Adding $\dot{V}_1$ to $\dot{V}_2$ yields:

$$\begin{aligned}\dot{V} \leq\ & b_1x(t)f(x(t)) - qx(t)f_1(x(t)) + b_1f(x(t))f_1(x(t)) \\ & + \frac{p}{q}(1-e^{-qh})f^2(x(t)) - q\left(\frac{pe^{-2qh}}{c^2(1-e^{-qh})} + 1\right)f_1^2(x(t)).\end{aligned}$$

If $b_1 < 0$ and $qc > 0$, then by applying the assumption $0 \le \frac{f(y)}{y} \le m$ one has:

$$\dot{V} \le \left[\frac{b_1}{m} + \frac{p}{q}(1 - e^{-qh})\right] f^2(x(t)) + \left[b_1 - \frac{q}{m}\right] f(x(t)) f_1(x(t))$$
$$- q\left[\frac{pe^{-2qh}}{c^2(1 - e^{-qh})} + 1\right] f_1^2(x(t)).$$

In order to satisfy the conditions $b_1 < 0$ and $qc > 0$, note the following:

$$b_1 = b + ce^{qh} < 0$$
$$\begin{cases} if \quad c > 0, b < 0 \qquad e^{qh} < -\dfrac{b}{c} \\ \qquad\qquad 0 < q < \dfrac{1}{h}\ln\left(-\dfrac{b}{c}\right) \quad that \quad |c| < |b|, \\ if \quad c < 0, b > 0 \qquad e^{qh} > -\dfrac{b}{c} \\ \qquad\qquad 0 > q > \dfrac{1}{h}\ln\left(-\dfrac{b}{c}\right) \quad that \quad |c| > |b|, \\ if \quad c < 0, b < 0 \qquad q \in \Re^{-}. \end{cases} \tag{4.28}$$

One of the results that is derived from the above inequalities is $b + c < 0$ (one of the assumptions of the problem). In order to convert $\dot{V}$ to the quadratic form in terms of $f$ or $f_1$, one should have the following inequalities:

$$i) \quad q\left[\frac{pe^{-2qh}}{c^2(1 - e^{-qh})}\right] + 1 \ \ge \ 0 \qquad\qquad p \ge c^2(e^{qh} - e^{2qh}),$$

$$ii) \quad \frac{b_1}{m} + \frac{p}{q}(1 - e^{-qh}) \ \le \ 0 \qquad\qquad p \le \frac{qb_1}{m(e^{-qh} - 1)}.$$

If inequality ($i$) is satisfied, then the following inequality should hold:

$$-q\left(\frac{pe^{-2qh}}{c^2(1 - e^{-qh})} + 1\right) f_1^2(x(t)) + \left(b_1 - \frac{q}{m}\right) f(x(t)) f_1(x(t))$$
$$\le \frac{\left(b_1 - \frac{q}{m}\right)^2}{4q\left(\frac{pe^{-2qh}}{c^2(1 - e^{-qh})} + 1\right)} f^2(x(t)),$$

Finally, $\dot{V}$ is obtained as follows:

$$\dot{V} \le \left[\frac{b_1}{m} + \frac{p}{q}(1 - e^{-qh}) + \frac{c^2(b_1 - \frac{q}{m})^2(1 - e^{qh})}{4q[pe^{-2qh} + c^2(1 - e^{-qh})]}\right] f^2(x(t)).$$

If $\dot{V}$ is negative semidefinite, (or the equilibrium state of the system (4.27) would be stable in the sense of Lyapunov), then:

$$\left[\frac{b_1}{m} + \frac{p}{q}(1 - e^{-qh}) + \frac{c^2(b_1 - \frac{q}{m})^2(1 - e^{qh})}{4q[pe^{-2qh} + c^2(1 - e^{-qh})]}\right] \le 0. \tag{4.29}$$

From (*i*) and (*ii*), there exists $p$ such that:

$$c^2(e^{qh} - e^{2qh}) \le p \le \frac{qb_1}{m(e^{-qh} - 1)}. \tag{4.30}$$

So, $q$ is derived properly from (4.28) and (4.39) initially and then $p$ will be determined from (4.30). Applying (4.30) in (4.29) yields:

$$c^2\left(e^{qh} - 1\right) + \frac{qb_1}{m(e^{qh} - 1)} - |c|\left(b_1 + \frac{q}{m}\right) \le 0. \tag{4.31}$$

The inequality (4.31) may be reduced to:

$$\begin{aligned} &(b + c)\left(-c + \frac{q}{m(e^{qh} - 1)}\right) \le 0, && \text{for} \quad c > 0, \\ &\left(c\left(e^{2qh} - 1\right) + b\right)\left(c + \frac{q}{m(e^{qh} - 1)}\right) \le 0, && \text{for} \quad c \le 0. \end{aligned} \tag{4.32}$$

The derivation of the stability conditions from (4.32) is obviously simpler than what follows from (4.29). For example, when $c > 0$, according to the constraint of the problem ($b + c < 0$), the inequality (4.29) would be converted to:

$$-c + \frac{q}{m(e^{qh} - 1)} \ge 0 \qquad \frac{q}{e^{qh} - 1} \ge mc,$$

If the above inequalities can be established with the selected $q$ from (4.28) and (4.30), the amount of delay, the coefficients of the system and a proper $m$, then the zero equilibrium state of the system (4.27) would be stable in the sense of Lyapunov. In other words, the amount of $m$ determines the boundary of stability region of the system.

**Remark 4.3:**

An important note for the proposed method is that when $c < 0$, one may derive delay-independent stability results for the system (4.27) by letting $q \to -\infty$. It is shown that if $b \le c$, then the system (4.27) is delay-independent Lyapunov stable.■

**Example 4.7:**

Consider the following dynamic equation:

$$\dot{x}(t) = x^3(t) - 20x^3(t-h), \qquad t \ge 0. \tag{4.33}$$

The initial conditions of the system are assumed to be as follows:

$$\phi(t) = 0.5e^{5t}, \qquad -h \le t \le 0.$$

The objective here is to investigate the effect of delay ($h$) on stability of system (4.33) (Figure 4.6) and compare the simulation results with the proposed method. By carrying out a simulation, one can observe that the system (4.33) would be stable for $h$, up to $h = 0.43$. For example, if $h = 0.2$, we choose $q = -0.1$ and obtain the range of $p$ as follows:

$$7.76 \le p \le 306.97.$$

Choosing $p = 50$, the Lypunov functional derivative is as follows:

$$\dot{V} \le \left[\frac{-18.604}{m} + 10 + 0.454\left(-18.604 + 0.1\frac{1}{m}\right)^2\right] f^2(x(t))$$

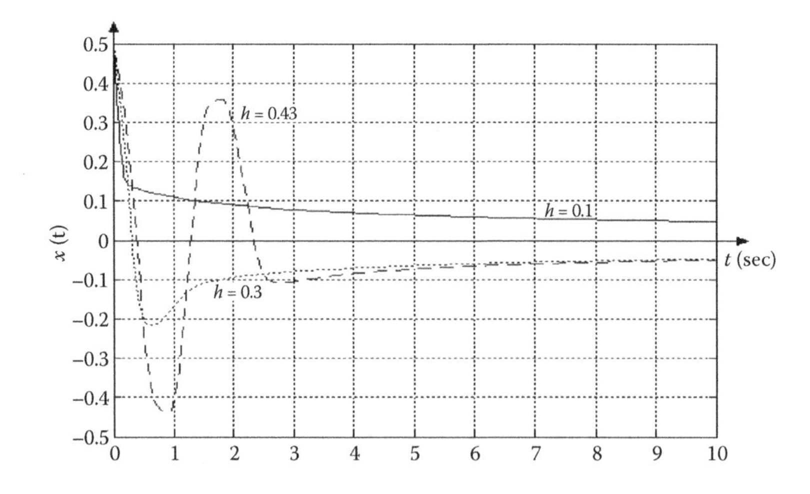


**FIGURE 4.6** The trajectories of the system (4.33) with $h \le 0.43$.

If the value of $m$ is within the interval $0 < m \le 0.3026$, then $\dot{V}$ is negative definite. In order to prove such a condition, the initial value of the state of the system (4.33) should be within the following range:

$$0 < \frac{x^3}{x} \le 0.3026 \qquad x(0) \le 0.56$$

The simulation results show that the zero equilibrium state of the system (4.33) is stable with the above preset values.

### Example 4.8:

Consider the following time-varying scalar system:

$$\dot{x}(t) = c(t)f(x(t-h)), \qquad t \ge 0. \tag{4.34}$$

In order to analyze the stability of this system, one has to set:

$$V_1(t, x_t) = \frac{1}{2}\left[x(t) + f_1(x(t))\right]^2,$$

where:

$$f_1(x(t)) = \int_{t-h}^{t} r(t,u)f(x(u))\,du = e^{-q(t-h)}\int_{t-h}^{t} c(u+h)e^{q(u)}f(x(u))\,du,$$

and the function $q(t)$is defined as follows:

$$q(t) = \alpha\int_{0}^{t+h} c(u+h)\,du, \quad c(t) < 0, \quad \alpha > 0.$$

Denote $B(t) \triangleq \int_{t}^{t+h} c(u+h)\,du$, then we have:

$$\dot{V}_1(t, x_t) = c(t+h)\Big[e^{\alpha B(t)}x(t)f(x(t)) - \alpha x(t)f_1(x(t)) + e^{\alpha B(t)}f(x(t))f_1(x(t)) - \alpha f_1^2(x(t))\Big].$$

Let:

$$V_2(t, x_t) = p\int_{-h}^{0}\int_{t+s}^{t} c(|s-u|+h)\ \ e^{-q(|s-u|-h)}c(u+h)\ e^{2q(u)-q(u-h)}f^2(x(u))\,du\,ds,$$

then:

$$\dot{V}_2(t,x_t) = p\left[f^2(x(t))c(t+h)\ e^{q(t)+\alpha B(t)}\int_{-h}^{0} c(|s-t|+h)e^{-q(|s-t|-h)}\,ds\right.$$
$$\left. - c(t+h)e^{-q(t-h)}\int_{-h}^{0} c(t+s+h)e^{2q(t+s)-q(t+s-h)}f^2(x(t+s))\,ds\right].$$

In other words:

$$\dot{V}_2(t,x_t) = pc(t+h)\left[f^2(x(t))e^{q(t)+\alpha B(t)}\int_{t}^{t+h} c(u+h)e^{-q(u-h)}\,du\right.$$
$$\left. + e^{-q(t-h)}\int_{t-h}^{t} |c(u+h)|e^{2q(u)-q(u-h)}f^2(x(u))\,du\right].$$

Applying the Schwarz integral inequality, one has:

$$pc(t+h)e^{-q(t-h)}\int_{t-h}^{t} |c(u+h)|e^{2q(u)-q(u-h)}f^2(x(u))\,du$$
$$\le\ pc(t+h)e^{-q(t-h)}\frac{\left[\int_{t-h}^{t} c(u+h)e^{q(u)}f(x(u))\,du\right]^2}{\int_{t-h}^{t} |c(u+h)|e^{q(u-h)}\,du}$$
$$\le\ pc(t+h)e^{-q(t-h)}\frac{\alpha}{e^{-\alpha B(t-h)}-1}f_1^2(x(t)),$$

then:

$$\dot{V}_2(t,x_t)\ \le\ pc(t+h)\left[\frac{1}{\alpha}e^{\alpha B(t)}\left(e^{\alpha B(t)}-1\right)f^2(x(t)) + \frac{\alpha}{e^{-\alpha B(t-h)}-1}f_1^2(x(t))\right]$$

Suppose that the constraint on $B(t)$ is as follows:

$$b_s \triangleq \underset{t>-h}{Sup}|B(t)| = \underset{t>-h}{Sup}\int_{t}^{t+h} |c(u+h)|\,du < \infty,$$

then:

$$\dot{V}(t,x_t)\ \le\ c(t+h)\left[e^{-\alpha b_s}x(t)f(x(t)) + \frac{p}{\alpha}e^{-\alpha b_s}\left(e^{-\alpha b_s}-1\right)f^2(x(t))\right.$$
$$\left. - \alpha x(t)f_1(x(t)) + e^{-\alpha b_s}f(x(t))f_1(x(t)) + \alpha\left(\frac{pe^{-\alpha b_s}}{1-e^{-\alpha b_s}}-1\right)f_1^2(x(t))\right].$$

Because the coefficients of $f$ and $f_1$ are negative, it is possible to apply the assumption $0 < \frac{f(y)}{y} \le m$ for the above relation, thus:

$$\dot{V}(t,x_t) \le c(t+h)\left[\left(\frac{e^{-\alpha b_s}}{m} + \frac{p}{\alpha}e^{-\alpha b_s}\left(e^{-\alpha b_s} - 1\right)\right)f^2(x(t)) + \left(e^{-\alpha b_s} - \alpha\right)f(x(t))f_1(x(t)) + \alpha\left(\frac{pe^{-\alpha b_s}}{1-e^{-\alpha b_s}} - 1\right)f_1^2(x(t))\right],$$

$$\dot{V}(t,x_t) \le |c(t+h)|\left[\left(\frac{p}{\alpha}e^{-\alpha b_s}\left(1-e^{-\alpha b_s}\right) - \frac{e^{-\alpha b_s}}{m}\right)f^2(x(t)) + \left(\alpha - e^{-\alpha b_s}\right)f(x(t))f_1(x(t)) - \alpha\left(\frac{pe^{-\alpha b_s}}{1-e^{-\alpha b_s}} - 1\right)f_1^2(x(t))\right].$$

To present $\dot{V}$ as a quadratic form in terms of $f$ or $f_1$, one should have the following inequalities:

$$i)\quad \alpha\left(\frac{pe^{-\alpha b_s}}{1-e^{-\alpha b_s}} - 1\right) \ge 0, \qquad p \ge e^{\alpha b_s} - 1,$$

$$ii)\quad \frac{p}{\alpha}e^{-\alpha b_s}(1-e^{-\alpha b_s}) - \frac{e^{-\alpha b_s}}{m} \le 0, \qquad p \le \frac{\alpha}{m(1-e^{-\alpha b_s})}, \tag{4.35}$$

similar to the last section, $\dot{V}$ is obtained as follows:

$$\dot{V}(t,x_t) \le |c(t+h)|\left[\frac{p}{\alpha}e^{-\alpha b_s}\left(1-e^{-\alpha b_s}\right) - \frac{e^{-\alpha b_s}}{m} + \frac{\left(\frac{\alpha}{m} - e^{-\alpha b_s}\right)^2\left(1-e^{-\alpha b_s}\right)}{4\alpha\left(pe^{-\alpha b_s} + e^{-\alpha b_s} - 1\right)}\right]f^2(x(t))$$

The system (4.34) is uniformly stable if:

$$\frac{p}{\alpha}e^{-\alpha b_s}\left(1-e^{-\alpha b_s}\right) - \frac{e^{-\alpha b_s}}{m} + \frac{\left(\frac{\alpha}{m} - e^{-\alpha b_s}\right)^2\left(1-e^{-\alpha b_s}\right)}{4\alpha\left(pe^{-\alpha b_s} + e^{-\alpha b_s} - 1\right)} \le 0. \tag{4.36}$$

From (i) and (ii), the range of $p$ can be obtained as follows:

$$e^{\alpha b_s} - 1 \le p \le \frac{\alpha}{m\left(1-e^{-\alpha b_s}\right)}. \tag{4.37}$$

So, at first, we select a proper positive $\alpha$ from (4.35) and then choose a proper $p$ from (4.37). By manipulating the relation (4.36), one has:

$$p^2 - \left(\frac{\alpha}{m(1-e^{-\alpha b_s})} + \left(e^{\alpha b_s} - 1\right)\right)p + \frac{1}{4}e^{2\alpha b_s}\left(\frac{\alpha}{m} + e^{-\alpha b_s}\right)^2 \leq 0. \tag{4.38}$$

By deriving more details from the above inequality, it is straightforward to show that the inequality (4.38) holds for some $p$ if $\alpha = \frac{m}{2}$. Using this relation, the inequality (4.37) can be simplified to the following form:

$$e^{2\alpha b_s} - \frac{5}{2}e^{\alpha b_s} + 1 \leq 0.$$

The above inequality holds if $0.5 \leq e^{\alpha b_s} \leq 2$. For example, if $e^{\alpha b_s} = 2$ then the range of $p$ can be obtained as follows:

$$e^{\alpha b_s} - 1 \leq p \leq \frac{\alpha}{m\left(1 - e^{-\alpha b_s}\right)} \qquad 1 \leq p \leq 2$$

Choosing $p = 1$, the inequality (4.36) will become as follows:

$$\begin{aligned}\dot{V} &\leq \frac{p}{\alpha}e^{-\alpha b_s}\left(1 - e^{-\alpha b_s}\right) - \frac{e^{-\alpha b_s}}{m} + \frac{\left(\frac{\alpha}{m} - e^{-\alpha b_s}\right)^2\left(1 - e^{-\alpha b_s}\right)}{4\alpha\left(pe^{-\alpha b_s} + e^{-\alpha b_s} - 1\right)} \\ &= \frac{1}{4\alpha} - \frac{1}{2m} + 0 = \frac{1}{2m} - \frac{1}{2m} = 0.\end{aligned}$$

Therefore, the equilibrium state of the system (4.34) is uniformly stable. ■

**Example 4.9:**

Consider the following dynamic equation:

$$\dot{x}(t) = -5e^{-t}x^3(t-h)\sin(x(t-h)), \qquad t \geq 0. \tag{4.39}$$

The initial conditions of the system are as follows:

$$\phi(t) = 5\sin(t), \qquad -h \leq t \leq 0.$$

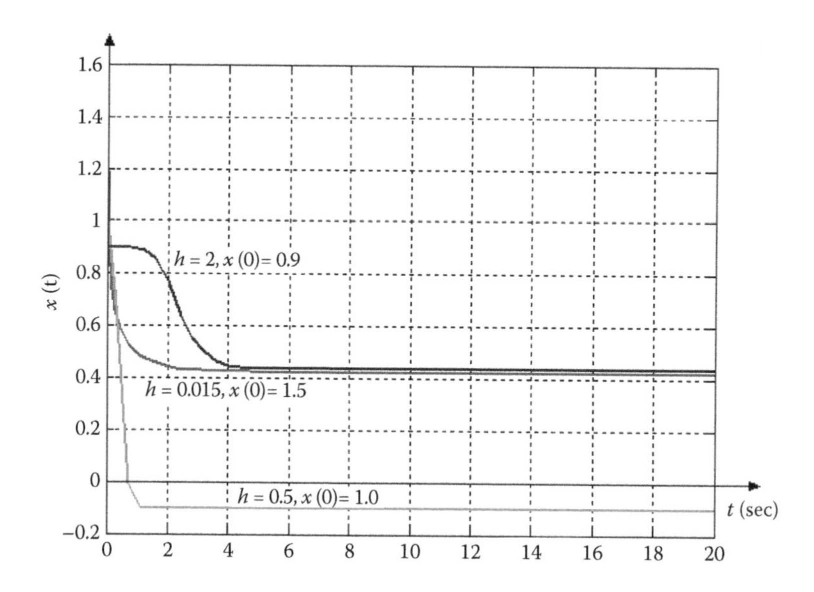


**FIGURE 4.7** The trajectories of the system (4.39) with $h \leq 0.43$.

Using the relations of the previous section, with $c(t) = -5e^{-t}$, $b_s$ is obtained as follows:

$$b_s = \underset{t>-h}{Sup} \int_t^{t+h} 5e^{-(u+h)} du = 5(1-e^{-h})$$

where $b_S$ is obviously dependent on delay ($h$). Using $\alpha = \frac{m}{2}$, $e^{\alpha b_s} = 2$, and $p = 1$ from the inequalities of the previous section, one has:

$$m = \frac{\ln 2}{2.5\left(1-e^{-h}\right)}. \tag{4.40}$$

The equation (4.40) shows that if $h$ increases then, $m$, which indicates the regain of stability of the system given in (4.34) will decrease. Establishing the above inequalities, the system given in (4.39) is uniformly stable. The simulation result verifies the analysis (Figure 4.7). ■

In this section, the Lyapunov–Krasovskii theorem is extended to some cases of nonlinear time-delay systems by deriving a new algorithm for generating proper Lyapunov–Krasovskii functionals. These functionals can be used to analyze the delay-dependent stability of the given systems. The conditions obtained are sufficient and could be used to analyze the local stability properties. The proposed results also give less conservative conditions. From the application of the proposed inequalities of the procedure to a typical system, better results have been obtained. Implementing a numerical example, the capability of the algorithm is demonstrated. Also, it is possible to derive delay-independent stability conditions in this case.

**PROBLEMS:**

**4.1:** Consider the following time-delayed linear system:

$$\dot{x}(t) = A_0\, x(t) + A_1\, x(t-\tau),$$

where $x \in R^n$ is the state vector, $\tau \in R^+$ is constant delay and $A_0$ and $A_1$ are constant matrices with appropriate dimensions. Using Razumikhin's theorem, show that the equilibrium state of the system is asymptotically stable if there exists symmetric positive definite constant matrix $Q$ and constant scalar $\beta > 1$, such that the following linear matrix inequality holds:

$$A_0 Q + Q\, A_0^T + A_1 Q\, A_1^T + \beta Q < 0.$$

**4.2:** Consider the following scalar time-delayed dynamical equation:

$$\dot{x}(t) = -b\, x(t-\tau).$$

Using the Leibnitz integral formula, transform the above dynamical equation to the following form:

$$\dot{x}(t) = -b\, x(t) + b^2 \int_{-2\tau}^{-\tau} x(t+\theta)\, d\theta.$$

then, employ the Razumikhin theorem to prove that the equilibrium state of the system is asymptotically stable for $\tau < \frac{1}{b}$.

**4.3:** A dynamical system is described by the following time-delayed equation:

$$\dot{x}(t) = A_0\, x(t) + A_1\, x(t-\tau).$$

Using the Leibnitz integral formula, rewrite the equation as follows:

$$\dot{x}(t) = (A_0 + A_1)\, x(t) - \int_{t-\tau}^{t} \dot{x}(\theta)\, d\theta$$

Let the Lyapunov–Krasovskii functional candidate be as follows:

$$V = x^T(t) P\, x(t) + \int_{t-\tau}^{t} x^T(\theta) Q x(\theta)\, d\theta + \int_{-\tau}^{0} \int_{t+\theta}^{t} \dot{x}^T(\lambda) Z\, \dot{x}(\lambda)\, d\lambda\, d\theta.$$

where $P$, $Q$, and $Z$ are symmetric positive definite matrices. Prove that the sufficient condition for asymptotic stability of this system is to exist symmetric positive definite matrices $P$, $Q$, and $Z$ such that the following matrix inequality holds:

$$\begin{bmatrix} (A_0 + A_1)^T P + P(A_0 + A_1) + Q + \tau A^T Z A & \tau A_0^T Z A_1 & \tau P A_1 \\ * & -Q + \tau A_1^T Z A_1 & 0 \\ * & * & -\tau Z \end{bmatrix} < 0.$$

where * stands for symmetric entries in the matrix. (Hint: Bind the cross terms by the completion of squares technique.)

**4.4:** For a linear time-delayed system described by the following equation:

$$\begin{bmatrix} \dot{x}_1(t) \\ \dot{x}_2(t) \end{bmatrix} = \begin{bmatrix} -1 & 0.5 \\ -0.5 & -1 \end{bmatrix} \begin{bmatrix} x_1(t) \\ x_2(t) \end{bmatrix} + \begin{bmatrix} -2 & 2 \\ -2 & -2 \end{bmatrix} \begin{bmatrix} x_1(t-\tau) \\ x_2(t-\tau) \end{bmatrix}$$

find the maximum allowable bound for $\tau$ for asymptotic stability of zero equilibrium state of the system by the approaches of Section 4.2 and Problem 4.3. Compare the results and justify your answer.

**4.5:** Consider the following scalar linear multiple-delay system:

$$\dot{x}(t) = -b_1\, x(t-\tau_1) - b_2\, x(t-\tau_2), \qquad b_i > 0,$$

showing that the equilibrium state of the system is stable for all delays $\tau_i \in [0,\, \tau_i^*]$, $i = 1,2$ if $b_1 \tau_1^* + b_2 \tau_2^* < 1$.

**4.6:** Consider the following linear multiple-delay system:

$$\dot{x}(t) = A\, x(t) + \sum_{i=1}^{r} A_i\, x(t-\tau_i)$$

(a) Show that if the following delay-independent condition is satisfied, then the above equilibrium state of the system is asymptotically stable:

$$\begin{bmatrix} -A^T P - P A - \sum_{i=1}^{r} S_i & P A_1 & \dots & P A_r \\ A_1^T P & S_1 & \dots & 0 \\ \vdots & & \ddots & \\ A_r^T P & 0 & \dots & S_r \end{bmatrix} > 0$$

where $P$ and $S_i,\, i = 1, \dots r$ are symmetric positive definite matrices. (Hint: Use the following Lyapunov–Krasovskii functional.):

$$V(x) = x^T(t) P x(t) + \sum_{i=1}^{r} \int_{-\tau_i}^{0} x^T(t+\theta)\, S_i\, x(t+\theta)\, d\theta$$

(b) Obtain a delay-dependent condition for stability analysis of the above system using the following Lyapunov–Krasovskii functional:

$$V(x) = x^T(t)P\,x(t) + \sum_{i=1}^{r}\int_{-\tau_i}^{0}\int_{t+\xi}^{t} \mathbf{x}^T(\xi)\,S_i\,x(\xi)d\xi d\theta$$

$$+\sum_{i=1}^{r}\sum_{j=1}^{r}\int_{-\tau_i-\tau_j}^{-\tau_j}\int_{t+\xi}^{t} x^T(\xi)\,S_{ij}\,x(\xi)\,d\xi d\theta.$$

**4.7:** Consider the linear time-delayed system as follows:

$$\dot{x}(t) = A_0\;x(t) + A_1\,x(t-\tau) + Bu(t).$$

Using Razumikhin's theorem, prove that if there exist symmetric matrices $\Lambda > 0$, $\Lambda_i > 0$, $i = 1,2,3$ and a matrix $\Gamma$ satisfying the following LMIs:

$$\begin{bmatrix} \Xi_1 + 2\tau\Lambda & \tau A_1(\mathbf{A}\Lambda + B\Gamma) & \tau A_1^2\,\Lambda & \tau A_1\,\Lambda \\ * & -\tau\Lambda_1 & 0 & 0 \\ * & * & -\tau\Lambda_2 & 0 \\ * & * & * & -\tau\Lambda_3 \end{bmatrix} < 0,$$

and

$$\Lambda_i - \Lambda \le 0\;,\; i = 1,2,3,$$

in which, $\Xi_1 = \Lambda(A + A_1)^T + (A + A_1)\Lambda + \Gamma^T B^T + B$, then the control law $u(t)$ = $K\,x(t)$ with $K = \Gamma\Lambda^{-1}$ will stabilize the equilibrium state of the system.

# 5 An Introduction to Stability Analysis of Linguistic Fuzzy Dynamic Systems

**Introduction:** For many real-world systems and processes, a mathematical description in the form of differential/difference equations or other conventional models is neither feasible nor practical due to the complexities involved, the time consideration, and the intrinsic nature of the data deficiency. To find a practical way, studies have been done to analyze the model free fuzzy control systems. The fuzzy models come in three main groups; TSK (Takagi-Sugeno-Kang), the linguistic, and the fuzzy relational models (FRMs). Tanaka [t1] proposed a method for stability analysis of the TSK model by finding a common Lyapunov function. Percup, Preitl, and Solyom [p5] used the center of manifold theory for a fuzzy system's stability analysis. Another study proposed a sufficient condition for stability analysis of a TSK fuzzy model [s4]; Farinawata and Linder worked separately on robust stability controller design. Many other efforts in the TSK stability analysis domain have been done [t1], in which the classical approaches for stability analysis of fuzzy systems have been used. Unfortunately, these approaches defy the simplicity object, which was the main aim of Zadeh when he proposed his fuzzy system approaches. For simplicity, in system analysis and humanity interface, a linguistic model is offered [z1]. Some authors have done some incomplete researches in this area.

Margaliot and Langholz [m11] have proposed some approaches for linguistic models of nonlinear systems stability analysis. The basic idea is to use the crisp equation of Lyapunov for the stability analysis in their approaches.

## 5.1 TSK FUZZY MODEL SYSTEM'S STABILITY ANALYSIS

In the following, first the switching system and its stability analysis will be considered, and then the results are used for TSK stability analysis. Consider the linear time-varying system:

$$\dot{x}(t) = A(t)x(t), \; x(0) = x_0, \tag{5.1}$$

where $x(t) \in R^n$ and the matrix $A(t)$ switches between asymptotic stable matrices belonging to set $\Omega = \{A_1, A_2, A_3, ....\}$. This system is referred to as the *switching system*. Narendra and Balakrishnan [n2], proved the following theorem.

---

**Theorem 5.1 [s4]:**

Consider the switching system (5.1) with $\Omega = \{A_1, A_2, A_3, ..., A_N\}$, where the matrices $A_i$, commuted pair-wise, are asymptotically stable (i.e., having eigenvalues with negative real parts), hence:

(i) The system is exponentially stable for any arbitrary switching sequence from the elements of $\Omega$.

(ii) Given a symmetrical positive definite initial matrix $P_0$, let $P_1, P_2, P_3, ..., P_N$ be the unique symmetrical positive definite solution for the following Lyapunov equation:

$$A_i^T P_i + P_i A_i = -P_{i-1}, \quad \text{i} = 1,2,3,\ldots,\text{N}.$$

Then the function $V(x) = x^T P_N x$ is a common Lyapunov function for each of the individual systems $\dot{x}(t) = A_i x(t)$ and hence the Lyapunov function for the switching system (5.1).

(iii) For a given choice of matrix $P_0$, the choice of matrices $A_1, A_2, A_3, ..., A_N$ in any order, yields the same solution $P_N$ in (ii).

(iv) The matrix $P_N$ can also be expressed in integral form as:

$$P_N = \int_0^\infty e^{A_N^T t_N} .... \left[ e^{A_2^T t_2} \left[ \int_0^\infty e^{A_1^T t_1} P_0 e^{A_1 t_1} dt_1 \right] e^{A_2 t_2} dt_2 \right] .... e^{A_N t_N} dt_N,$$

where, as in part (iii), the order in which the matrices $A_1, A_2, A_3, ..., A_N$ appear can be replaced by any permutation.

**Proof:**

See Suratgar and Nikravesh [s4]. ■

In the following, this theorem is used and a sufficient condition for stability analysis of fuzzy TSK system is derived [s4].

Consider a TSK fuzzy model as follows:

$$\textit{Rule ith: if } x_1(t) \textit{ is } B_1^i \textit{ and...and } x_N(t) \textit{ is } B_n^i \textit{ then}:$$

$$\left[\dot{x}_1(t)......\dot{x}_N(t)\right]^T = A_i \left[x_1(t)......x_N(t)\right]^T, \tag{5.2}$$

where $x(t) = [x_1(t)......x_N(t)]^T$ is the state vector and $B_j^i$ is the fuzzy set that belongs to $i$th rule and $j$th state. With the Center of Area (CoA) defuzzification procedure, the crisp dynamic equation of the state of the fuzzy system is obtained as:

$$\dot{x}(t) = \frac{\sum_{i=1}^{M} (Dof_p)_i A_i x(t)}{\sum_{i=1}^{M} (Dof_p)_i} = \frac{\sum_{i=1}^{M} (Dof_p)_i A_i}{\sum_{i=1}^{M} (Dof_p)_i} x(t), \tag{5.3}$$

where $M$ is the total number of rules. $(Dof_p)_i$ is the degree of firing of the $i$th rule. $(Dof_p)_i$ is defined as:

$$(Dof_p)_i = \mu_{B_1^l}(x_1^P(t)) \times \mu_{B_2^l}(x_2^P(t)) \times ..... \times \mu_{B_N^l}(x_N^P(t)), \tag{5.4}$$

where the value of membership function is $\mu_{B_k^l}(x_k^P(t))$ for the fuzzy set $B_k^i$ for the $k$th element of state for the $p$th pair of data and for the $i$th rule. $Z_P$ and $W_i$ are defined as follows

$$W_i = \frac{(Dof_p)_i}{\sum_{i=1}^{M} (Dof_p)_i}, \tag{5.5}$$

$$Z_P = \sum_{i=1}^{M} W_i A_i . \tag{5.6}$$

It should be noted that the $W_i$ is selected from the interval [0, 1] in such a way that

$$\sum_i W_i = 1.$$

Since $W_i$, i = 1,...,M are positive real scalars, for a diagonal matrix with negative real part entries, the $Z_p$, p = 1,2,... would be diagonal matrices whose eigenvalues have negative real parts.

In the fuzzy TSK system inference, for each linguistic value of $x(t)$, the dynamical equation is as follows:

$$\dot{x}(t) = Z_p x(t)$$

where $Z_p$ for different linguistic value of $x(t)$ switches between their values that are matrices with negative real part eigenvalues. Theorem 5.1 concludes that the above fuzzy system is exponentially stable. So, the following theorem is adapted:

**Theorem 5.2 [s4]:**

Consider the TSK fuzzy model of a system as follows:

$$Rule \text{ ith}: if \quad x_i(t) is\ B_1^i\ and...and\ x_N(t)\ is\ B_n^i\ then:$$

$$\left[\dot{x}(t)......\dot{x}_N(t)\right]^T = A_i\left[x_1(t)......x_N(t)\right]^T.$$

The sufficient condition for exponential stability of a TSK fuzzy model is that $A_i$ be diagonal matrices whose eigenvalues have negative real parts. ■

This theorem can be expanded to the cases where $A_i$ is either upper or lower triangular matrix.

We believe these approaches defy the simplicity objective, which is the cornerstone of fuzzy logic methodologies. As is obvious, the linguistic model is offered for simplifying the system analysis and human interface considerations [z1]. Some authors have carried out further research in these topics, but the stability analysis of a linguistic model is still an open problem. We propose here a new approach for linguistic fuzzy modeling [s5–s12], which is suitable for stability analysis of such systems.

## 5.2 LINGUISTIC FUZZY STABILITY ANALYSIS USING A FUZZY PETRI NET [s5]

This section presents a new approach for linguistic fuzzy system's modeling, which is also suitable for stability analysis of linguistic models. An approach is presented first which is called *infinite place* model, using modified fuzzy Petri net, and a new *place* definition based on the concept of physical infinity state. The method has some practical difficulties in its primary presentation, which are taken care of in the second approach called the *variation* model. This section elaborates the two models using some definitions and stating *a necessary and a sufficient condition* for stability of a class of linguistic fuzzy systems. This stability analysis is verified using some benchmark systems simulations.

Hasewaga and Furuhashi [h6] proposed a definition for the equilibrium state in a fuzzy linguistic model. There seems to be some flaws in their definition.

A new definition of fuzzy equilibrium state will be presented here. This section uses fuzzy Petri net (FPN) and presents two linguistic models, suitable for stability analysis of linguistic fuzzy systems modeling. The Petri net and fuzzy Petri net are presented in what follows.

### 5.2.1 Review of a Petri Net and Fuzzy Petri Net

The generic components of Petri nets include a finite set of places *(P)* and finite set of transitions *(T)*. The Petri nets are finite bipartite graphs where places are linked by the transitions, which in turn are connected to the output places. Equivalently, there are input and output sets of transitions defined for a given place. Distribution of tokens over places is called *marking of the net*. Generally, a transition may fire when each of its input places contains at least one token. The transition firing results in removing tokens from input places and adding them to output places.

The transition produces a new marking. Therefore, a marking represents the state of a system and changes when a transition is fired. Formally, a Petri net can be described as the following structure:

$$N_p = (P,T,F,W,M_0), \tag{5.7}$$

where:

$P = \{p_1, p_2, p_3, \ldots, p_n\}$ is a finite set of places, $T = \{trans_1, trans_2, \ldots, trans_m\}$ is a finite set of transitions, $F \quad (P \times T) \bigcup (T \times P)$ is a set of arcs, $W : F \rightarrow \{1,2,3,..\}$ is a weight function, and $M_0 : P \rightarrow \{0,1,2,3,..\}$ is an initial marking.

In the fuzzy Petri net (FPN) each place may or may not contain a token associated with the truth value of a proposition; this value is a quantity between zero and one. Additionally, each transition is associated with a certainty factor between zero and one. Formally, this model of FPN is the following topple:

$$N_f = (P,T,D,I,O,\alpha,\beta), \tag{5.8}$$

where $P$ and $T$ are defined above, and $D$ is finite set of propositions (linguistic values), $I : T \rightarrow P$ is an input mapping, $O$ and $f$ are the output and association mappings, respectively, $O : T \rightarrow P$ and $f : T \rightarrow [0,1]$ respectively. In addition, $\alpha : P \rightarrow [0, 1]$ and $\beta : P \rightarrow D$. In $N_f$ token value in place $p_i, p_i \in P$, is denoted by $\alpha(p_i)$ *where*, $\alpha(p_i) \in [0, 1]$. A transition "$trans_i$" is enabled if for all $p_j \in I(trans_i)$ *and* $\alpha(p_j) \geq \lambda$, where $\lambda$ is a threshold value, $\lambda \in [0, 1]$. If this transitions fires, tokens are removed from their input places and each token is deposited into each of its output places. For instance, the following fuzzy production rule:

$$If\ d_j\ then\ d_k\ (CF = \mu_i), \tag{5.9}$$

can be modeled by FPN as shown in Figure 5.1, where the truth values $y_j$ and $\mu_i$ are aggregated through the algebraic product, $y_k = y_j \mu_i$.

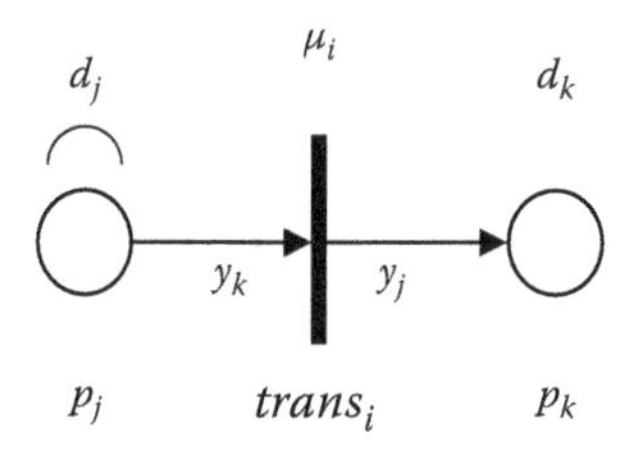


**FIGURE 5.1** The fuzzy Petri net model of the fuzzy rule.

For the case where the "*if part*" consists of more than one statement, its aggregation is as follows:

$$If\ d_{j1}\ and...and\ d_{jn}\ then\ d_k\ (CF = \mu_i)\ y_k = Min(y_{j1},...,y_{jn})\mu_i. \tag{5.10}$$

In the following section, FPN is used with some proposed modifications for dynamical linguistic fuzzy modeling.

### 5.2.2 Appropriate Models for Linguistic Stability Analysis

Two models are proposed in this section [s4,s5].

(a) The infinite place model.
(b) The variation model.

These models use a Petri net for linguistic fuzzy modeling and are suitable for stability analysis of linguistic models. In these models, each place is associated with a linguistic label and each transition is associated with a given rule of the rule base. Of course, in infinite place model, there is an additive place, defined as follows:

#### 5.2.2.1 The Infinite Place Model

Consider the following nonlinear system:

$$y(k) = f(u(k), y(k-1)). \tag{5.11}$$

Let the rule base of a linguistic model of this system be described as follows:

$$\begin{aligned} R^i &: If\ u(k)\ is\ Ling_i 1\ and\ y(k-1)\ is\ Ling_i 2 \\ &then\ y(k)\ is\ Ling_i 3, \quad i = 1,...,m. \end{aligned} \tag{5.12}$$

where $m$ is the total number of rules, $Ling_i j, j = 1,...,n$ describes linguistic values, that is, $Ling_i j \in \{Zero, Small, Medium, ...\}$, $j$ is the index of the linguistic set, $i$ is the rule number, and $n$ is the total number of linguistic labels. The linguistic model is described by a new FPN. In this Petri net, infinity place $P_\infty$ is defined, which is useful for stability analysis.

---

**Definition 5.1: The Infinity Place**

A place in a fuzzy Petri net is defined to be an *infinity place* iff it satisfies the following conditions:

(a) If a token belongs to a place, then the membership value of that place is infinity.
(b) If there is no token in a place, then the membership value of that place is zero. ■

Therefore, the membership value of $P_\infty$ belongs to $\{0,\infty\}$ a set which consists of only two elements. The infinity place has a physical meaning; if a token belongs to $P_\infty$, this means that its associated variable has no definite value, so its membership value in $P_\infty$ is indefinite, that is, is infinity. Similarly, if there is no token in $P_\infty$, this means that its associated variable has definite value, then its membership value in $P_\infty$ is zero. As an example, in the case of a control valve, if the opening percentage is more than 100%, it is physically infinite, and thus its associated opening percentage $P_\infty$ has a token. If the opening percentage is under 100%, it is physically finite, and thus its associated opening percentage $P_\infty$ has no token.

For presenting a linguistic model with FPN, consider the system that is given in Equation (5.12). This rule base could be modeled using modified FPN as shown in Figure 5.2. Note that the system's expert operator determines the infinite place rule. Now this approach together with the defined stable and unstable systems are used in the sense of BIBO (Bounded Input Bounded Output) stability.

#### 5.2.2.2 The BIBO Stability in the Infinite Place Model

If, after some transitions firing with finite inputs, infinite place associated with system output earns a token and does not lose the token at least for an interval of time, then the system is unstable in the sense of BIBO stability, otherwise it is BIBO stable. For example in Figure 5.2, the system is unstable, because for some *"small"* value of *u(k)* and *"medium"* value of *y(k-1)*, the output is infinite. This method suffers from the following shortcoming. If there is no expertise for system behavior, in order to model an open-loop unstable system, then the system output must reach physical infinity. Therefore, this approach is impractical. To overcome this problem, the following method is proposed. The internal stability is not considered here.

#### 5.2.2.3 The Variation Model

The main advantage of this approach is its capability for practical modeling of the open-loop unstable systems. This method describes the relations among the output variations with respect to the input variations, previous output variations, previous

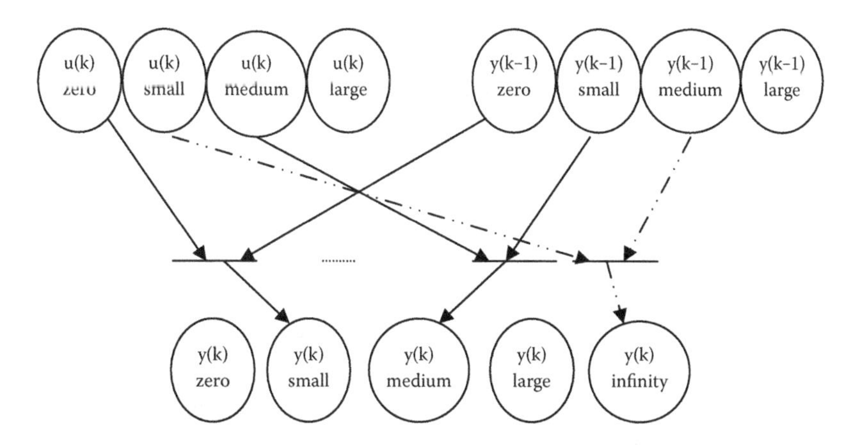


**FIGURE 5.2** The fuzzy Petri net description of an infinite place model.

inputs and outputs. For the simple system given in (5.11), the variation model's rule base is as follows:

$$R^i : If\ u(k) is\ Ling_i 1 and\ \ y(k)\ is\ Ling_i 2\ and|\ \ y(k)|\ is$$
$$Ling_i 3\ v\ \ u(k+1) is\ Ling_i 4\ then \tag{5.13}$$
$$|\ \ y(k+1)|\ is\ Ling_i 6, \quad i = 1,...,m,$$

where $\Delta y(k) \triangleq y(k) - y(k-1)$ and $\Delta u(k) \triangleq u(k) - u(k-1)$. In order to describe the variation model systematically, we use FPN. Figure 5.3 is drawn to describe (5.13), where $Ling_i j \in \{Zero, Small, Medium, Large\}$.

For the system stability analysis, the incremental behavior of the system output can be used. For introducing the procedure, the following definitions are needed:

---

**Definition 5.2: The Linguistic Measure**

Consider $\Omega = \{D_1, D_2, D_3, ..., D_n\}$, as a set of the fuzzy sets. The fuzzy sets in $\Omega$ are labeled by the linguistic values in $\Pi = \{Ling1, Ling2, ...., Lingn\}$, respectively. The fuzzy linguistic measure of variable $\chi$ is *LingM* iff:

$$\underset{i=1,...,n}{Max}(\mu_{D_i}(\chi)) = \mu_{D_M}(\chi), \tag{5.14}$$

where $\mu_{D_i}(\chi)$ is the membership value of $\chi$ in the $D_i$. ■

---

**Definition 5.3: The Linguistic Index**

Consider $\Omega = \{D_1, D_2, D_3, ..., D_n\}$ as a set of fuzzy sets. The fuzzy sets in $\Omega$ are labeled by linguistic values in $\Pi = \{Ling1, Ling2, ...., Lingn\}$, respectively. Assume that they are sorted from zero to very large. The fuzzy linguistic index of *LingM* is defined as:

$$|lingM|_{ling} = M.$$

■

### 5.2.3 The Necessary and Sufficient Condition for Stability Analysis of a First-Order Linear System Using Variation Models

To elaborate the general linguistic stability concept, let us start with a low-order system. Consider a relaxed, first-order, open-loop system with the transfer function $G(s) = \frac{1}{s+a}$, which is the input-output relation of this system.

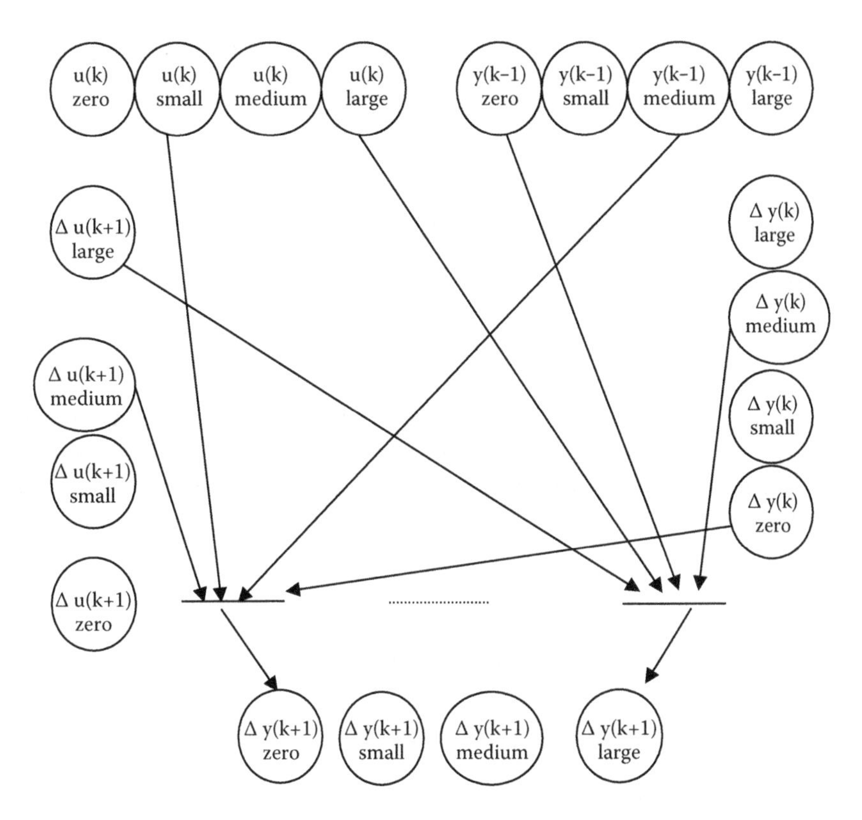


**FIGURE 5.3** The fuzzy Petri net description of the variation mode.

---

**Theorem 5.3: (Suratgar–Nikravesh Theorem 1)**

Consider the rules of the variation model that describe the step response of the first-order LTI system, as in (5.13), then this system is stable iff:

$$\forall i \in Z^{+} : \left| \; y(k+1) \right|_{ling}^{i} \leq \left| \; y(k) \right|_{ling}^{i},$$

and,

$$i \in Z^{+} : \left| \; y(k+1) \right|_{ling}^{i} \neq \left| \; y(k) \right|_{ling}^{i}, \tag{5.15}$$

where $\left| \; y(k) \right|_{ling}^{i}$ is the linguistic index associated with $| \; y(k) |$ in the $i$th rule. ■

**Proof:**

Consider the first-order system with the transfer function $\frac{1}{s+a}$. Using the classical approach, the system impulse response is as follows:

$$y(t) = exp(-at). \tag{5.16}$$

For $a > 0$, the system is stable and for $a \leq 0$, the system is unstable. $\Delta y(k)$ is defined as:

$$\Delta y(k) \triangleq y(kT) - y((k-1)T), \tag{5.17}$$

where for the first-order system, we have:

$$y(k) = \exp(-akT)\left[1 - \exp(aT)\right], \tag{5.18}$$

similarly,

$$y(k+1) \triangleq y((k+1)T) - y((k)T) = \exp(-akT)\left[\exp(-aT) - 1\right]. \tag{5.19}$$

Obviously, for all positive $T$, if $a > 0$, then $| \ y(k+1)| < | \ y(k)|$. If $A < 0$, then $| \ y(k+1)| > | \ y(k)|$ and if $a = 0$, then $| \ y(k+1)| = | \ y(k)|$. Therefore, if the variation model describes the system behavior close enough (the fitness), the stability concept is as follows:

$$\text{If } a > 0, \text{ then } \forall i : \left| \ y(k+1)\right|_{ling}^{i} \leq \left| \ y(k)\right|_{ling}^{i},$$

$$\text{If } a < 0, \text{ then } \forall i : \left| \ y(k+1)\right|_{ling}^{i} \geq \left| \ y(k)\right|_{ling}^{i},$$

$$\text{If } a = 0, \text{ then } \forall i : \left| \ y(k+1)\right|_{ling}^{i} = \left| \ y(k)\right|_{ling}^{i}.$$

Thus, the system is stable iff:

$$\forall i \in Z^{+} : \left| \ y(k+1)\right|_{ling}^{i} \leq \left| \ y(k)\right|_{ling}^{i}, \tag{5.20}$$ ■

and

$$i \in Z^{+} : \left| \ y(k+1)\right|_{ling}^{i} \neq \left| \ y(k)\right|_{ling}^{i}.$$

In the next section, a stability criterion is established by using Theorem 5.3.

### 5.2.4 Stability Criterion

In this subsection, a stability criterion is proposed by presenting the following theorem. Some definitions are proposed first. In order to find the stability criterion for the variation model, the fuzzy Petri net (FPN) description of the model (i.e., [5.13]) as described in Figure 5.3 is used. The $D_{\ y(k)}$ is defined as follows:

$$D_{\ y(k)} = \begin{bmatrix} d_{11} & \ldots & d_{1m} \\ & \vdots & \\ d_{n1} & & d_{nm} \end{bmatrix}, \tag{5.21}$$

in which:

$$d_{ij} = \begin{cases} 1\, if\, |\; y(k)|\; is\; ling_j\; for\; i\text{th}\; rule, \\ 0 \qquad\qquad\qquad otherwise, \end{cases} \tag{5.22}$$

where $m$ is the total number of linguistic labels and $n$ is the total number of rules.

---

**Definition 5.4: The Difference Matrix**

Consider the fuzzy variation model defined in Section 5.2.2.3 with its fuzzy Petri net (FPN) description of (5.13). The difference matrix denoted by *Diff(S)* is defined as follows:

$$Diff(S(k)) = D_{\;y(k)} - D_{\;y(k+1)}. \tag{5.23}$$

■

---

**Definition 5.5: The Characteristic Vector**

Consider the scalar matrix *A;* the *characteristic vector* of *A* which is denoted by *Char(A)* is defined as:

$$Char(A) = A \times V_2 \times .... \times V_m \times \begin{bmatrix} 1 & ... & 1 \end{bmatrix}_{1\times m}^{T} \tag{5.24}$$

where:

$$V_i = I_{m\times m} + M_i,\; \forall i = 2,...,m \tag{5.25}$$

and:

$$M_i(p,q) - \begin{cases} 1 & p < i, q = i, \\ 0 & otherwise, \end{cases} \tag{5.26}$$

where $M_i(p,q)$ is the $p,q$ element of $M_i$. ■

---

**Definition 5.6: The Equal Replacing Vector**

Consider the scalar-valued vector *V.* To obtain the equal replacing vector of *V,* denoted by $E_r(V)$, replace all zero elements of *V,* if any, by its nonzero elements arbitrarily. Note that if all entries of *V* are zero, then *V* will not be changed by the equal replacing procedure. ■

**Theorem 5.4 [s4,s5]: (Suratgar–Nikravesh Theorem 2)**

The linguistic model of a low-order LTI system is BIBO stable, iff:

$$\sum_{i=1}^{n} sign(Er(Char(Diff(S))) = -n, \tag{5.27}$$

where $n$ is the total number of rules, which are fired in the system's step response, in the system's rule base. ■

**Proof:**

Consider the fuzzy Petri net (FPN) description of the variation model (i.e., [5.13]) as described in Figure 5.3. Consider *Diff(S)* matrix that is, (5.23). If in each row of *Diff(S)*, which contains only -1 and/or +1, and the negative element is on the left-hand side of the positive one, the stability conditions of Theorem 5.3 are satisfied. Therefore, if $W = Char(Diff(S))$, it is obvious that:

(i) If a row in $W$, which is the $n \times 1$ vector, is negative, then in the rule associated with this row, one has:

$$| \; y(k+1)|_{ling}^{i} < | \; y(k)|_{ling}^{i} .$$

(ii) If a row in $W$ is positive, then in the rule associated with this row, one has:

$$| \; y(k+1)|_{ling}^{i} > | \; y(k)|_{ling}^{i} .$$

(iii) If a row in $W$ is zero, then in the rule associated with this row, one has:

$$| \; y(k+1)|_{ling}^{i} = | \; y(k)|_{ling}^{i} .$$

Hence, none of the elements of $W$ neither are positive nor zero iff:

$$\forall i \in Z^{+} : | \; y(k+1)|_{ling}^{i} \leq | \; y(k)|_{ling}^{i}$$

and:

$$j: \; | \; y(k)|_{ling}^{j} \neq | \; y(k+1)|_{ling}^{j}$$

Now, in such a case $\forall j \in \{1,2,\ldots,n\}; | \; y(k)|_{ling}^{j} = | \; y(k+1)|_{ling}^{j} = M \neq 0$ and we may end up, for example, with an integrator that is obviously an unstable system.

Let $Z = Er(W)$ be the $Er$ operation acting on $W$ to generate vector $Z$, therefore, using Theorem 5.3, the system model is stable iff:

$$\sum_{i=1}^{n} sign(Z_i) = \sum_{i=1}^{n} sign(Er(Char(Diff(S))) = -n \tag{5.28}$$

where $Z_i$ is the $i$th element of $Z$ vector. ■

## 5.3 LINGUISTIC MODEL STABILITY ANALYSIS

In this section, the stability analysis method is extended to fuzzy linguistic systems associated with a class of applied nonlinear systems. Some preliminary definitions are given as follows [z1]:

### 5.3.1 Definitions in Linguistic Calculus*

The following proposed definitions are needed to define equilibrium state and its stability concepts described in the next section.

---

**Definition 5.7: Center of Linguistic Value**

Consider for instance, the following membership functions (Figure 5.4):

The linguistic values are $\{Ling1, Ling2, Ling3, Ling4, \ldots.\}$ associated with {very small, small, medium, big, very big,...}. The $x_{cent}$ is the center of linguistic value *LingM* if;

$$X_{cent} = \left\{ x \,\middle|\, \max_{x} \mu_{\,lingM}(x) = \mu_{lingM}(x) \right\}. \tag{5.29}$$

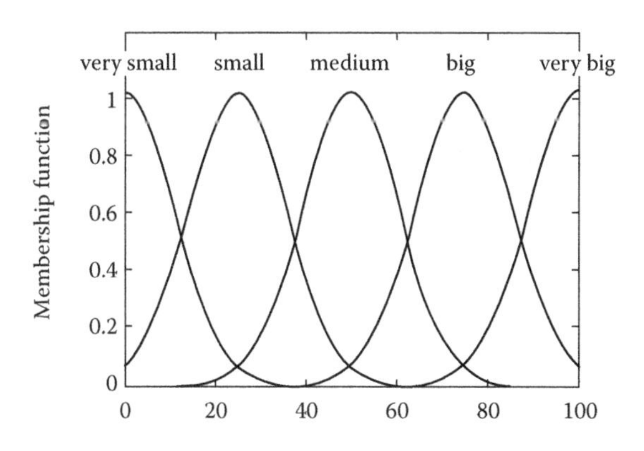


**FIGURE 5.4** Gaussian membership functions.

---

* Such a calculus is known as *computation with word*.

**Remark 5.1:**

If there is finite interval (i.e., $[a,b]$) where $\mu_{lingM}(x)=1$, then the mean point of the interval is defined as the center of linguistic value. ■

**Remark 5.2:**

If there is an open infinite interval (i.e., $[a,\infty)$) or $(-\infty,a]$ where $\mu_{lingM}(x)=1$, then the finite boundary of interval is defined as the center of linguistic value ■

**Definition 5.8: Linguistic Neighborhood**

Consider a set of fuzzy sets as:

$$Set=\{A,B,C,...\}, \tag{5.30}$$

where $A,B,C,...$ are the fuzzy sets. Sort the center of linguistic values of these fuzzy sets increasingly. Associate the sorted sequence with $N_m^1$.* Thus, the first term is "1" and its associated fuzzy set is labeled by *ling*1, the second term is "2" and its associated fuzzy set is labeled by *ling*2,.... The $\zeta \in Z^+$ neighborhood of *lingR* is denoted by:

$$\underset{\zeta}{\Omega}^{L.N.}(lingR)=\{lingj \mid R-\zeta \le j \le R+\zeta\}. \tag{5.31}$$

■

**Example 5.1:**

Consider the fuzzy sets with the membership functions given in Figure 5.5, then:

$$\underset{\zeta=1}{\Omega}^{L.N.}(ling3)=\{ling2, ling3, ling4\}. \tag{5.32}$$

■

**Definition 5.9: Linguistic Deleted Neighborhood**

The linguistic neighborhood of *lingR* which does not contain *lingR*, is called the *linguistic deleted neighborhood* of *lingR* and is denoted as:

$$\underset{\zeta}{\Omega}^{L.D.N..}(lingR)=\left\{lingj \mid R-\zeta_1 \le j \le R+\zeta_2, j \ne R\right\}. \tag{5.33}$$

If $\zeta_1=\zeta_2=\zeta$, then the linguistic deleted neighborhood is called the *symmetrical deleted neighborhood.* ■

* $N_m$ is the first *m* term of natural number sequence.

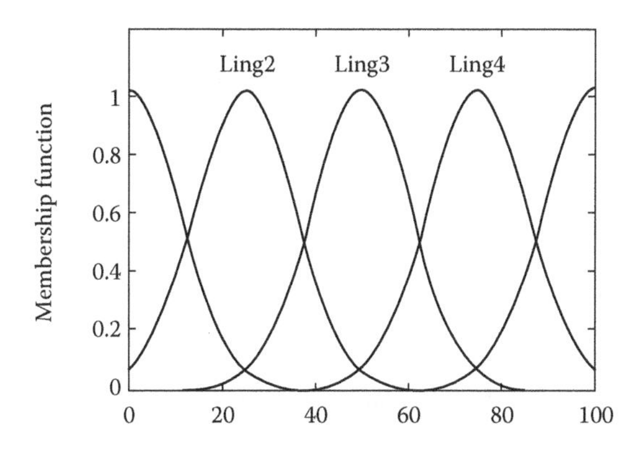


**FIGURE 5.5** A sample of membership function.

---

**Definition 5.10: Fuzzy Equilibrium Subset**

Consider a rule base describing the dynamic relation of the state variables of a system without exogenous input as follows:

$$R^i: \textit{If } x(k)\, \textit{is Ling}_i H \quad \textit{then } x(k+1)\, \textit{is Ling}_i E,\ i = 1,\ldots,m. \tag{5.34a}$$

If there is a rule in which $Ling_i H$ and $Ling_i E$ are linguistically equal, then $Ling_i H \equiv Ling_i E \equiv Ling_i eq$ is called a *fuzzy equilibrium subset*. ■

---

**Definition 5.11: Equilibrium Halo**

Consider $P = \{lingP_1, lingP_2, .., lingP_N\}$, as a set of all fuzzy equilibrium subsets. The $lingP_i$ is called the equilibrium halo if for an initial condition $x(k_0) = lingP_i$ the following rule:

$$\textit{if } x(k) \textit{ is } lingP_i \textit{ then } x(k+1) \textit{ is } lingP_i, \tag{5.34b}$$

is fired with the highest degree of firing among the other rules, for all $k > k_0$. ■

### 5.3.2 A Necessary and Sufficient Condition for Stability Analysis of a Class of Applied Mechanical Systems

This subsection reviews some definitions about the potential energy and conservative forces in pure mechanical systems. It has been observed that the work done against a gravitational or an elastic force depends only on the net change of position and not on the particular path followed in reaching the new position. The force fields with this characteristic are called *conservative force fields*, which possess an important mathematical property and will be illustrated as follows:

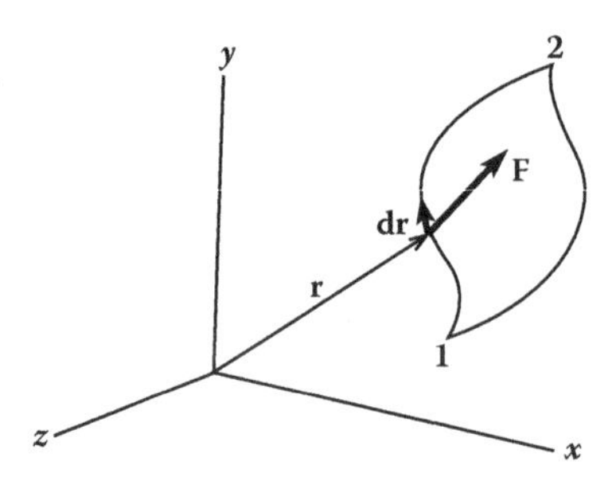


**FIGURE 5.6** The sample path for describing the potential energy.

Consider a force field as shown in Figure 5.6, where the force $F$ is a function of its coordinates. The work done by the force $F$ in the displacement $dr$ of its point of application is:

$$dU = F.dr. \tag{5.35}$$

The total work done along its path form point 1 to point 2 is:

$$U = \int_1^2 F.dr = \int_1^2 (F_x\,dx + F_y\,dy + F_z\,dz). \tag{5.36}$$

The integral of $F$, that is, $\int F.dr$, is a line integral depending, in general, on the particular path followed between any two points 1 and 2 in space. However, if $F.dr$ is an exact differential of some scalar function $V$ of the coordinates, that is, $-dV$, then:

$$U = \int_{V_1}^{V_2} -dV = -(V_2 - V_1), \tag{5.37}$$

depends only on the end points of the motion and is *independent* of the path followed. The minus sign in front of $dV$ is arbitrary but is chosen to agree with the customary designation of the sign of potential energy changes in the earth's gravity field. If $V$ exists, the differential change in $V$ becomes:

$$dV = \frac{\partial V}{\partial x}dx + \frac{\partial V}{\partial y}dy + \frac{\partial V}{\partial z}dz. \tag{5.38}$$

Comparing it with:

$$-dV = F.dr = F_x dx + F_y dy + F_z dz. \tag{5.39}$$

implies:

$$F_x = -\frac{\partial V}{\partial x},\ F_y = -\frac{\partial V}{\partial y} \text{ and } F_z = -\frac{\partial V}{\partial z}, \tag{5.40}$$

therefore, the force may also be written as the following vector form of:

$$F = -\quad V, \tag{5.41}$$

where the symbol "∇" stands for the vector operator "del" defined as follows:

$$= i\frac{\partial}{\partial x} + j\frac{\partial}{\partial y} + k\frac{\partial}{\partial z}. \tag{5.42}$$

The quantity $V$ is known as the *potential function*, and the expression $\nabla V$ is known as the *gradient of the potential function.* When the force components are differentiable from a potential function as described above, the force is said to be *conservative* and the work done by the force between any two points is independent of the path followed.

In the following, necessary and sufficient conditions are proposed for asymptotic stability of an equilibrium state of some applied mechanical systems.

---

**Theorem 5.5: (Suratgar–Nikravesh Theorem 3)**

Consider a mechanical system, which is perturbed with a given force and starts to move at some initial time. There is no exogenous input. Assume all forces are conservative except the frictional torque. Let $x_e$ be the system's isolated equilibrium state. The equilibrium state $x_e$ is asymptotic stable iff there is a symmetrical deleted neighborhood $\Omega_{SDNE}$ of $x_e$ such that:

$$\forall x \in \Omega_{SDNE} : \frac{dE_{potential}(x)}{dx_i} \times \frac{|x_i|}{x_i} > 0,$$

and:

$$\frac{dE_{potential}(x_{eq})}{dx_i} = 0 \qquad \mathrm{i} = 1,2,3, \tag{5.43}$$

where the $E_{potential}$ is the potential energy of the system. ■

---

**Remark 5.3:**

Without loss of generality, $x_e$ is chosen as the origin of coordinates and is assumed to be the reference of a system potential energy, that is,

$$E_{potential}(x_e) = 0. \tag{5.44}$$

■

---

**Remark 5.4:**

In Newtonian dynamic systems, the set of two numbers, the position and the velocity at time $t_0$, are qualified to be called the *state vector of the system at time* $t_0$. In other words, $[\ x\ \ y\ \ z\ \ \dot{x}\ \ \dot{y}\ \ \dot{z}\ ]^T$ is the state vector. This concept is used in the theorem's proof. ■

---

**Remark 5.5:**

It is assumed that $E_{potential}$ is directional derivable on $\Omega_{SDNE}$. ■

---

**Remark 5.6:**

It is assumed that the perturbed system at its initial condition has potential energy and does not have kinetic energy. ■

**Proof:**

See Suratgar and Nikravesh [s9]. ■

### 5.3.3 A Necessary and Sufficient Condition for Stability Analysis of a Class of Linguistic Fuzzy Models

This subsection uses Theorem 5.5 and proposes a necessary and sufficient condition for stability analysis of the linguistic models associated with a class of linear and/or nonlinear mechanical systems.

---

**Theorem 5.6: (Suratgar–Nikravesh Theorem 4)**

Consider the unforced system with the assumption that all applied forces are conservative forces except the frictional forces. An equilibrium halo is asymptotic stable iff there is a symmetrical linguistic deleted neighborhood of equilibrium halo such that the potential energy of the system is greater than or equal to the potential energy of equilibrium halo. ■

**Proof:**

In the previous section, in Theorem 5.5, the necessary and sufficient condition for asymptotic stability for an equilibrium state for the given system was considered.

The following relations are the same:

1. $\Omega_{SDNE}; \forall x \in \Omega_{SDNE}: \dfrac{x_i}{|x_i|}\dfrac{dE_{potential}(x)}{dx_i} > 0.$
2. A local minimum of $E_{potential}(x)$ is $E_{potential}(x_{eq})$.

Consider an arbitrary direction $x_i$. If the minimum potential energy is at $x_{eq}$ then in a given neighborhood of $x_{eq}$, one has:

$$\text{If } 0 = x_{eq} > x_i \text{ then } E_{potential}(x_{eq}) < E_{potential}(x_i).$$

$$\text{If } 0 = x_{eq} < x_i \text{ then } E_{potential}(x_{eq}) < E_{potential}(x_i), \text{ thus:}$$

$$\frac{\partial E_{potential}(x)}{\partial x_i}\Big|_{x>x_{eq}=0} > 0, \tag{5.45}$$

$$\frac{\partial E_{potential}(x)}{\partial x_i}\Big|_{x<x_{eq}=0} < 0, \tag{5.46}$$

therefore:

$$x_i > x_{eq} = 0 \qquad \frac{\partial E_{potential}(x)}{\partial x_i}\frac{x_i}{|x_i|} > 0,$$

$$x_i < x_{eq} = 0 \qquad \frac{\partial E_{potential}(x)}{\partial x_i}\frac{x_i}{|x_i|} > 0.$$

It means that, if a local minimum of $E_{potential}(x)$ is $E_{potential}(x_{eq})$, then $\Omega_{SDNE}$; $\forall x \in \Omega_{SDNE}: \frac{x_i}{|x_i|}\frac{dE_{potential}(x)}{dx_i} > 0$. On the other hand, if $\Omega_{SDNE}$ ;$\forall x \in \Omega_{SDNE}$: $\frac{x_i}{|x_i|}\frac{dE_{potential}(x)}{dx_i} > 0$, it is obvious that a local minimum of $E_{potential}(x)$ is $E_{potential}(x_{eq})$.

Having the above statement, Theorem 5.5 can be restated as follows:

The equilibrium state of the system given in Theorem 5.5 is asymptotically stable iff a local minimum of potential energy is in $x_{eq}$.

However, the crisp model concepts can be extended in the fuzzy model. The equilibrium state in the crisp model is associated with equilibrium halo in the fuzzy model; the neighborhood in crisp is associated with the linguistic neighborhood in the fuzzy model and so on. ■

Using the above idea, it can be concluded that:

In the system given in Theorem 5.5, an equilibrium halo is asymptotic stable iff there is a symmetrical linguistic deleted neighborhood of equilibrium halo such that potential energy of the system there, is greater than or equal to the potential energy of equilibrium halo. ■

**Example 5.2:**

Consider the pendulum system given in Figure 5.7. The rules of equilibrium subsets are as follows:

$$X(k)=\begin{bmatrix} Ling19 \\ Ling4 \\ Ling13 \\ Ling14 \\ Ling11 \\ Ling10 \\ Ling9 \\ Ling8 \\ Ling10 \\ Ling11 \\ Ling9 \\ Ling1 \\ Ling16 \\ Ling7 \\ Ling6 \\ Ling12 \end{bmatrix} \quad Z(k)=\begin{bmatrix} Ling10 \\ Ling10 \\ Ling10 \\ Ling10 \\ Ling11 \\ Ling11 \\ Ling11 \\ Ling10 \\ Ling10 \\ Ling10 \\ Ling10 \\ Ling10 \\ Ling10 \\ Ling10 \\ Ling10 \\ Ling10 \end{bmatrix}$$

The equilibrium halo of the system is the given state obtained from the linguistic fuzzy model that is derived from a modified one pass method [s11], and is obtained as follows, where the $Z$ and $X$ are the coordinates of the system:

$$X(k) = Ling10 \text{ and } Z(k) = Ling10.$$

The system's potential energy rule base of the equilibrium halo is as follows:

*If $X(k)$ is Ling10 and $Z(k)$ is Ling10 then $E_{Potential}$ is Ling10.*

Using the system's potential energy rule base, the deleted neighborhood's potential energy is greater than or equal to the equilibrium halo's potential energy.

*If $X(k)$ is Ling9 and $Z(k)$ is Ling11 then $E_{Potential}$ is Ling11.*

*If $X(k)$ is Ling9 and $Z(k)$ is Ling10 then $E_{Potential}$ is Ling10.*

*If $X(k)$ is Ling11 and $Z(k)$ is Ling10 then $E_{Potential}$ is Ling10.*

*If $X(k)$ is Ling11 and $Z(k)$ is Ling11 then $E_{Potential}$ is Ling11.*

where the Ling *i*th is defined in Section 5.3.1. Using Theorem 5.6, the equilibrium halo of this system is asymptotic stable. This analysis coincides with the classic stability analysis given in the literature. ■

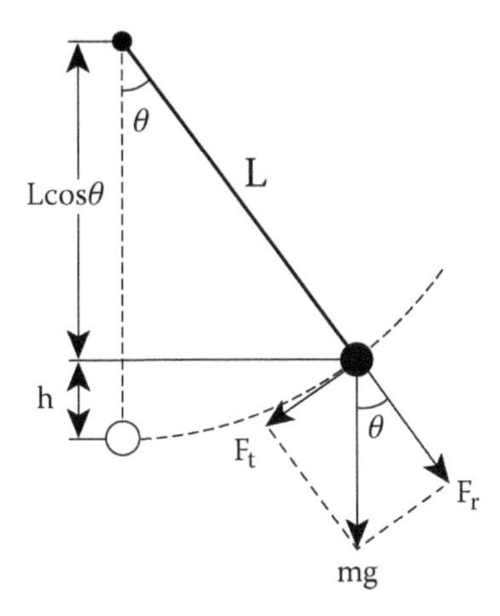


**FIGURE 5.7** A pendulum system.

### Example 5.3:

Consider a system without exogenous inputs where all the applied forces are conservative forces except the frictions forces. The fuzzy linguistic model rule base of the system is as follows:

*If $x_1(k)$ is ling1 and $x_2(k)$ is ling2 then $x_1(k+1)$ is ling2 and $x_2(k+1)$ is ling 3*

*If $x_1(k)$ is ling2 and $x_2(k)$ is ling1 then $x_1(k+1)$ is ling5 and $x_2(k+1)$ is ling1*

*If $x_1(k)$ is ling7 and $x_2(k)$ is ling7 then $x_1(k+1)$ is ling8 and $x_2(k+1)$ is ling8*

*If $x_1(k)$ is ling3 and $x_2(k)$ is ling4 then $x_1(k+1)$ is ling3 and $x_2(k+1)$ is ling4*

*If $x_1(k)$ is ling5 and $x_2(k)$ is ling6 then $x_1(k+1)$ is ling3 and $x_2(k+1)$ is ling5*

If $x_1(k)$ is ling2 and $x_2(k)$ is ling4 then $x_1(k+1)$ is ling5 and $x_2(k+1)$ is ling5

The potential energy of the system is described by the following rule base:

*If $x_1(k)$ is ling1 and $x_2(k)$ is ling3 then $E_{potential}$ is ling2*

*If $x_1(k)$ is ling2 and $x_2(k)$ is ling3 then $E_{potential}$ is ling3*

*If $x_1(k)$ is ling3 and $x_2(k)$ is ling4 then $E_{potential}$ is ling1*

*If $x_1(k)$ is ling3 and $x_2(k)$ is ling3 then $E_{potential}$ is ling4*

*If $x_1(k)$ is ling4 and $x_2(k)$ is ling3 then $E_{potential}$ is ling5*

*If $x_1(k)$ is ling2 and $x_2(k)$ is ling4 then $E_{potential}$ is ling2*

*If $x_1(k)$ is ling4 and $x_2(k)$ is ling4 then $E_{potential}$ is ling3*

*If $x_1(k)$ is ling2 and $x_2(k)$ is ling5 then $E_{potential}$ is ling2*

*If $x_1(k)$ is ling3 and $x_2(k)$ is ling5 then $E_{potential}$ is ling4*

*If $x_1(k)$ is ling4 and $x_2(k)$ is ling5 then $E_{potential}$ is ling2*

(a) Find the equilibrium halo of the system?
(a) Is the equilibrium halo of the system stable?

**Solution:**

Using Definition 5.9 from the system rule base, the equilibrium halo is:

$$\begin{bmatrix} x_{1eq} \\ x_{2eq} \end{bmatrix} = \begin{bmatrix} ling3 \\ ling4 \end{bmatrix}$$

Using Theorem 5.6 from the potential energy of the system rule base, we can conclude that the above equilibrium halo of the system is stable. ■

## 5.4 STABILITY ANALYSIS OF FUZZY RELATIONAL DYNAMIC SYSTEMS

The focus of this section and the remaining parts of this chapter is on the investigation of the stability of the equilibrium state of Fuzzy Relational Models (FRMs). It is worth mentioning that FRMs can be considered extensions of the fuzzy linguistic models, as their concept suggest. In fact, in FRMs, a degree of truthfulness is assigned to every rule that can be constituted. The collection of such degrees is gathered in a matrix which is representative of the rule base in the model and is called a *relational matrix*. The FRMs provide more systematic frameworks for fuzzy modeling. These models can use different types of fuzzy compositions and have been used to model static and dynamic systems, see [a9,a10] and [c2]. The materials of Section 5.4 through Section 5.6 concerning the stability issue are essentially taken from Aghili Ashtiani [a5], Aghili Ashtiani and Nikravesh [a6], Aghili Ashtiani and Nikravesh [a7], and Aghili Ashtiani, Nikravesh, and Raja [a8]. Other useful materials in this regard can be found in Chen and Chen [c2] and Kandel, Luo, and Zhang [k6].

A fuzzy relational dynamic system (FRDS) is in fact a fuzzy relational model (FRM) with feedback, and can be categorized as a class of linguistic fuzzy systems. Many works are available in function approximation and system model identification using FRMs. An introduction to FRM based on the fuzzy relational equations (FREs) was introduced in 1976 by Sanchez [s14]. A typical FRE is written as $b = a \circ R$, in which $a$ and $b$ are the fuzzy input and output vectors, respectively. This equation models a function $f$, which can be used to model a dynamic system. Figure 5.8 shows a model of a dynamic system without any external input.

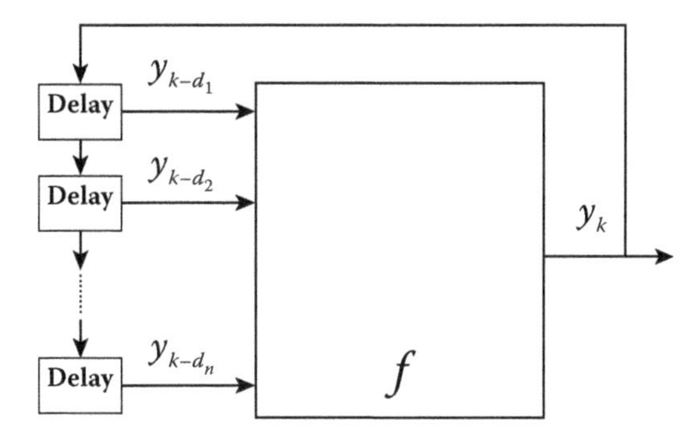


**FIGURE 5.8** Model of a dynamic system with feedback and without any external input.

If the actual system is not too complex, then we may use only one of the past values of the output variable. In this case, the FRM can be described by $b(k) = b(k-d) \circ R$, where $R$ is a $p \times q$ fuzzy relational matrix. For simplicity one may let $p = q$. Such a model can be called a *first-order FRM*. The variable represents the proper value of a constant delay to give the best modeling result. It is obvious that the modeling capability can be increased to some extent by increasing the number of linguistic terms.

### 5.4.1 Model Representation and Configuration

A typical FRM consists of three main parts, a fuzzifier, a defuzzifier, and a relational matrix with its appropriate fuzzy composition operator. The general scheme of an FRM is shown in Figure 5.9. This figure is in fact the implementation of the FRE $b(k) = b(k-d) \circ R$. In general, the appropriate type for each part is selected by the model designer.

Note that the fuzzy relational composition operator is of the form s-t as in general FRMs, that is, s-norm over t-norm; for example, "max"-"min", "max"-"prod", "Sum"-"prod" and, and so forth. Furthermore, some specific types of FRM parts are considered. The fuzzifier is made up of triangular membership functions with the property of sum-normality, that is, the sum of the membership function values for every nonfuzzy value is one. The parameter is the number of linguistic terms, and hence, the number of membership functions. For further references, such a fuzzifier is called a *standard fuzzifier*. Figure 5.10 represents the standard fuzzifier for $q = 3$. Note that the centers of membership functions do not need to be spaced equally in the nonfuzzy universe of discourse, here [–1,1].

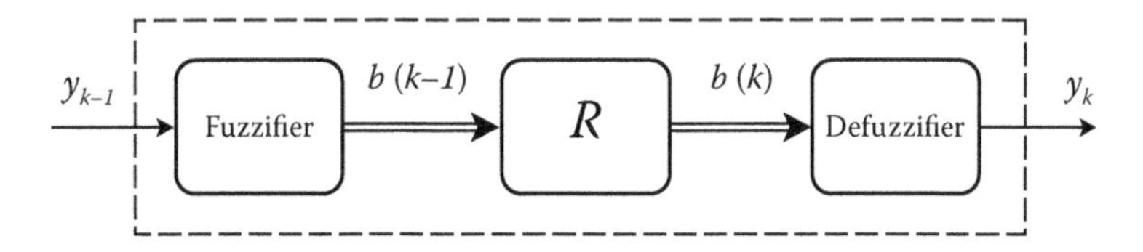


**FIGURE 5.9** A simple FRDS (a FRM plus a unit delay feedback). General block diagram of an FRM.

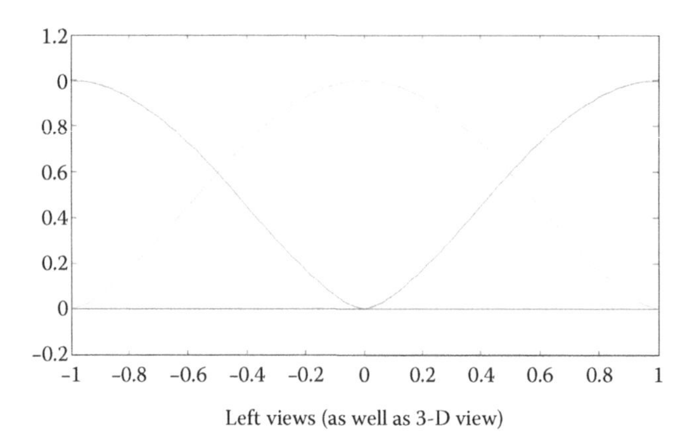


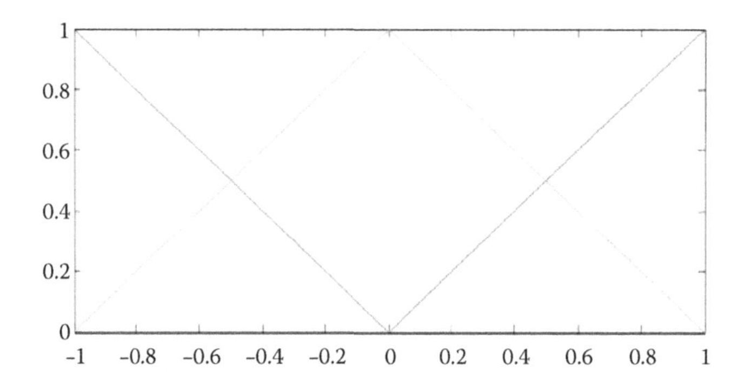


**FIGURE 5.10** The standard fuzzifier with three membership functions.

The defuzzifier is of a *weighted average* type and is simply expressed by

$$y = \sum_{i=1}^{q} b_i c_i \,/\, \sum_{i=1}^{q} b_i, \tag{5.47}$$

where $b_i$'s are the elements of the fuzzy vector $b$ and $c_i$'s are the centers of the defuzzifiers. Therefore, the general scheme of the FRM is as follows:

1. Fuzzifier: standard fuzzifier (mentioned above),
2. Fuzzy Relational Composition: algebraic s-t composition,
3. Defuzzifier: weighted average defuzzifier.

---

**Remark 5.7:**

An algebraic FRDS or a "sum"-"prod" FRDS, means an FRDS in which the fuzzy composition is of the form s-t, the s-norm is the algebraic sum, and the t-norm is the algebraic product. ■

---

**Definition 5.12:**

In Figure 5.9, if the fuzzifier is a standard fuzzifier, its "R" matrix is a unit-row relational matrix, its defuzzifier is a weighted average defuzzifier and its fuzzy relational composition is of algebric type composition then, the resulted FRDS is called *SUWA-FRDS.* ■

### 5.4.2 Stability in an FRDS: An Analytical Glance

This section serves as a motivating and inspiring preparatory study, which leads to the approaches of stability analysis taken in the sequel. First, some helpful definitions are introduced.

---

**Definition 5.13:**

The nonfuzzy scalar value of every quantity is converted to a vector of length $p$ through the fuzzifier, where $p$ is the number of linguistic terms defined for the quantity. This vector is called the *fuzzy vector.* It is also called the *possibility vector* in the literature. ■

---

**Definition 5.14:**

The set of all possible fuzzy vectors in the $p$-dimensional coordinate system constitute the hypercube $[0, 1]^p$ when the membership functions are normal. This hypercube is called the *fuzzy space* and is denoted by $\tilde{X}$ in sequel. ■

---

**Definition 5.15:**

Let be the nonfuzzy input space on which the nonfuzzy variable is defined. The value of is converted to a fuzzy vector through the fuzzifier function $\Phi: X \rightarrow \tilde{X}$ where $\tilde{X}$ is the relating fuzzy space. The fuzzification curve $\tilde{\phi}$ is defined as the image of the nonfuzzy input space $X$, that is $\tilde{\phi} = \Phi(X)$. In other words, the fuzzification curve is the range of the vector-valued fuzzifier function $\Phi$ while $\tilde{X}$ is its codomain, that is $\tilde{\phi} \quad \tilde{X}$. ■

**Example 5.4:**

Assume that the position of a robot arm is in the range that is normalized to simplify the formulation and generalization of the modeling process. Consider also three linguistic terms $N$, $Z$, and $P$ defined for that variable. Then, the dimension of the resulting fuzzy space is $p = 3$ and so three membership functions should be defined on the nonfuzzy input space $X$. Our fuzzifier is a simple, yet effective standard fuzzifier as depicted in Figure 5.10. Therefore, the fuzzy space in this example is the cube $[0,1]^3$. The fuzzification curve is depicted in the 3-D diagram of Figure 5.11. Different views of the curve are depicted as well as its perspective view to prevent imagination confusion.

Note that the fuzzifier diagram of Figure 5.10, in fact shows the one-to-one mapping between all points of and corresponding points in $\tilde{X}$. For instance, the point in the actual nonfuzzy space is mapped by the fuzzifier to the point in the virtual fuzzy space. The fuzzification

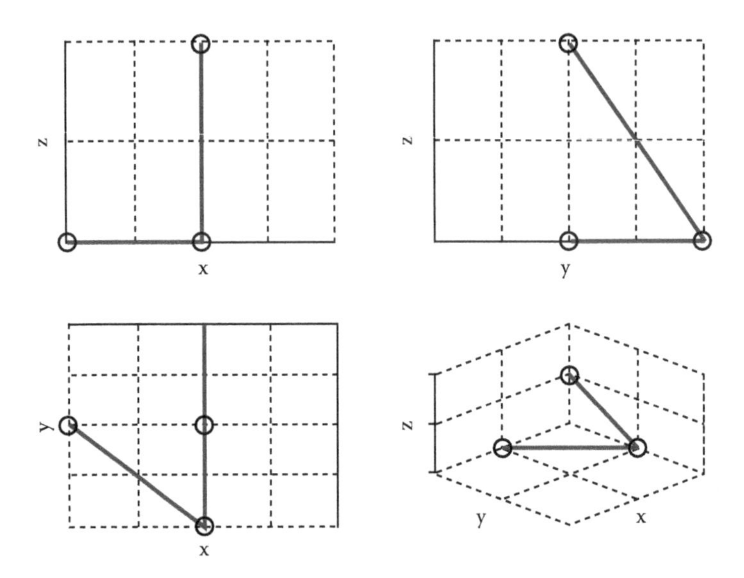


**FIGURE 5.11** Fuzzification curve for the standard fuzzifier of Figure 5.10. (Front-Up-Left views as well as 3-D view.)

curve of Figure 5.11 shows the set of all possible fuzzy points at the output of the fuzzifier. See Figure 5.12 as another example, which shows a smooth fuzzifier and the relating fuzzification curve. Of course, again $\tilde{\phi}$ is an one-dimensional manifold in the fuzzy space $\tilde{X}$.

Obviously, the dimension and the form of the fuzzification curve in each problem depends on the dimension of the fuzzifier, that is, the shape and other characteristics of the membership functions.

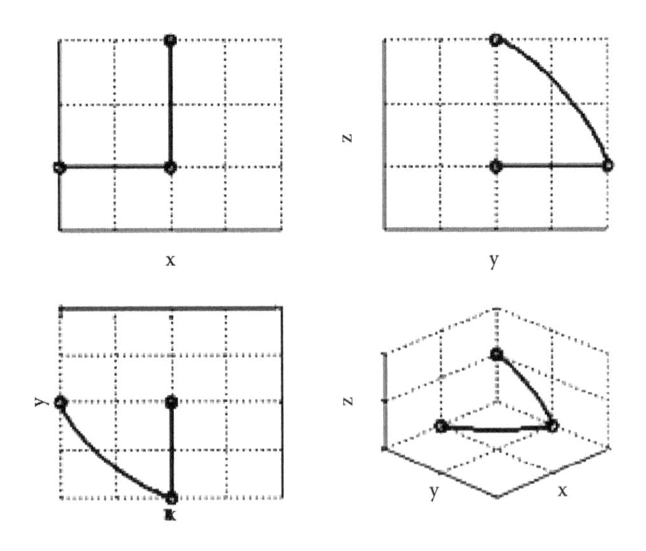


**FIGURE 5.12** A smooth fuzzifier and the related fuzzification curve.

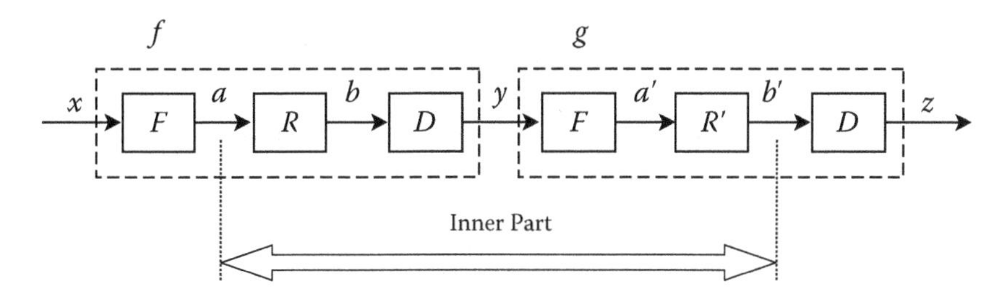


**FIGURE 5.13** Two FRMs cascaded to model the composite function.

The output evolution in our FRM is modeled by $y(k) = f(y(k-1))$. Hence, between the fuzzy interfaces, fuzzifier and defuzzifier, one has $b(k) = b(k-1) \circ R$. But there is an important point here in this model. The output fuzzy vector leads to an equivalent nonfuzzy value using a defuzzifier. However, defuzzifiers are not one-to-one mappings in general.

Hence, several different fuzzy trajectories may lead to a specified output trajectory. This complicates the study of the system's stability via trajectory analysis. If a one-to-one defuzzifier could be chosen, then this problem would have been resolved, but as will be seen later, this is not possible because in such a case the fuzziness of the model becomes trivial.

Two major factors in choosing a defuzzifier are the nature of the problem at hand and the ease of calculation. Now, choose a one-to-one defuzzifier, such a simplifying choice for a one-to-one defuzzifier is the exact inverse of the fuzzifier.

Let $f$ and $g$ be two functions and $h = g \circ f$. In Figure 5.13, the fuzzy relational implementation of the composition $z = g(f(x))$ is illustrated. Suppose $R$, $R'^{1}$,* and $Q$ are the fuzzy relational matrices of $f$, $g$, and $h$, respectively, that is,

$$a = \left[a_1,\ldots,a_i,\ldots,a_p\right],\ b = \left[b_1,\ldots,b_j,\ldots,b_q\right],\ R = \left[r_{ij}\right]_{p\times q},\ R' = \left[r'_{ij}\right]_{p\times q},\ c = a \circ Q$$

The question that arises is whether or not $Q = R \circ R'$. The answer, in general, is obviously not. However, a close look at the problem would lead to the selection of a proper type of fuzzy composition to establish later the useful conditions and assumptions.

For the two relational matrices to become equal, the first step is to pick optimal fuzzy interfaces, that is, choosing a fuzzifier and a defuzzifier that are inverse of each other. So, assume $= \Phi^{-1}$.

Then

$$a = [a_1,\ldots,a_i,\ldots,a_p],\quad b = [b_1,\ldots,b_j,\ldots,b_q],$$
$$R = [r_{ij}]_{p\times q},\quad R' = [r'_{ij}]_{p\times q},$$

and

$c = a \circ Q$. Therefore, the inner part of Figure 5.13 reduces to Figure 5.14.

* There is no mathematical relationship between R and R'.

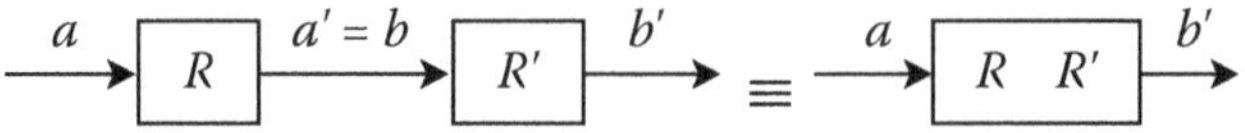

**FIGURE 5.14** The inner part of Figure 5.13 with the assumption of optimal interfaces.

Performing some simple mathematical manipulations leads to a specific type of fuzzy composition as follows:

$$\begin{cases} b_i = \underset{i}{s}\, t(a_i, r_{ij}) \\ b'_{i'} = \underset{i'}{s^t}(a'_{i'}, r'_{i'j'}) \overset{a' \equiv b}{\phantom{=}} \boxed{b'_{j'}} = \underset{i'}{s^t}(b_{i'}, r'_{i'j'}) = \underset{i'}{s^t}(\underset{i}{s^t}(a_i, r_{ii'}), r'_{i'j'}) = \underset{j}{s^t}(\underset{i}{s^t}(a, r_{ij}), r'_{i'j'}) \end{cases}$$

By definition one has:

$$\boxed{c_i} = \underset{i}{s^t}(a_i, q_{ij})$$

Note that:

$$q_{ij} = s_k^t(r_{ik}, r'_{kj})$$

using $q_{ij}$ and $c_j$, one has:

$$c_j = (a_i, s_k^t(r_{ik}, r'_{kj})) \qquad c \neq b' \qquad Q \neq R^0 R'.$$

To make $c = b'$, in the second step, satisfy the equality by selecting the proper fuzzy norms. In this regard, an algebraic product is choosen as t-norm. This is also known as *probabilistic t-norm*, $t(a, b) = ab$.

Now, choose the well-known Lukasiewicz s-norm which is defined as. $s(a,b) = min((a+b),1) \triangleq (a+b)\wedge 1$. Note that this is a nonlinear function due to the min operator. But, if the summation of the arguments of the function is less than or equal to one, then it will serve as a linear function. Note that this condition can be assured by putting some constraints on the FRM. This issue will be discussed and used in the following sections.

$$\begin{cases} b'_j = \sum_k \left( r'_{kj} \sum_i (a_i r_{ik}) \right) \\ c_j = \sum_i \left( a_i \sum_k \left( r'_{kj} r'_{ik} \right) \right) \end{cases} \qquad \begin{cases} b'_j = \sum_k \sum_i \left( r'_{kj} a_i r_{ik} \right) \\ c_j = \sum_i \sum_k \left( a_i r'_{kj} r_{ik} \right) \end{cases}$$

$$c = b'; \forall = 1,\ldots,q \qquad c = b'.$$

The above material is summarized in Theorem 5.7, after introducing a definition.

**Definition 5.16:**

A fuzzy composition is an algebraic fuzzy composition in which the s-norm and the t-norm are both algebraic types. ■

Note that the algebraic s-norm and the algebraic t-norm are also known as the *Lukasiewicz s-norm*, that is, $s(a, b) = (a + b) \wedge 1$, and probabilistic t-norm, that is, $t(a,b) = ab$, respectively.

**Theorem 5.7:**

In the scheme represented in this subsection (Section 5.4.2), iff:

1. The defuzzifier is the inverse of the fuzzifier.
2. The fuzzy composition operator is the algebraic fuzzy composition.

Now return to the FRM of a dynamic system, $X_{k+1} = X_k \circ R$. From this relational equation, one has $X_2 = X_1 \circ R$ and so $X_2 = X_1 \circ R = (X_0 \circ R) \circ R$. Under the assumption of Theorem 5.7, the above formula can be rewritten as $X_2 = X_1 R = X_0 \circ R^2$. Furthermore, the result of operating the relational matrix on any point on the fuzzification curve should stay on the same curve, or simply, the fuzzification curve should be invariant under the operation "o $R$", that is,

$$\forall X_2 \in, = X_{k+1} = X_k o R = X_k \;\; R \in$$

Toward this end, some constraints may be set on the relational matrix. This reduces the modeling capability of a specified fuzzy relational structure to some extent.

Consider the triple $(s,t,\phi)$ in which $s,t$ are s-norm and t-norm, respectively in the fuzzy s-t composition and $\phi$ stands for the fuzzification curve. The issue is to find a triple $(s,t,\phi)$ such that some constraints can be put on the relational matrix R to obtain $u \circ R \in \psi, \forall u \in \psi$, and yet some degree of freedom remains in the relational matrix. There is unfortunately no such triple $(s,t,\phi)$ that meets the above condition. In fact, the space of fuzzy trajectories cannot be confined to a 1-D manifold in the fuzzy space, because there would be only one degree of freedom, and hence, no fuzziness.

On the other hand, the behavior of the fuzzy trajectories of the system is very complicated when the whole fuzzy space is considered as a valid region for fuzzy vectors. In such cases, the stability analysis of the system and even finding the whole set of equilibrium states is not an easy task. Therefore, it would be a good idea to confine the fuzzy space to a specific subspace of it, to achieve a more analyzable model. For example, by implementing some constraints on the relational matrix $R$, one can confine the fuzzy trajectory of the system to a hyper-surface in the fuzzy space. This idea is used in the next section.

## 5.5 ASYMPTOTIC STABILITY IN A SUM-PROD FRDS

In this section an FRDS based on sum-prod fuzzy composition is considered. A set of sufficient conditions for asymptotic convergence to the center of a linguistic term is obtained and then the possibility of extending this result to other kinds of fuzzy composition, that is, the class of s-prod composition, is contemplated. The concepts of fuzzy intraplanarity and unit-row matrix are introduced to obtain the proposed conditions.

---

**Definition 5.17:**

An FRM described by $b(k+1) = b(k) \circ R$ is defined to have the property of intra-planarity on the hyperplane (5.48), if the output fuzzy vector of the model always lies on that hyperplane. In other words, the hyperplane (5.48) is invariant under the sum-prod fuzzy composition when the FRM is intraplanar.

$$\Gamma_f : \sum_{j=1}^{q} d_J = 1. \tag{5.48}$$

■

---

**Definition 5.18:**

The matrix $R$ is defined as the unit-row matrix if the sum of the elements of each row of the matrix is equal to one, that is,

$$\sum_{j=1}^{q} r_{ij} = 1 \quad \forall i \in \{1, \dots, p\} \tag{5.49}$$

■

---

**Lemma 5.1:**

Assume an FRDS, or equally an FRM is described by $b(k+1) = b(k) \circ R$, with sum-prod fuzzy composition and standard fuzzifier. The model has the property of fuzzy intraplanarity on the hyperplane iff the relational matrix is the unit-row matrix. ■

**Proof:**

The lemma states that when (5.49) is true then, $\sum_{i=1}^{p} a_i = 1$ leads to $\sum_{j=1}^{q} b_j = 1$ and $b \in \Gamma_f, \forall a \in \Gamma_f$. Also, if the property of *fuzzy* intraplanarity is satisfied, then (5.49) can be concluded. To prove the lemma, consider the relational matrix $R = [r_{ij}]$.

According to the sum-prod composition, the $j$th element of the output fuzzy vector can be written as:

$$b_j = \overset{p}{\underset{i=1}{S}} t(a_i, r_{ij}) = \left( \sum_{i=1}^{p} a_i r_{ij} \right) \wedge 1.$$

(a) Sufficient condition:

$$\begin{cases} a \in \Gamma_f & \sum_{i=1}^{p} a_i = 1 \\ \sum_{j=1}^{q} r_{ij} = 1 & \end{cases} \qquad \sum_{i=1}^{p} a_i r_{ij} \leq 1 \qquad b_j = \sum_{i=1}^{p} a_i r_{ij}$$

$$\sum_{j=1}^{q} b_j = \sum_{j=1}^{q} \sum_{i=1}^{p} a_i r_{ij} = \sum_{i=1}^{p} a_i \sum_{j=1}^{q} r_{ij} = \sum_{i=1}^{p} a_i = 1.$$

(b) Necessary condition:

$$\begin{cases} a \in \Gamma_f & \sum_{i=1}^{p} a_i = 1 \\ b \in \Gamma_f & \sum_{j=1}^{q} b_j = 1 \end{cases} \qquad \forall j \quad b_j \leq 1 \qquad b_j = \sum_{i=1}^{p} a_i r_{ij}$$

$$\sum_{j=1}^{q} b_j = \sum_{j=1}^{q} \sum_{i=1}^{p} a_i r_{ij} = \sum_{i=1}^{p} \sum_{j=1}^{q} a_i r_{ij} = \sum_{i=1}^{p} a_i \sum_{j=1}^{q} r_{ij} = 1.$$

$$\sum_{i=1}^{p} a_i \sum_{j=1}^{q} r_{ij} = \sum_{i=1}^{p} a_i = 1.$$

The variables $r_{ij}$ *and* $a_i$ belong to the unit interval [0, 1], thus $\sum_{j=1}^{q} r_{ij} = 1$, and the proof is completed. ■

---

**Corollary 5.1:**

For an FRDS described by $b(k+1) = b(k) \circ R$, with sum-prod fuzzy composition and standard fuzzifier, when the relational matrix is the unit-row matrix, then defuzzifier (5.47) can be simplified as

$$y = \sum_{j=1}^{q} b_j c_j$$

■

---

**Theorem 5.8:**

In an FRDS described by $b(k+1) = b(k) \circ R$, with the unit-row relational matrix and standard fuzzifier, the actual output converges globally asymptotically to the center of the $l$th membership function if:

1. $ru = 1$.
2. $r_{il} \neq 0 \ \forall i$.

**Proof:**

The fuzzy output of the model evolves according to the following FRE: $X(k+1) = X(k) \circ R$. The $j$th element of this fuzzy vector is obtained as:

$$x_j(k+1) = \underset{i}{S}\, t(x_i(k), r_{ij}) = \sum_i x_i(k) r_{ij},$$

where $i$ belongs to $\{1,\ldots,p\}$ and specifically for the $l$th element of this fuzzy vector, it becomes as follows:

$$r_{ll} = 1 \qquad x_1(k) r_{ll} = x_l(k) \qquad x_l(k+1) = x_l(k) + \sum_{i=l} x_i(k) r_{il}.$$

Since $x_i$ and $r_{ij}$ are nonnegative, the inequality (5.50) holds as follows:

$$x_l(k+1) \geq x_l(k). \tag{5.50}$$

This inequality shows that the value of $x_l(k)$ grows up or remains constant as time passes. The equality occurs only when $\sum_{i \neq l} x_i(k) r_{il} = 0$ and then $x_l(k)$ remains unchanged. Using the second condition of the theorem, $r_{il} \neq 0$, yields the following:

$$x_l(k+1) = x_l(k) \qquad \sum_{i \neq l} x_i(k) r_{il} = 0 \qquad x_i(k) r_{il} = 0 \quad \forall i \neq l.$$

Hence:

$$x_i(k) = 0 \quad \forall i \neq l. \tag{5.51}$$

Combining (5.50) and (5.51), the $l$th element of the fuzzy output vector grows unless all other elements of the same vector vanish. Thus, it is now sufficient to prove that for every $j$ other than $l$, $x_j(k)$, vanishes as the time passes.

Note that $R$ is the unit-row matrix and hence the fuzzy vector $x(k)$ always ends on the hyperplane $\Gamma_f$:

$$\sum_{j=1}^{q} x_j(k) = 1. \tag{5.52}$$

Rewriting (5.52) yields:

$$\sum_{j=1}^{q} x_j(k) = x_l(k) + \sum_{\substack{j=1 \\ j\neq l}}^{q} x_j(k) = 1 \quad \sum_{\substack{j=1 \\ j\neq l}}^{q} x_j(k) = 1 - x_l(k). \tag{5.53}$$

From (5.50) and (5.51), it is found that $x_l(k)$ emerges to one, unless $\sum_{j\neq l} x_j(k) r_{jl}$ vanishes. However, according to (5.52), $\sum_{j\neq l} x_j(k) r_{jl}$ does not vanish unless $x_l(k) = 1$. Thus, the proof is completed and the final result can be written as follows for a given $l$:

$$\lim_{k\to\infty} x_i(k) = \begin{cases} 1 & j = l, \\ 0 & j \neq l. \end{cases}$$

This means that the actual output variable converges to the center of the $l$th membership function. ■

---

**Definition 5.19:**

Consider the surface $S_f$ defined in the $q$-D fuzzy space by an s-norm $S$ as:

$$S_f : \underset{j=1}{\overset{q}{S}}\; b_j = 1.$$

When the fuzzy output of an FRM stays on the surface $S_f$, that is, it always satisfies the above condition, the model can be said to have the property of *intrasurficiality in* $S_f$, or simply the FRM is *intrasurficial in* $S_f$. ■

Note that $S_{j=1}^{q} b_j$ means operating the s-norm on all elements of $b_j$ for from 1 to $q$, where $q$ is the number of linguistic terms. Applying the s-norm on more than two elements is sequentially possible due to the associativity of the s-norm operator. Also, note that the form or the shape of the surface depends on the type of s-norm $S$ and hence on the type of fuzzy composition.

---

**Lemma 5.2:**

If there exist a scalar $\alpha$ such that $s(\alpha a, \alpha b) = \alpha s(a, b)$, then $s(\alpha b) = \alpha s(b)$ where $b = [b_1 ... bq]$ and $s(b) = s(b_1, ..., b_2)$.

**Proof:**

Based on the associativity of the s-norm and according to the assumption $s(\alpha b_1, \alpha b_2) = \alpha s(b_1, b_2)$ for $q = 3$ we have:

$s(\alpha a_1, \alpha b_2, \alpha b_3) = s(\alpha b_1, s(\alpha b_2, \alpha b_3)) = s(\alpha b_1, \alpha s(b_2, b_3)) = \alpha s(b_1, s(b_2, b_3)) = \alpha s(b_1, b_2 b_3)$, and so the lemma could be proved by induction for any other values of $q$. ■

---

**Definition 5.20:**

Having a set $A \subset R$, the multi-input function $s$ has the property of homogeneity for $A$, or $s$ is homogeneous for $A$, when $s(\alpha a_1, \alpha b_2) = \alpha s(b_1, b_2)$, for every $\alpha \in A$. ■

**Example 5.5:**

Zadeh's s-norm," max" function, is homogeneous for $[0, \infty]$, for example, $0.2 \vee 0.4 \vee 0.9 = 2(0.1 \vee 0.2 \vee 0.45)$. Thus, it has the property of special homogeneity for $A = [0, \infty]$. ■

---

**Lemma 5.3:**

For a strictly monotone s-norm S, if $\underset{j=1}{\overset{q}{S}} b_j = 1$ then $j : b_j = 1$. ■

**Proof:**

It is sufficient to prove that if $b_j < 1$, for every $j \in \{1, ..., q\}$, then $\underset{j=1}{\overset{q}{S}} b_j < 1$. First, consider the case $q = 2$. Since is strictly monotone, one has $s(b_1, b_2) < s(1, b_2)$. But $s(1, b_2) = 1$, by the boundary condition of s-norm, and hence, $s(b_1, b_2) < 1$. Similarly, for $q = 3$, one can write $s(b_1, b_2, b_3) = s(b_1, s(b_2, b_3)) < s(1, s(b_2, b_3)) = 1$, and so the lemma could be proved by induction for any other values of $q$. ■

---

**Lemma 5.4:**

For an FR system with s-t fuzzy composition in which the s-norm is homogeneous for [0, 1], the t-norm is algebraic product, and the fuzzification curve lies on $S_f : \underset{j}{S} x_j = 1$ the following two statements are true:

1. If $\underset{j=1}{\overset{q}{S}} r_{ij} = 1 \ \forall i \in \{1,...,p\}$, then the FRM is intrasurficial in $S_f$, or simply $b \in S_f \quad \forall \alpha \in S_f$, (sufficient condition).
2. If the FRM is intrasurficial in $S_f$, then $i \in \{1,...,p\}, \underset{j=1}{\overset{q}{S}} r_{ij} = 1$, (necessary condition).

**Proof:**

Denote the *ith* row and the *j*th column of the matrix $R = \lfloor r_{ij} \rfloor$, respectively by $r_i$ and $C_j$. Any element of the fuzzy output vector can be written as:

$$b_j = \underset{i=1}{\overset{p}{S}} t(a_i, r_{ij}) = \underset{i=1}{\overset{p}{S}} a_i r_{ij}.$$

Hence:

$$\underset{j=1}{\overset{p}{S}} b_j = \underset{j=1}{\overset{q}{S}} \underset{i=1}{\overset{p}{S}} a_i r_{ij} = \underset{i=1}{\overset{p}{S}} \underset{j=1}{\overset{q}{S}} a_i r_{ij} = \underset{i=1}{\overset{p}{S}} a_i \underset{j=1}{\overset{q}{S}} r_{ij}, \tag{5.54}$$

since the s-norm is commutative and homogeneous for [0, 1]. The proof for each case follows separately. ■

**Proof of Statement 1:**

(The sufficient condition): Equation (5.54) and the assumption $\underset{j=1}{\overset{q}{S}} r_{ij} = 1$ lead to $\underset{j=1}{\overset{q}{S}} b_j = \underset{i=1}{\overset{q}{S}} a_i = 1$, which means both $a_i$ and $b_i$ lie on $S_f$, that is, $a$, $b \in S_f$. ■

**Proof of Statement 2:**

(The necessary condition): Having $\underset{i=1}{\overset{q}{S}} a_i = 1$ and $\underset{j=1}{\overset{a}{S}} b_j = 1$ is sufficient to show that $\underset{j=1}{\overset{q}{S}} r_{ij} = 1$. Rewrite (5.54) with a change of variable $p_i = \underset{j=1}{\overset{q}{S}} r_{ij}$ as $\underset{j=1}{\overset{q}{S}} b_j = \underset{i=1}{\overset{q}{S}} a_i p_i$, which yields $\underset{i=1}{\overset{p}{S}} a_i p_i = 1$. Therefore, by Lemma 5.3 at least one of the terms $a_i\ p_i$ is one, that is, $i : a_i p_i = 1$. Hence $i : (a_i = 1, p_i = 1)$, since $a_i p_i \in [0,1]$. Thus, at least one of the variables is one, which means:

$$i \mid \underset{j=1}{\overset{q}{S}} r_{ij} = 1.$$

■

---

**Lemma 5.5:**

For every strictly monotone s-norm $s, s(a,b) > a$ if $b \neq 0$. ■

**Proof:**

For a strictly monotone s-norm, one has: $b > c \quad s(a,b) > s(a,c)$. Since $b > 0$, putting in the property, the result is obtained. ■

---

**Corollary 5.2:**

When the operation of a strictly monotone s-norm results in zero, then all of its arguments are zero. ■

---

**Theorem 5.9:**

In a first-order s-t FRDS with strictly monotone fuzzy norms and with the property of intrasurficiality (in $S_f : \underset{j}{S}\, x_j = 1$), the center of the $lth$ linguistic term is a globally asymptotically stable equilibrium state if the following two conditions hold:

1. $r_{ll} = 1$,
2. $r_{il} \neq 0 \ \forall i$.

**Proof:**

Using $x(k+1) = x(k) \circ R$ we have and hence:

$$x_j(k+1) = \underset{i}{S}\, t(x_i(k), r_{ij})$$

$$x_l(k+1) = \underset{i}{S}\, t(x_i(k), r_{il}) = s(t(x_l(k), r_{ll}), \underset{i \neq l}{S}\, t(x_i(k), r_{il})).$$

$$r_{ll} = 1 \quad t(x_l(k), r_{ll}) = x_i(k) \quad x_l(k+1) = s(x_1(k), \underset{i \neq l}{S}\, t(x_i(k), r_{il})).$$

Thus:

$$x_l(k+1) \geq x_l(k).$$

This indicates that $x_l(k)$ grows or remains constant as time passes. However, since the s-norm is strictly monotone, according to Lemma 5.3 the equality in (5.50) holds only when $\underset{i \neq l}{S}\, t(x_i(k), r_{il}) = 0$. In this case $x_l(k)$ remains constant.

$$x_l(k+1) = x_l(k) \quad \underset{i \neq l}{S}\, t(x_i(k), r_{il}) = 0 \quad t(x_i(k), r_{il}) = 0 \ \forall i \neq l.$$

But $r_{il} \neq 0$ by assumption, and therefore:

$$x_l(k+1) = x_l(k) \quad x_i(k) = 0 \ \forall i \neq l.$$

Finally, having $\underset{j}{S}\, x_j(k)=1,$ according to the intrasurficiality property of the model in $S_f$: $\underset{j}{S}\, x_j=1$, the proof is completed by the following expansion of $\underset{j}{S}\, x_j(k)$:

$$\underset{j}{S}\, x_j(k)=x_l(k)+\underset{j\neq l}{S}\, x_j(k)=1 \qquad \underset{j\neq l}{S}\, x_j(k)=1-x_l(k).$$

$$k\rightarrow\infty \qquad \underset{j\neq l}{S}\, x_j(k)\rightarrow 0 \qquad x_j(k)\rightarrow 0 \;\; \forall j\neq l.$$

Thus, the result can be written as $\lim_{k\rightarrow\infty} x_j(k)=\begin{cases}1 & j=l\\ 0 & j=l\end{cases}$, which means that the output variable approaches the center of the *lth* linguistic term. ■

Now focus on the assumed properties of the s-norm in Theorem 5.9, that is, strict monotonicity and intrasurficiality. Consider the algebraic product as the t-norm, then, according to Lemma 5.4, the model is intrasurficial in $S_f$ when the s-norm is homogeneous for [0,1]. So, the following two properties for the s-norm are considered in this analysis, "strict monotonicity" and "homogeneity for [0,1]". Strict monotonicity holds for many types of s-norms, but, the question that arises here is whether there exists an s-norm that satisfies both properties simultaneously. The answer is apparently "no." An example of an s-norm that is homogeneous for [0,1] is the well-known Zadeh norm, max function, which is not strictly monotone.

Now, let us obtain a less conservative set of conditions. Remember that the general relation for the s-t composition is:

$$x_j(k+1)=\underset{i}{S}\, t(x_i(k),r_{ij}).$$

The above iterative equation is very simple to understand. This equation means that all elements of the fuzzy vector in the previous instant contribute in determining the value of the *j*th element of the current fuzzy vector. When the element $r_{ij}$ is zero, then there is no direct dependency between the *i*th element of the previous fuzzy vector and the *j*th element of the current one. Nevertheless, there may be an indirect dependency between them, for example, the *j*th element depends on the *i*th one and the *j*th element depends on the *i*th one. In such a case, it takes two instant of times for the *i*th element to affect the value of the *j*th element.

---

**Definition 5.21:**

When the *j*th element of the fuzzy vector depends directly on the *i*th element of it through iterative operation of the relational matrix, it can be said that there is a route of length one from the *i*th element to the *j*th element, since it takes one instant of time for the *i*th element to affect the value of the *i*th element. Similarly, by a route of length $l_{path}$, it means that the value of $x_j(k)$ depends on the value of $x_i(k-l_{path})$. ■

**Example 5.6:**

Consider the following relational matrix. The dependency of the fifth element of the fuzzy vector of the correspondent FRM to the first element is investigated.

$$R = \begin{bmatrix} 0.5 & 0.5 & 0 & 0 & 0 \\ 0 & 0.5 & 0.5 & 0 & 0 \\ 0 & 0 & 0.5 & 0.5 & 0 \\ 0 & 0 & 0 & 0.5 & 0.5 \\ 0 & 0 & 0 & 0 & 1 \end{bmatrix}.$$

Obviously, the fifth element does not depend directly on the first element since $r_{15} = 0$. Hence, there is no route of length one from 1 to 5. However, to see if there is an indirect route from the first element to the fifth, the Boolean mapping is defined. ■

---

**Definition 5.22:**

A Boolean matrix is defined as a matrix with the same size as the relational matrix with only zero and one as its elements such that each element is zero (one) when the corresponding element in the relational matrix is zero (nonzero). The transformation that maps a relational matrix to its Boolean matrix is called Boolean mapping and is denoted with $B(R)$, where the relational matrix is. ■

More clearly the Boolean matrix $B$ is obtained as follows:

$$B = B(R), \quad B = [b_{ij}], \quad b_{ij} = \begin{cases} 0 & r_{ij} = 0, \\ 1 & r_{ij} \neq 0. \end{cases}$$

---

**Remark 5.8:**

Obviously the elemental value of 1 (0) in the Boolean matrix indicates the existence (nonexistence) of direct dependence of the $j$th element on the $i$-th element. ■

---

**Convention:**

The rows and columns of a matrix respectively are hereby denoted by $r_i(B)$ and $c_j(B)$. So the matrix itself can be expressed as:

$$B = [r_i(B)] = [C_j(B)]$$ ■

---

**Lemma 5.6:**

There exists at least one route of length two from the *i*th element of the fuzzy vector to the *j*th one iff:

$$r_i(B) \underset{zadeh}{\circ} c_j(B) = 1$$

where $\underset{zadeh}{\circ}$ means fuzzy composition with Zadeh norms, that is, max-min composition. ■

**Proof:**

The existence of a route of length two from $i$ to $j$ means that the *j*th element of the current fuzzy vector depends on the *i*th element of the fuzzy vector of two instants ago and this means that the *j*th element depends directly on an intermediate element and that intermediate element depends directly on the *i*th element. Hence, two direct dependencies should be verified. This is done by the “min” operator. The intermediate element can be any of the $q$ elements of the fuzzy vector. Therefore, the route can be made through any or some of them and so we conclude the existence of the route by the “max” operator. The inverse is obviously true using a similar reasoning. ■

---

**Remark 5.9:**

Remember that in Boolean algebra, the “max” (“min”) operator is equivalent to logical “or” (“and”) and hence, $r_i(B) \underset{zadeh}{\circ} c_j(B) = 1$ means that there exists at least one intermediate element on which the *jth* element depends and that one itself depends on the *ith* element. ■

---

**Lemma 5.7:**

The number of the routes of length two, $n_{path}$, is determined by:

$$n_{path} = r_i(B) \underset{algebraic}{\circ} c_j(B),$$

where $\underset{algebraic}{\circ}$ is algebraic fuzzy composition, that is, sum-prod composition. ■

**Proof:**

A similar argument as in the previous lemma is used to prove this lemma. Note that by using the sum operator instead of the “max” operator here, the number of the existing routes is calculated. Furthermore, the “prod” operator can be replaced with the “min” operator. ■

**Definition 5.23:**

By the length-two route map, we mean a matrix in which each element $(i, j)$ represents the existence of at least one route of length two from the $i$th element to the $j$th element. In a similar way, the length-$l_{path}$ route map is defined for investigating the existence of routes of length $l_{path}$, where 1 (0) indicates the existence (nonexistence) of routes. A matrix whose elements represent the numbers of existing routes is called *length-$l_{path}$ routes map*. Obviously, these elements are nonnegative integers. ■

**Lemma 5.8:**

The length-two route map and the length-two routes map are determined by $B \underset{Zadeh}{\circ} B$ and $B \underset{algebraic}{\circ} B$, respectively.

**Proof:**

The proof is obvious using Lemmas 5.6 and 5.7. ■

**Example 5.7:**

Consider the following relational matrix $R$ and study its direct routes as well as indirect routes of length two.

$$R = \begin{bmatrix} 0.1 & 0.5 & 0.3 & 0 & 0 \\ 0 & 1.2 & 0.5 & 0.2 & 0 \\ 0 & 0 & 1 & 0.1 & 0.4 \\ 0 & 0 & 0.8 & 0.2 & 0.1 \\ 0 & 0 & 0 & 0.3 & 0.1 \end{bmatrix} \quad B = \begin{bmatrix} 1 & 1 & 1 & 0 & 0 \\ 0 & 1 & 1 & 1 & 0 \\ 0 & 0 & 1 & 1 & 1 \\ 0 & 0 & 1 & 1 & 1 \\ 0 & 0 & 0 & 1 & 1 \end{bmatrix},$$

$$B \underset{zadeh}{\circ} B = \begin{bmatrix} 1 & 1 & 1 & 0 & 0 \\ 0 & 1 & 1 & 1 & 0 \\ 0 & 0 & 1 & 1 & 1 \\ 0 & 0 & 1 & 1 & 1 \\ 0 & 0 & 0 & 1 & 1 \end{bmatrix} \underset{zadeh}{\circ} \begin{bmatrix} 1 & 1 & 1 & 0 & 0 \\ 0 & 1 & 1 & 1 & 0 \\ 0 & 0 & 1 & 1 & 1 \\ 0 & 0 & 1 & 1 & 1 \\ 0 & 0 & 0 & 1 & 1 \end{bmatrix} = \begin{bmatrix} 1 & 1 & 1 & \boxed{1} & \boxed{1} \\ 0 & 1 & 1 & 1 & \boxed{1} \\ 0 & 0 & 1 & 1 & 1 \\ 0 & 0 & 1 & 1 & 1 \\ 0 & 0 & \boxed{1} & 1 & 1 \end{bmatrix},$$

$$B \underset{algebraic}{\circ} B = \begin{bmatrix} 1 & 1 & 1 & 0 & 0 \\ 0 & 1 & 1 & 1 & 0 \\ 0 & 0 & 1 & 1 & 1 \\ 0 & 0 & 1 & 1 & 1 \\ 0 & 0 & 0 & 1 & 1 \end{bmatrix} \underset{algebraic}{\circ} \begin{bmatrix} 1 & 1 & 1 & 0 & 0 \\ 0 & 1 & 1 & 1 & 0 \\ 0 & 0 & 1 & 1 & 1 \\ 0 & 0 & 1 & 1 & 1 \\ 0 & 0 & 0 & 1 & 1 \end{bmatrix} = \begin{bmatrix} 1 & 2 & 3 & 2 & 1 \\ 0 & 1 & 3 & 3 & 2 \\ 0 & 0 & 2 & 3 & 3 \\ 0 & 0 & 2 & 3 & 3 \\ 0 & 0 & 1 & 2 & 2 \end{bmatrix}$$

For example, consider the element (1,4). It is seen from the maps that there are two indirect routes (of length two) from the first element to the fourth element of the fuzzy vector. These routes are through the second and third elements. ■

Thus, an index for the existence and the number of length-two routes is obtained. However, whether it is possible to have an index for the intensity of the effect of the $i$th element on the $j$th element by length-two routes remains open. A good index in this respect can be $R \circ R$, where the fuzzy composition here is the same as the fuzzy composition used in constructing the FRM.

**Example 5.8:**

For the relational matrix of Example 5.7, calculate the inter-effect map of linguistic terms for two cases, "max"-"min" composition, and "sum"-"prod" composition.

$$R \circ R = \begin{bmatrix} 0.1 & 0.5 & 0.3 & 0 & 0 \\ 0 & 0.2 & 0.5 & 0.2 & 0 \\ 0 & 0 & 1 & 0.1 & 0.4 \\ 0 & 0 & 0.8 & 0.2 & 0.1 \\ 0 & 0 & 0 & 0.3 & 0.1 \end{bmatrix} \circ \begin{bmatrix} 0.1 & 0.5 & 0.3 & 0 & 0 \\ 0 & 0.2 & 0.5 & 0.2 & 0 \\ 0 & 0 & 1 & 0.1 & 0.4 \\ 0 & 0 & 0.8 & 0.2 & 0.1 \\ 0 & 0 & 0 & 0.3 & 0.1 \end{bmatrix},$$

$$R \underset{max-min}{\circ} R = \begin{bmatrix} 0.1 & 0.2 & 0.5 & 0.2 & 0.3 \\ 0 & 0.2 & 0.5 & 0.2 & 0.4 \\ 0 & 0 & 1 & 0.3 & 0.4 \\ 0 & 0 & 0.8 & 0.2 & 0.4 \\ 0 & 0 & 0.3 & 0.2 & 0.1 \end{bmatrix},$$

$$R \underset{sum-prod}{\circ} R = \begin{bmatrix} 0.01 & 0.15 & 0.58 & 0.13 & 0.12 \\ 0 & 0.04 & 0.76 & 0.13 & 0.22 \\ 0 & 0 & 1 & 0.24 & 0.45 \\ 0 & 0 & 0.96 & 0.15 & 0.35 \\ 0 & 0 & 0.24 & 0.09 & 0.04 \end{bmatrix},$$

■

---

**Remark 5.10:**

Using Definition 5.22, it can easily be shown that $B(R \circ R) = B \underset{zadeh}{\circ} B$. ■

**Lemma 5.9:**

For a $n \times n$ relational matrix, the length of the longest route is $n - 1$. ■

**Proof:**

For the $n \times n$ relational matrix, the problem can be restated that there exist points and one wants to go from any arbitrary source point to any arbitrary destination point such that none of the points are met more than once. Hence, the longest route consists of $n - 1$ segments, and so it is said that the route length is $n - 1$. ■

---

**Lemma 5.10:**

The length-$l_{path}$ route map and the length-$l_{path}$ routes map are obtained respectively by $B_{zadeh}^{l_{path}}$ and $B_{algebraic}^{l_{path}}$ as follows for a relational matrix $R_{p \times q}$:

$$B_{zadeh}^{l_{path}} \cong \overbrace{B \underset{zadeh}{\circ} \cdots \underset{zadeh}{\circ} B}^{l_{path}}, \quad B_{algebraic}^{l_{path}} \cong \overbrace{B \underset{algebraic}{\circ} \cdots \underset{algebraic}{\circ} B}^{l_{path}},$$

where

$$l_{path} \leq n-1, n = p \wedge q.$$ ■

**Proof:**

It is a simple generalization of Lemma 5.8. ■

---

**Definition 5.24:**

The complete route map is a matrix in which the element $(i, j)$ indicates the existence of a route of any length from the $i$th element of the fuzzy vector to the $j$th one; 1 (0) for existence (nonexistence). If the number of all routes of any length from the $i$th element of the fuzzy vector to the $j$th one is shown as the element $(i, j)$ of a matrix, then this matrix is called a *complete routes map* matrix. ■

---

**Corollary 5.3:**

The complete route map matrix and the complete routes map matrix of a relational matrix are obtained as given by Table 5.1. ■

**TABLE 5.1**
**Obtaining the Complete Route (s) Maps Matrix of an FRM**

| The Complete Route | The Complete Routes Map |
|---|---|
| $\bigvee_{i=1}^{l_{path}} B_{zadeh}^{i}$ | $\sum_{i=1}^{l_{path}} B_{algebraic}^{i}$ |

**Theorem 5.10: (Aghili–Nikravesh Theorem 1)**

In a SUWA FRDS described by $b(k+1) = b(k)_0 R$, the output (of the defuzzifier) converges globally asymptotically to the center of the $l$th membership function if the following two conditions hold:

1. $r_{ll} = 1 \psi\, r_{ij} < \forall (i, j) \neq (l, l)$.
2. All the elements of the $l$th column in the complete route map of $R$ be nonzero.

**Proof:**

The fuzzy output of the model evolves according to the FRE $X(k+1) = X(k) \circ R$. The $j$th element of this fuzzy vector is obtained as $x_j\,(k+1) = \underset{i}{S}\,t\,(x_i(k), r_{ij}) = \sum_i x_i\,(k) r_{ij}$, where $i \in \{1,..., p\}$ and specifically, for the $l$-th element it becomes as follows:

$$r_{ll} = 1 \quad x_l\,(k+1) = x_l\,(k) + \sum_{i \neq l} x_i\,(k) r_{il}.$$

Since $x_i$ and $r_{ij}$ are nonnegative, the inequality (5.55) holds as follows:

$$x_l\,(k+1) \geq x_l\,(k). \tag{5.55}$$

This inequality shows that the value of $x_l(k)$ grows up or remains constant as time passes. The equality holds only when $\sum_{i \neq l} x_i\,(k) r_{il} = 0$ and then $x_l\,(k)$ remains unchanged.

It is known by the second assumption of the theorem that all elements of the $l$th column in the complete route map of $R$ are nonzero, hence there is at least one non-zero element in the $l$th column, other than $r_{ll}$. Since $r_{il}$ can be zero for some values of $i$, define $I = \{i : r_{il} \neq 0, i \neq 0\}$, thus, $\forall i \in I : x_i(k) = 0$, since $r_{il} \neq 0$. For $i \notin I$, a little more work is needed. It is known by assumption that all elements of the $l$th column in the complete route map of $R$ are nonzero. Hence, for every $i \notin I$ there is an index $i' \notin I$ such that $x_{i'}(k)$ depends on $x_i(k)$, that is, there exists at least one route from $i$ to $i'$. On the other hand, it is known that this dependency is through the s-norm operator. Hence

$$x_i(k) = 0 \text{ since } x_{i'}(k) = 0.$$

Thus,

$$\forall i \notin I : x_i(k) = 0.$$

Therefore:

$$x_i(k) = 0 \;\; \forall i \neq l. \tag{5.56}$$

Combining (5.55) and (5.56), we conclude that the $l$th element of the fuzzy output vector grows unless all other elements of the same vector vanish. Thus, it is now sufficient to prove that for every $j$ other than $l$, $x_k(k)$, vanishes as time passes.

Note that $R$ is a unit-row matrix and hence the fuzzy vector $x(k)$ always ends on the hyperplane $\Gamma_f$, that is:

$$\sum_{j=1}^{q} x_j(k) = 1. \tag{5.57}$$

Rewriting (5.57) yields:

$$\sum_{j=1}^{q} x_j(k) = x_l(k) + \sum_{\substack{j=1 \\ j\neq 1}}^{q} x_j(k) = 1 \quad \sum_{\substack{j=1 \\ j\neq 1}}^{q} x_j(k) = 1 - x_l(k).$$

From (5.55) and (5.56), it is found that $x_l(k)$ approaches toward one, unless $\sum_{j\neq l} x_j(k) r_{jl}$ vanishes. However, according to (5.57), $\sum_{j\neq l} x_j(k) r_{jl}$ does not vanish unless $x_l(k) = 1$. This completes the proof and can be summarized as follows:

$$\lim_{k\to\infty} x_j(k) = \begin{cases} 1 & j = l \\ 0 & j \neq l \end{cases}.$$

This means that the output converges to the center of the $l$th membership function. ■

**Example 5.9:**

Consider the FRDS described by $b(k+1) = b(k) o R$ with "sum"-"prod" fuzzy composition and standard fuzzifier. The relational matrix is as follows:

$$R = \begin{bmatrix} 0.8 & 0.2 & 0 & 0 & 0 \\ 0 & 0.8 & 0.2 & 0 & 0 \\ 0 & 0 & 1 & 0 & 0 \\ 0 & 0 & 0.2 & 0.8 & 0 \\ 0 & 0 & 0 & 0.2 & 0.8 \end{bmatrix}.$$

It is clear that five linguistic values are considered to describe the fuzzy variable, see Figure 5.15. The nonfuzzy universe of discourse is [–2,2].

Both conditions of Theorem 5.10 hold here for the third linguistic term. Hence, it can be said that the actual output of the dynamic system certainly converges to the center of the third membership function, that is, 0, here, from every arbitrary initial point in the nonfuzzy universe of discourse. The set of the centers of the membership functions is {–2.–1,0,1,2}.

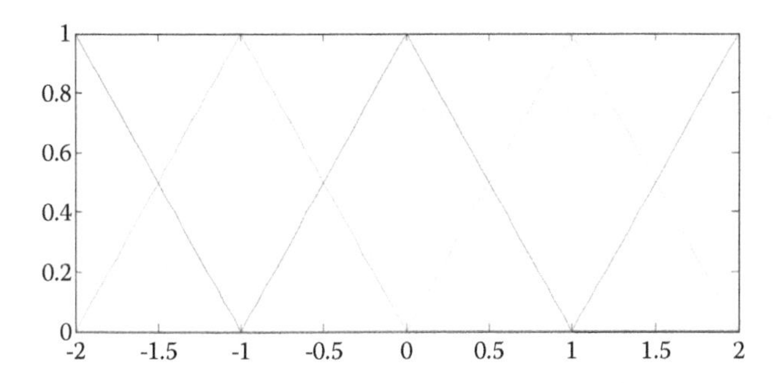


**FIGURE 5.15** The membership functions for the Example 5.9.

Figure 5.16 shows the trajectory of the output of the model initialized from $y(0) = -2$. A simple exponential function $e^{-\alpha t}$ is fitted to the output trajectory to see how close the trajectory of this specific example is to that of a first-order nonfuzzy dynamic system. Here $\alpha = -15$.

Note that, as expected, in Table 5.2, the sum of the elements of each row, that is, the fuzzy vector, is one. It means that having a unit-row relational matrix in the FRDS $b(k+1) = b(k) \circ R$, makes the hyperplane (5.48). ■

## 5.6 ASYMPTOTIC CONVERGENCE TO THE EQUILIBRIUM STATE

In this section, the same problem of the previous section is considered; however, a different approach is taken to solve the problem. The mathematical foundation of this approach is based on a special notion of symmetry for the relational matrix

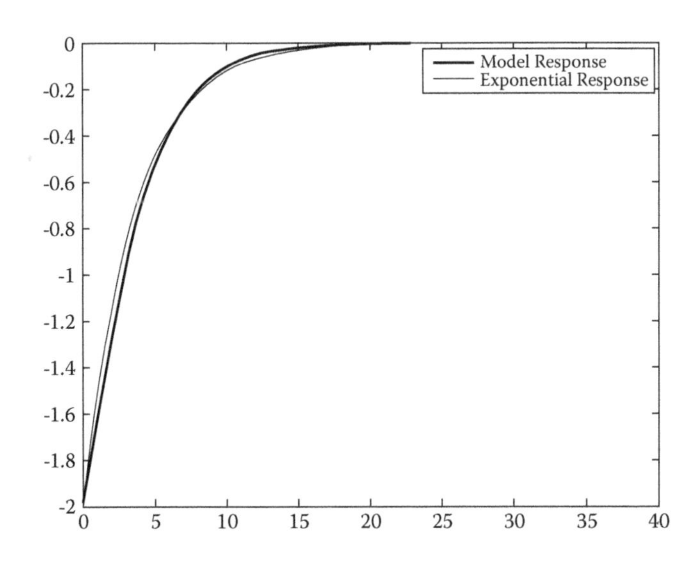


**FIGURE 5.16** Trajectory of the nonfuzzy output and its exponential fitting function.

**TABLE 5.2**
**Evolution of the Output Fuzzy Vector from [1,0,0,0,0] to [0,0,1,0,0]**

| $(k)$ | $\underline{b}(k)$ | | | | | |
|---|---|---|---|---|---|---|
| 0 | 1 | $\circ R$ | 0 | 0 | 0 | 0 |
| 1 | 0.8 | | 0.2 | 0 | 0 | 0 |
| 2 | 0.64 | | 0.32 | 0.04 | 0 | 0 |
| 3 | 0.512 | | 0.384 | 0.104 | 0 | 0 |
| 4 | 0.4096 | | 0.4096 | 0.1808 | 0 | 0 |
| 5 | 0.3277 | | 0.4096 | 0.2627 | 0 | 0 |
| ⋮ | ⋮ | | ⋮ | ⋮ | ⋮ | ⋮ |
| 100 | 0 | | 0 | 1 | 1 | 0 |

introduced in Appendix A5. A symmetric approach to the problem is then taken and a sufficient condition for the stability of the equilibrium state based on the location of the eigenvalues of the fuzzy relational matrix is obtained. The non-fuzzy universe of discourse is normalized to [0,1] and the fuzzifier is symmetric with respect to the given membership function. The defuzzifier is a simple weighted average.

First, some useful definitions from Appendix A5 are briefly restated here.

**Definition 5.25:**

The square matrix $R$ is Centrally Symmetric (CS) when

$$r(i,j) = r(n+1-i, n+1-j). \qquad \blacksquare$$

**Definition 5.26:**

The matrix $R$ is Row-Wise (Column-Wise) Symmetric (RWS/CWS) when:

$$r(i,j) = r(n+1-i, j) \quad (r(i,j) = r(i, n+1-j)). \qquad \blacksquare$$

**Definition 5.27:**

The matrix R is Row-Wise (Column-Wise) Negative Symmetric (RWNS/CWNS) when:

$$r(i,j) = -r(n+1-i, j) \quad (r(i,j) = -r(i, n+1-j)). \qquad \blacksquare$$

**Definition 5.28:**

The square matrix $R$ is Plus Symmetric (PS) when:

$$r(i,j)=r(n+1-i,j)=r(i,n+1-j)=r(n+1-i,n+1-j). \quad \blacksquare$$

**Remark 5.11:**
None of the above-mentioned definitions are limited to square matrices, except the CS and the PS matrices. For example, an RWS column vector is meaningful and is used hereafter. ■

**Example 5.10:**

The type of the following matrices and vectors are determined using Definition 5.25 through Definition 5.28.

$$\begin{pmatrix} 0.5 & 0.1 & 1 \\ 0.7 & 0.9 & 0.7 \\ 1 & 0.1 & 0.5 \end{pmatrix}: CS, \quad \begin{pmatrix} -0.5 & 0.1 & -1 \\ 0 & 0 & 0 \\ 0.5 & -0.1 & 1 \end{pmatrix}: RWNS,$$

$$\begin{pmatrix} 0.5 & 0.1 & 1 \\ 0.7 & 0.9 & 0.7 \\ 0.5 & 0.1 & 1 \end{pmatrix}: RWS, \quad \begin{pmatrix} 0.5 \\ 0.7 \\ 0.1 \end{pmatrix}: RWS,$$

$$\begin{pmatrix} 0.5 & -0.3 & 0.5 \\ 0.7 & 0.9 & 0.7 \\ 0.5 & -0.3 & 0.5 \end{pmatrix}: PS, \quad \begin{pmatrix} 0.1 \\ 0 \\ -0.1 \end{pmatrix}: RWNS,$$

**Lemma 5.11:**

Assume a symmetric weighted average defuzzifier as described in Section 5.4.1 with $2m+1$ linguistic terms, each associated to a center of a membership function. Let: $c=[-c_m \cdots -c_1\ 0\ c_1 \cdots c_m]$ denote the vector of those center values. If the fuzzy vector $b$, which is a row vector here, is CWS, then the nonfuzzy output of the defuzzifier is zero. ■

**Proof:**

The defuzzifier is:

$$y=\sum_{j=1}^{q} b_j c_j \Big/ \sum_{j=1}^{q} b_j, \quad c_0=0.$$

The fuzzy row vector $b$ is CWS, and so it can be written as:

$$b = [b_i \cdots b_1 \;\; b_0 \; b_1 \cdots b_i].$$

Hence, the numerator of the defuzzifier vanishes, thus:

$$b_i c_i + \cdots + b_1 c_1 + b_0 (0) - b_1 c_1 - \cdots - b_i c_i = 0, \text{ and so } y = 0.$$

■

**Theorem 5.11: (Theorem 4 in Aghili Ashtiani [a5])**

Let $R$ be an $n \times n$ CS matrix that has $n$ distinct eigenvalues. Then $R^k$ tends to a PS matrix if all eigenvalues associated with RWNS eigenvectors are located in the unit circle. ■

**Proof:**

See Appendix A5. ■

**Remark 5.12:**

Note that the condition of eigenvalues in Theorem 5.11 is not a hard restriction and is true for most cases as simulation results show. ■

**Theorem 5.12: (Aghili–Nikravesh Theorem 2)**

In a SUWA FRDS described by $b = (k+1) = b(k)_0 R$ in which the $q \times q$ relational matrix $R$ is CS and has $q$ distinct eigenvalues, the output of the symmetric defuzzifier converges globally asymptotically to the equilibrium state from every arbitrary initial point, if all eigenvalues associated with the RWNS eigenvectors are located in the unit circle. ■

**Proof:**

The FRE of the model can be written as:

$$b(k) = b(k-1)R = b(0)R^k,$$

since the fuzzy composition is "sum"-"prod" and the relational matrix is a unit-row matrix. Note that all vectors here are row vectors. For every element, one has

$b_j(k) = b(0)c_j(R^k), \forall_j = \{1,...,q\}$ where $c_j(R^k)$ is the $j$th column of $R^k$. By Theorem 5.11, $R^k$ tends toward a PS matrix as $k \to \infty$ and every PS matrix is both an RWS matrix and a CWS matrix. Therefore, $c_j(R^k) \to c_{q+1-j}(R^k)$ as $k \to \infty$ and so is for $b_j(k) \to b_{q+1-j}(k)$ as $k \to \infty$, for all $j = \{1,...,q\}$. In this way Lemma 5.11 would lead to the final conclusion, that is, $y \to 0$ as $k \to \infty$. ■

Theorem 5.12 indicates that when the conditions are met, the origin is a globally stable equilibrium state of the system; so when the system is somehow perturbed from the origin, the system comes back to the equilibrium state if the external perturbation is removed.

### Example 5.11:

Consider an FRDS described by $b(k+1) = b(k) \circ R$ and configured as the specific configuration in Section 5.4.1 with $q = 5$. To verify Theorem 5.12, a random CS unit-row relational matrix can be generated and can be used to run the model from any arbitrary initial condition. Such a relational matrix is as follows:

$$R = \begin{pmatrix} 0.3684 & 0.3684 & 0.0789 & 0.0789 & 0.1053 \\ 0.2889 & 0.2000 & 0.2444 & 0.1111 & 0.1556 \\ 0.2167 & 0.1500 & 0.2667 & 0.1500 & 0.2167 \\ 0.1556 & 0.1111 & 0.2444 & 0.2000 & 0.2889 \\ 0.1053 & 0.0789 & 0.0789 & 0.3684 & 0.3684 \end{pmatrix}.$$

The eigenvectors associated with RWNS eigenvectors should be checked. The modal matrix for which its columns are the eigenvectors of $R$ and the corresponding eigenvalues are as follows:

$$M = \begin{pmatrix} 0.2279 & 0.1583 & 0.1138 & 0.3514 & 0.1486 \\ 0.1857 & 0.4047 & -0.5903 & 0.3367 & -0.3367 \\ 0.1728 & -1.1259 & 0.9531 & 0 & 0 \\ 0.1857 & 0.4047 & -0.5903 & -0.3367 & 0.3367 \\ 0.2279 & 0.1583 & 0.1138 & -0.3514 & -0.1486 \end{pmatrix}.$$

The eigenvalues associated with the above eigenvectors are as follows, respectively:

$$[1.0000 \;\; 0.0688 \;\; -0.0173 \;\; 0.3909 \;\; -0.0389].$$

Note that the conditions of Theorem 5.12 are satisfied. It is obvious that one of the eigenvalues is 1 and thus, it is not in the unit circle; but according to the theorem, this is not a problem since it is not related to any of the RWNS eigenvectors. Hence, we expect that the output of the system converges to the origin from any arbitrary initial condition.

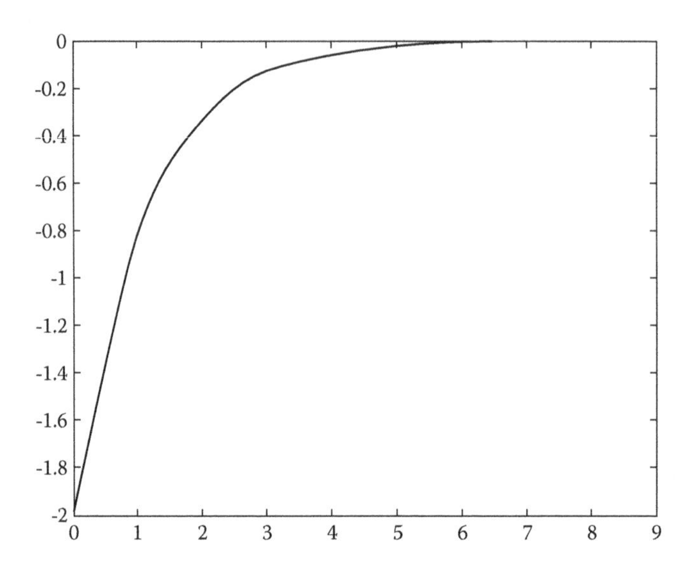


**FIGURE 5.17** Convergence of the output trajectory to the origin.

Consider, for example, the system with initial condition $y(0) = -2$, which is equivalent to the fuzzy vector [1 0 0 0 0] according to the specified fuzzifier. Figure 5.17 shows the nonfuzzy and fuzzy output in some instances of time. It is seen that the fuzzy vector tends toward a CWS row vector and the nonfuzzy output tends toward zero. Also, the output trajectory is depicted in Figure 5.17 (Table 5.3). ■

**TABLE 5.3**
**Evolution of the System Output and its Fuzzy Vector**

| $k$ | $y(k)$ | $b(k)$ | | | | |
|---|---|---|---|---|---|---|
| 0 | -2.000 | 1.0000 | 0 | 0 | 0 | 0 |
| 1 | -0.816 | 0.3684 | 0.3684 | 0.0789 | 0.0789 | 0.1053 |
| 2 | -0.318 | 0.2826 | 0.2383 | 0.1678 | 0.1364 | 0.1748 |
| 3 | -0.124 | 0.2490 | 0.2059 | 0.1725 | 0.1656 | 0.2070 |
| 4 | -0.049 | 0.2361 | 0.1935 | 0.1728 | 0.1778 | 0.2197 |
| 8 | -0.001 | 0.2281 | 0.1858 | 0.1728 | 0.1855 | 0.2277 |
| 9 | -0.000 | 0.2280 | 0.1857 | 0.1728 | 0.1856 | 0.2279 |

**PROBLEMS:**

**5.1:** Consider the FRDS with standard fuzzifier, weighted-average defuzzifier, and algebraic fuzzy relational composition. For which of the following relational matrices, global convergence to a unique equilibrium point can be implied? What is that equilibrium point?

$$R_1 = \begin{bmatrix} 0.1 & 0.3 & 0.6 \\ 0.0 & 1.0 & 0.0 \\ 0.4 & 0.3 & 0.3 \end{bmatrix}$$

$$R_2 = \begin{bmatrix} 0.1 & 0.3 & 0.6 \\ 0.0 & 1.0 & 0.0 \\ 0.4 & 0.0 & 0.3 \end{bmatrix}$$

$$R_3 = \begin{bmatrix} 0.1 & 0.3 & 0.6 \\ 0.0 & 1.0 & 0.0 \\ 0.4 & 0.0 & 0.0 \end{bmatrix}$$

$$R_4 = \begin{bmatrix} 0.1 & 0.3 & 0.6 \\ 0.0 & 1.0 & 0.0 \\ 1.0 & 0.0 & 0.0 \end{bmatrix}$$

$$R_5 = \begin{bmatrix} 0.2 & 0.3 & 0.3 & 0.2 \\ 0.2 & 0.3 & 0.4 & 0.1 \\ 0.1 & 0.4 & 0.3 & 0.2 \\ 0.2 & 0.5 & 0.2 & 0.1 \end{bmatrix}$$

$$R_6 = \begin{bmatrix} 0.2 & 0.2 & 0.2 & 0.2 & 0.2 \\ 0.1 & 0.2 & 0.4 & 0.2 & 0.1 \\ 0.0 & 0.0 & 0.1 & 0.0 & 0.0 \\ 0.2 & 0.1 & 0.4 & 0.1 & 0.2 \\ 0.2 & 0.2 & 0.2 & 0.2 & 0.2 \end{bmatrix}$$

$$R_7 = \begin{bmatrix} 0.2 & 0.2 & 0.2 & 0.2 & 0.2 \\ 0.1 & 0.2 & 0.4 & 0.2 & 0.1 \\ 0.1 & 0.3 & 0.2 & 0.3 & 0.1 \\ 0.1 & 0.2 & 0.4 & 0.2 & 0.1 \\ 0.2 & 0.2 & 0.2 & 0.2 & 0.2 \end{bmatrix}$$

$$R_8 = \begin{bmatrix} 0.2 & 0.3 & 0.0 & 0.3 & 0.2 \\ 0.1 & 0.2 & 0.4 & 0.2 & 0.1 \\ 0.0 & 0.0 & 1.0 & 0.0 & 0.0 \\ 0.6 & 0.0 & 0.0 & 0.3 & 0.1 \\ 0.7 & 0.0 & 0.0 & 0.2 & 0.1 \end{bmatrix}$$

$$R_9 = \begin{bmatrix} 0.1 & 0.2 & 0.3 & 0.1 & 0.2 & 0.1 \\ 0.2 & 0.1 & 0.1 & 0.3 & 0.1 & 0.2 \\ 0.3 & 0.1 & 0.0 & 0.1 & 0.2 & 0.3 \\ 0.3 & 0.2 & 0.1 & 0.0 & 0.1 & 0.3 \\ 0.2 & 0.1 & 0.3 & 0.1 & 0.1 & 0.2 \\ 0.1 & 0.2 & 0.1 & 0.3 & 0.2 & 0.1 \end{bmatrix}$$

$$R_9 = \begin{bmatrix} 0.0 & 0.0 & 0.5 & 0.0 & 0.0 & 0.5 \\ 0.5 & 0.0 & 0.5 & 0.0 & 0.0 & 0.0 \\ 0.0 & 0.0 & 0.5 & 0.5 & 0.0 & 0.0 \\ 0.0 & 0.5 & 0.5 & 0.0 & 0.0 & 0.0 \\ 0.0 & 0.0 & 0.0 & 0.0 & 1.0 & 0.0 \\ 0.0 & 0.0 & 0.5 & 0.0 & 0.5 & 0.0 \end{bmatrix}$$

**5.2:** Regarding the theorems in this section:

(a) Is it possible to achieve some necessary conditions for the existence of a unique stable equilibrium point?

(b) Concerning the sufficient conditions, can you extend the results to cover the case in which the existence of a unique stable equilibrium point somewhere between the centers of the linguistic terms is implied?

(c) Concerning the sufficient conditions, can you extend the results to cover the case in which the existence of more than one equilibrium point is implied?

**5.3:** What is your idea about the fuzzy relational models of order higher than one? How do you define the equilibrium point for such models? Is it possible to obtain some conditions for the existence of any stable equilibrium points?

# References

## a


1. Anderson, L.R. and Leighton, W. "Liapunov functions for autonomous systems of second order." *Journal of Mathematical Analysis and Application* vol. 23, no. 3, September 1968.
2. Aeyels, D. and Peuteman, J. "On exponential stability of nonlinear time-variant differential equations." *Automatica* 35(6) 1091–1100, 1999.
3. Aeyels, D. "Asymptotic stability of non-autonomous systems by Liapunov's direct method." *Systems and Control Letters* 25, no. 4:273–280, 1995.
4. Alaviani, S. Sh, "Stabilization of Special Class of Linear Time Varying Time Delay Systems." MSc Thesis, Tehran, Iran: Amirkabir University of Technology (AUT), (in Farsi) January 2006.
5. Aghili Ashtiani, A. "Fuzzy Relational Equations, Extension to Dynamic Systems and Investigating their Stability." MSc Thesis, Tehran, Iran: Amirkabir University of Technology (AUT), (in Farsi) 2007.
6. Aghili Ashtiani, A. and Nikravesh, S.K.Y. "A new approach to stability analysis in fuzzy relational dynamic systems." *Iranian Journal of Fuzzy Systems*, vol.9, no.1 pp. 39–48, 2012.
7. Aghili Ashtiani, A. and Nikravesh, S.K.Y. "Fuzzy relational stability in fuzzy relational dynamic systems *Iranian Journal of Fuzzy Systems*."
8. Aghili Ashtiani, A., Nikravesh, S.K.Y., and Raja, P. "Centrally Symmetric, Plus Symmetric, and Row-Wise Symmetric Matrices and Their Properties." Proc. 5th Seminar on Linear Algebra, UMZ, Iran, October 2009.
9. Aghili Ashtiani, A. and Menhaj, M.B. "Introducing a New Pair of Differentiable Fuzzy Norms and its Application to Fuzzy Relational Function Approximation." Proc. 10th Joint Conference on Information Sciences. Salt Lake City, United States: World Scientific Publishing Co. pp. 1329–1336, 2007.
10. Aghili Ashtiani, A. and Menhaj, M.B. "Numerical solution of fuzzy relational equations based on smooth fuzzy norms." *Soft Computing*, Springer, 2009. DOI:10.1007/s00500-009–0425–1.
11. Aggarwal, J.K. and Vidyasagar, M. *Nonlinear Systems Stability Analysis*. John Wiley, New York, 1977.


## b


1. Barnett, S. and Storey, C. *Matrix Method in Stability Theory*. New York: Nelson, 1970.
2. Butz, A. "Higher order derivatives of Liapunov functions." *IEEE Trans. on Automatic Control (Correspondence)* 14: 111–112, 1969.
3. Boyd, S., Ghaoui, E., Feron, E., and Balakrishnan, V. "Linear matrix inequalities in system and control theory." *SIAM Studies in Applied Mathematics* vol. 15, Philadelphia, PA: SIAM, 1994
4. Bianchini, R.M. and Stefani G. "Grated applications and controllability along a trajectory" *SIAM J. Control and Optimization* (SICON) 28(4):903–924, 1990.
5. Bacciotti, A. and Roiser, L. *Liapunov Functions and Stability in Control Theory*. The Netherlands: Springer, 2010.

### c

1. Chellaboina, V. and Haddad, W. Teaching Time-Varying Stability Theory Using Autonomous Partial Stability Theory." Proceedings of the 40th IEEE Conference on Decision and Control. Orlando, Florida, pp. 3230–3235, 2001.
2. Chen, J. and. Chen, L. "Study on stability of fuzzy closed-loop control systems." *Fuzzy Sets and Systems* 57, 2:159–168, 1993.

### d

1. Dehghani, M. and Nikravesh, S.K.Y. "State space model parameter identification in large-scale power systems." *IEEE Trans. On Power Systems* vol. 23, no. 3:1449–1457, August 2008.
2. Dehghani, M. and Nikravesh, S.K.Y. "Estimation of synchronous generator parameters in a large-scale power systems." *International Journal of Innovative Computing, Information and Control* vol. 5, no. 8, August 2009.

### e

1. Eslami, M. *Theory of Sensitivity in Dynamic Systems, an Introduction.* Heidelberg, Germany: Springer-Verlag, 1994.

### f

1. Fathabadi, H. "New Method for Stability Analysis of Nonlinear Dynamical Systems." PhD Dissertation, Tehran, Iran: Electrical Eng. Dept., Amirkabir University, (in Farsi) 2002.
2. Fathabadi, H. and Nikravesh, S.K.Y. "A theoretical method for design and realization or fixed amplitude sinusoidal oscillators." *Analog Integrated Circuits and Signal Processing* 39: 123–130, 2004.

### g

1. Gibson, G.E. *Nonlinear Automatic Control.* New York: McGraw-Hill, 1963.
2. Gurel, O. and Lapidus, L. "A guide to the generation of Liapunov functions." *Industrial and Engineering Chemistry* 61(3): 1969.

### h

1. Hsu, J.C. and Meyer, A.V. *Modern Control Principles and Applications.* New York: McGraw-Hill, 1968.
2. Hermit, J.R. and Story, C. "Numerical application of Szego's method for constructing Liapunov functions." *IEEE Trans. Automatic Control* vol. AC-14, February 1969.
3. Hang, C.C. and Chang, J.A. "An algorithm for constructing Liapunov functions based on the variable gradient method." *IEEE Trans. Automatic Control (Correspondence)* vol. AC–15:510–512, August 1970,.
4. Hale, J. and Lunel, S. *Introduction to Functional Differential Equations.* New York: Springer Verlag, 1993.
5. Hardy, G.H., Littlewood, J.E., and Polya, G. *Inequalities*, 2nd ed., Cambridge: Cambridge University Press, 1952.

6. Hasegawa, T. and Furuhashi, T. "Stability analysis of fuzzy control systems simplified as a discrete system." *Control and Cybernetics* (27):565–577, 1998.
7. Haddad, W.M. and Chellaboina, V. *Nonlinear Dynamical Systems and Control: A Lyapunov Based Approach*. New Jersey: Princeton University Press, 2008.

## i

1. Ingwerson, D.R. "A Modified Liapunov Method for Nonlinear Stability Problems." Doctoral Dissertation, Stanford, California: Stanford University, 1960.
2. Infante, E.F. and Clark, L.G. "A method for the determination of the domain of stability of second-order nonlinear autonomous system." *ASME, J. of App. Mech*: Vol 31. 315–320, 1964.

## j

1. Jankovic, M. "Extension of Control Lyapunov Functions to Time-Delay System." In Proceedings of the 3rd IEEE Conference on Decision and Control, Sydney, Australia, 2000.

## k

1. Khalil, H.K. *Nonlinear Systems*, 3rd ed., Upper Saddle River, New Jersey: Prentice Hall, 2002.
2. Ku, Y.H. and Puri, N.N. "On Liapunov functions of higher order nonlinear systems." *J. Franklin Inst.* 276: 339–346, 1963.
3. Kinnen, S.E. and Chen, C. "Liapunov Functions for a Class of *n*th-Order Nonlinear Differential Equations." NASA CR-687, January 1967 DOI: 19700027500_1970027500.
4. Kinnen, S.E. and Chen, C. "Liapunov Functions from Auxiliary Exact Differential Equations." NASA CR-799, 1967. Also in *Automatica* vol. 4:195–204.
5. Kiet, T.T., Vu, N.G., and Phat, C. "Lyapunov stability of nonlinear time-varying differential equation." *ACTA mathematic A. Veitnamica*: 231–249, 2006.
6. Kandel, A., Luo Y., and Zhang, Y.Q. "Stability analysis of fuzzy control systems." *Fuzzy Sets and Systems* 105:33–48, 1999.
7. Kwon, W.H., Lee, Y.S., and Han, S.H. "Receding Horizon Predictive Control for Nonlinear Time-Delay Systems." In Proceedings of the 6th IFAC Symposium on Dynamics and Control of Process Systems, 2001 (DYCOPS-6), 277–282.

## l

1. Leondes, C.T. *Advances in Control Systems Theory and Applications* Vol. 2, Academic Press, pp. 1–64, 1965.
2. LaSalle, J. and Lefschetz, S. *Stability by Liapunov's Direct Method with Applications*. New York: Academic Press, USA, 1961.
3. Lefferts, E.J. "A Guide for the Application of the Liapunov Direct Method to Flight Control Systems." NASA CR-209, DOI: 19650011559_1965011559 1965.
4. Leighton, W. "On the construction of Liapunov functions for certain autonomous nonlinear differential equations." *Contributions to Differential Equations* vol. 2, no. 1–4, 1963.
5. LaSalle, J.P. "Some extensions of Liapunov's second method." *IRE Trans. on Circuit Theory*, 520–527, December 1960.
6. Lakshmikantham, V., Matrosov, V.M., and Sivasundaram, S. *Vector Lyapunov Functions and Stability Analysis of Nonlinear Systems*. The Netherlands: Springer, 2010.

## m

1. Margolis, S.G. and Vogt, W.G. "Control engineering applications of V.I. Zubov's construction procedure for Liapunov function." *IEEE Trans. Automatic Control* vol. AC-8:104–113, April 1963.
2. Meigoli, V. and Nikravesh, S.K.Y. "Extension of higher order derivatives of Liapunov functions in stability analysis of nonlinear systems." *Amirkabir International Journal of MISC* 41(1): 25–33, 2009.
3. Meigoli, V. "Stability Analysis of Homogeneous Nonlinear Systems." PhD Dissertation, Tehran, Iran: Dept. of Electrical Eng., Amirkabir University of Technology, May 2009.
4. Meigoli, V. and Nikravesh, S.K.Y. "A new theorem on higher-order derivatives of Liapunov functions." *ISA Trans.* 48: 173–179, 2009. doi:10.1016/j.isatra.2009.01.001.
5. Meigoli, V. and Nikravesh, S.K.Y. "Higher order derivatives of a Liapunov function approach for stability analysis of nonlinear systems." (Submitted in 2008 to *Systems and Control Letters.*)
6. Meigoli, V. and Nikravesh, S.K.Y. "Application of Higher Order Derivatives of Liapunov Functions in Stability Analysis of Nonlinear Homogeneous Systems." IAENG International Conference on Control and Automation ICCA 2009, Hong Kong, 18–20 March 2009.
7. Martynyuk, A.A. "Stability analysis by comparison technique." *Nonlinear Analysis* 62: 629–641, 2005.
8. Mahboobi Esfanjani, R., Reble, M., Muenz, U., Nikravesh, S.K.Y., and Allgower, F. "Model Predictive Control of Constrained Nonlinear Time-Delay Systems." Proceedings of the IEEE Conference on Decision and Control, Shanghai, December 2009.
9. Mahboobi Esfanjani, R. and Nikravesh, S.K.Y. "Stabilizing predictive control of nonlinear time-delay systems using control Lyapunov–Krasovskii functionals." *IET Control Theory and Applications* 3(10): 1395–1400, 2009.
10. Mahboobi Esfanjani, R. and Nikravesh, S.K.Y. "Predictive control for a class of distributed delay systems using Chebyshev polynomials." *International Journal of Computational Mathematics*, June 2009.
11. Margaliot M. and Langholz G. *New Approaches to Fuzzy Modeling and Control Design.* World Scientific Press, 2000.
12. Menhaj, M.B., Suratgar, A.A., and Karrari, M. "A Modified Dynamic Non-Singleton Fuzzy Logic System for Nonlinear Modeling." IEEE International Joint Conference on Neural Networks, Washington DC, 10 July-16 July 1999.
13. Meigoli, V. and Nikravesh, S.K.Y. "Stability analysis of nonlinear systems using higher order derivatives of Lyapunov function candidates."

## n

1. Nikravesh, S.K.Y. and Hoft, R.D. "Survey of analytical method of generating Lyapunov functions." *NSF, GK* 3441 ox, 1973.
2. Narendra, K.S. and Balakrishnan, J. "A common Liapunov function for stable LTI system with commuting A-matrices." *IEEE Trans. On Automatic Control* vol. 39: 2469–2471, 1994.

## o

1. Ogata, K. *State Space Analysis of Control Systems.* New Jersey: Prentice-Hall, Inc., 1967.

## p

1. Puri, N.N. and Weygandt, C.N. "Second method of Lyapunov and Routh's canonical form." *J. Franklin Inst.* 276: 365–383, 1963.

2. Palosenski, O. and Stern, P. "Comments on an Energy Metric Algorithm for the Generation of Lyapunov Functions." *IEEE Trans. Automatic Control (Correspondence)* vol. AC-14:110–111, April 1969.
3. Peuteman, J, and Aeyels, D. "Averaging results and the study of uniform asymptotic stability of homogeneous differential equations that are not fast time-varying." *SIAM J. Control and Optimization* 37(4):997–1010, 1999.
4. Phat Vu, N. and Niamsup, P. " Stability of linear time-varying delay systems and applications to control problems." *Computational and Applied Mathematics* vol. 194:343–356, 2006.
5. Percup, R.E., Preitl, S., and Solyom, S. "Center of Manifold Theory Approach to the Stability Analysis of Fuzzy Control Systems." In Proceedings of EUFIT, pp. 382–390, 1999.
6. Pedrycz, W. "An identification algorithm in fuzzy relational systems." *Fuzzy Sets and Systems* 13:153–167, 1984.

## r

1. Rekasius, Z.V. and Gibson, J.E. "Stability analysis of nonlinear control system by the second method of Liapunov." *IRE Trans. on Automatic Control*, January 1962.
2. Rodden, J.J. "Applications of Lyapunov Stability Theory." PhD Dissertation, Stanford, California: Stanford University, 1964.
3. Rosenbrock, H.H. "A Lyapunov function with applications to some nonlinear physical systems." *Automatica* 1: 31–53, 1962.
4. Reiss, R. and Geiss, G. "The construction of Lyapunov function." *IEEE Trans. Automatic Control* vol. AC-8:382–383 October 1963.
5. Royden, H. L. *Real Analysis*. New York, NY: Macmillan, 1988.
6. Refaa, T., Angrick, C., Findeisen, R., Kim, J.S., and Allgower, F. "Model Predictive Control for Nonlinear Time-Delay System." In Proc. of the 7th IFAC Symposium on Nonlinear Systems, 2007. CSIR International Convention Center, South Africa, (22–24), August, 2007.

## s

1. Slotine, J.J.E. and Li, W. *Applied Nonlinear Control*. New Jersey: Prentice-Hall, 1991.
2. Szego, G.P. "A Contribution to Liapunov's Second Method, Nonlinear Autonomous Systems." *ASME Ser D*. December 1962. (Also, abbreviated version is found in *International Symposium in Nonlinear Differential Equations and Nonlinear Mechanics*. J.P. LaSalle and S. Lefschetz, eds., Academic Press New York, 1963.
3. Schultz, D.G. and Gibson, J.E. "The Variable Gradient Method for Generating Liapunov Functions." *AIEE Trans.* Part II, vol. 81:203–210, 1962.
4. Suratgar, A.A. and Nikravesh, S.K.Y. "A New Sufficient Condition For Stability of Fuzzy System." ICEE 2002, Tabriz, Iran, May, 2002.
5. Suratgar, A.A. and Nikravesh, S.K.Y. "Two new methods for linguistic modeling with necessary and sufficient condition for their stability analysis." *Daneshvar Journal* no. 43:55–66, Tehran, Iran, 2003.
6. Suratgar A. A. and Nikravesh S.K.Y. "Potential energy based stability analysis of fuzzy linguistic systems." *Iranian Journal of Fuzzy Systems*, 65–74 2005.
7. Suratgar, A.A. and Nikravesh, S.K.Y. "A new method for linguistic fuzzy modeling with stability analysis and applications." *Autosoft Journal*, vol 15. no 3. pp. 329–342, 2004.
8. Suratgar, A.A. and Nikravesh, S.K.Y. "Two New Approaches for Linguistic Fuzzy Modeling and Introduction to Their Stability Analysis." IEEE World Congress Computational Intelligence (in Fuzzy Systems), Honolulu, HI, 2002.
9. Suratgar, A.A. and Nikravesh, S.K.Y. "Variation Model: Perception and their Stability Analysis and Its Application in Electrical Device and Aerospace." ICEE, Shiraz, Iran, May, 2003.

10. Suratgar, A.A. and Nikravesh, S.K.Y. "Variation Model: The Concept and Stability Analysis." IEEE World Congress Computational May 2003 Intelligence in Fuzzy Systems, St. Louis, MO., 2003.
11. Suratgar, A.A. and Nikravesh, S.K.Y. "Variation model for symbolic modeling and its stability analysis: Application to active suspension system." Iran: *ISME,* 2003.
12. Suratgar, A.A. and Nikravesh, S.K.Y. "Necessary and Sufficient Conditions for Asymptotic Stability of a Class of Applied Nonlinear, Dynamical Systems." IEEE Circuit and System Conference, Sharjah, United Arab Emirates, August, 2003.
13. Sanchez, E. "Resolution of Composite Fuzzy Relation Equations." *Information and Control* 30 (1): 38–48, 1976.
14. Sastry, S. *Systems: Analysis, Stability and Control* (1): Springer, New York, 2010.

## t

1. Tanaka, K. "Stability and stabiliziability of fuzzy neural-linear control systems." *IEEE Trans. On Fuzzy Systems* 3: 438–447, 1995.

## v

1. Vidyasagar, M. *Nonlinear Systems Analysis,* 2nd ed., Englewood Cliffs, New Jersey: Prentice-Hall Inc., 1993.
2. Vali, A.R. and Nikravesh, S.K.Y. "Delay-dependent stability analysis of nonlinear time-invariant time delay systems." *Amirkabir J. of Science & Technology,* 16(61A), EBC:39–46, Tehran, Iran, Spring 2005.
3. Vali, A.R. and Nikravesh, S.K.Y. " Lyapunov–Krasovskii approach for delay-dependent stability analysis of nonlinear time-delay systems." *Scientica Iranica* vol. 14. no.6:586–590, December 2007.

## w

1. Walker, J.A. and Clark, L.G. "An Integral Method of Lyapunov Function Generation for Nonlinear Autonomous Systems." *ASME J. of App. Mech.*, 32, 569–575, 1965.
2. Wall, E.T. and Moe, M.L. "An energy metric algorithm for the generation Liapunov functions." *IEEE Trans. Automatic Control (Correspondence)* vol. AC-13:121–122, February 1968.

## y

1. Yoshizawa, T. *Stability Theory by Lyapunov's Second Method.* Tokyo: Gakuyutsutosho Printing Company, 1966.

## z

1. Zadeh, L.A. "Outline of a new approach to analysis of complex systems and decision process." *IEEE Trans. On System Man and Cybernetics* vol. SMC-3:28–44, 1973.
2. Zubov, V.L. *Methods of A.M. Lyapunov and Their Application.* Groningen, The Netherlands: P. Noordhoff, 1964.
3. Zubov, V.L. *The Dynamics of Controlled Systems.* Moscow: Vysshaya, Shikola, 1982.

# Appendix A1: Application of VLF in Nonlinear Power System Stabilization

Consider the following smooth nonlinear large-scale system ($\Sigma_N$), which consists of $N$ affine nonlinear subsystems such as:

$$\begin{aligned} \Sigma_N : \dot{x}_i &= f_i(x_i) + g_i(x_i)u_i + h_i(x) + k_i(x_i)d_i, \\ y_i &= z_i(x_i). \end{aligned} \tag{A1.1}$$

where $x_i \in R^n, u_i \in R^m, d_i \in R^q$ *and* $y_i \in R^p$. $x_i$, $u_i$, and $y_i$ are the state vector, the input, and the output of each subsystem, respectively. $f_i(x_i)$, $g_i(x_i)$, $z_i(x_i)$, and $h_i(x)$ are smooth functions with appropriate dimensions and $d_i$ is the disturbance. It is assumed that $f_i(x_{0i}) = 0$, $h_i(x_0) = 0$ and $z_i(x_{0i}) = 0$.

Defining the system states, inputs, outputs, and disturbances as follows:

$$x = \begin{bmatrix} x_1 \\ x_2 \\ \vdots \\ x_N \end{bmatrix}, u = \begin{bmatrix} u_1 \\ u_2 \\ \vdots \\ u_N \end{bmatrix}, y = \begin{bmatrix} y_1 \\ y_2 \\ \vdots \\ y_N \end{bmatrix} \text{ and } d = \begin{bmatrix} d_1 \\ d_2 \\ \vdots \\ d_N \end{bmatrix}; \tag{A1.2}$$

the augmented large-scale system model is given by:

$$\begin{aligned} \dot{x} &= f(x) + g(x)u + k(x)d, \\ y &= z(x). \end{aligned} \tag{A1.3}$$

where:

$$f(x) = \begin{bmatrix} f_1(x_1) + h_1(x) \\ f_2(x_2) + h_2(x) \\ \vdots \\ f_N(x_N) + h_N(x) \end{bmatrix}, \quad g(x) = \begin{bmatrix} g_1(x_1) \\ g_2(x_2) \\ \vdots \\ g_N(x_N) \end{bmatrix},$$

$$z(x) = \begin{bmatrix} z_1(x_1) \\ z_2(x_2) \\ \vdots \\ z_N(x_N) \end{bmatrix} \text{ and } k(x) = \begin{bmatrix} k_1(x_1) \\ k_2(x_2) \\ \vdots \\ k_N(x_N) \end{bmatrix}. \tag{A1.4}$$

The objective is to design control laws "$u_i$" in each subsystem, such that:

$$\int_0^T \left(\|y_i(t)\|^2 + \|u_i(t)\|^2\right) dt \le \gamma^2 \int_0^T \|d_i(t)\|^2 \, dt, \forall T \ge 0 \quad \text{and} \quad d_i \in L_2(0,T).$$

In the following, the idea of nonlinear $H_\infty$ controller design in small-scale systems is extended to find decentralized controllers for each subsystem in large-scale systems. To solve the problem, three theorems are given.

---

**Theorem A1.1 [d1]:**

Consider the system model in (A1.1), if there exists a smooth positive definite Lyapunov function $V_i(x_i): R^{n_i} \to R^+$ for each subsystem such that $V_i(x_{0i}) = 0$ and the following Hamilton–Jacobi inequality is satisfied:

$$\begin{aligned} &\frac{\partial V_i}{\partial x_i}(x_i) f_i(x_i) + \frac{\partial V_i}{\partial x_i}(x_i) h_i(x) + \frac{1}{2} z_i^T(x_i) z_i(x_i) \\ &+ \frac{1}{2}\frac{\partial V_i}{\partial x_i}(x_i)\left[\frac{1}{\gamma^2} k_i(x_i) k_i^T(x_i) - g_i(x_i) g_i^T(x_i)\right]\frac{\partial^T V_i}{\partial x_i}(x_i) \le 0. \end{aligned} \qquad \text{(A1.5)}$$

Then, the equilibrium state of the closed-loop system with the following state feedback controller is stable and has $L_2$ gain less than $\gamma$.

$$u_i(x_i) = -g_i^T(x_i)\frac{\partial^T V_i}{\partial x_i}(x_i). \qquad \text{(A1.6)}$$

■

This theorem has some shortcomings, which will be mentioned in the following.

The main problem with this theorem is that the condition is rarely satisfied. If the following notation is used:

$$\begin{aligned} O_i(x_i) = &\frac{\partial V_i}{\partial x_i}(x_i) f_i(x_i) + \frac{1}{2} z_i^T(x_i) z_i(x_i) \\ &+ \frac{1}{2}\frac{\partial V_i}{\partial x_i}(x_i)\left[\frac{1}{\gamma^2} k_i(x_i) k_i^T(x_i) - g_i(x_i) g_i^T(x_i)\right]\frac{\partial^T V_i}{\partial x_i}(x_i), \end{aligned}$$

then the condition (A1.5) simplifies to:

$$\frac{\partial V_i}{\partial x_i}(x_i) h_i(x) \le -O_i(x_i).$$

This condition states that a function of a subsystem interaction term should be bounded by some function of the states of the same subsystem. In general, one cannot expect this condition to be satisfied. On the other hand, Theorem A1.1 just guarantees the stability of the equilibrium state of each subsystem. The large-scale system stability could not be concluded from Theorem A1.1, so it is not practical to use this theorem.

Another suggestion is to assume the Lyapunov function of a subsystem as a function of the states of that subsystem and all other subsystems that have interaction with it, that is, $V_i(x_i, x_j): R^{n_i} \to R^+$. In this case, in the Hamilton–Jacobi equation of' '*i*th' subsystem, the control of the other subsystems will be seen, too. As we cannot control the other subsystems in subsystem "*i*th," this Lyapunov function is not useful.

In the next theorem, a useful Lyapunov function is introduced that also overcomes the above-mentioned problems.

---

**Theorem A1.2:**

Consider the system model in (A1.1). Assume that there exists a smooth positive definite Lyapunov function $V_i(x_i): R^{n_i} \to R^+$ for each subsystem such that $V_i(x_{0i}) = 0$. Define: $V(x) \triangleq \sum_{i=1}^{N} V_i(x_i)$. $N$ is the total number of all subsystems and $V(x)$ is the Lyapunov function of the large-scale system, which is the summation of the Lyapunov functions of all subsystems.

If the Lyapunov functions satisfy the Hamilton–Jacobi inequality as follows:

$$\sum_{i=1}^{N} \frac{\partial V_i}{\partial x_i}(x_i) f_i(x_i) + \sum_{i=1}^{N} \frac{\partial V_i}{\partial x_i}(x_i) h_i(x) + \frac{1}{2}\sum_{i=1}^{N} z_i^T(x_i) z_i(x_i)$$
$$+\frac{1}{2}\sum_{i=1}^{N} \frac{\partial V_i}{\partial x_i}(x_i)\left[\frac{1}{\gamma^2} k_i(x_i) k_i^T(x_i) - g_i(x_i) g_i^T(x_i)\right]\frac{\partial^T V_i}{\partial x_i}(x_i) \le 0, \tag{A1.7}$$

then the equilibrium state of the closed-loop system with the following state feedback controller is stable and it has $L_2$ gain less than $\gamma$.

$$u_i(x_i) = -g_i^T(x_i)\frac{\partial^T V_i}{\partial x_i}(x_i). \tag{A1.8}$$

**Proof:**

Calculate the derivative of the Lyapunov function as follows:

$$\frac{dV}{dt} = \sum_{i=1}^{N} \frac{\partial V_i}{\partial x_i}(x_i) \frac{\partial x_i}{\partial t}$$
$$= \sum_{i=1}^{N} \left(\frac{\partial V_i}{\partial x_i}(x_i) f_i(x_i) + \frac{\partial V_i}{\partial x_i}(x_i) h_i(x)\right)$$
$$+ \sum_{i=1}^{N} \left(\frac{\partial V_i}{\partial x_i}(x_i) g_i(x_i) u_i + \frac{\partial V_i}{\partial x_i}(x_i) k_i(x_i) d_i\right).$$

The terms $\frac{\partial V_i}{\partial x_i}(x_i)g_i(x_i)u_i$ and $\frac{\partial V_i}{\partial x_i}(x_i)k_i(x_i)d_i$ are scalar, so they are equal to their transpose. Moreover, one can add to and at the same time subtract the following terms from the right-hand side of the above equation.

$$\frac{1}{2}\sum_{i=1}^{N}\frac{\partial V_i}{\partial x_i}(x_i)g_i(x_i)g_i^T(x_i)\frac{\partial^T V_i}{\partial x_i}(x_i).$$

$$\frac{1}{2\gamma^2}\sum_{i=1}^{N}\frac{\partial V_i}{\partial x_i}(x_i)k_i(x_i)k_i^T(x_i)\frac{\partial^T V_i}{\partial x_i}(x_i).$$

$$\frac{1}{2}\sum_{i=1}^{N}\|u_i\|^2.$$

$$\frac{1}{2}\gamma^2\sum_{i=1}^{N}\|d_i\|^2.$$

Finally, it yields:

$$\begin{aligned}\frac{dV}{dt} &= \sum_{i=1}^{N}\frac{\partial V_i}{\partial x_i}(x_i)f_i(x_i)+\sum_{i=1}^{N}\frac{\partial V_i}{\partial x_i}(x_i)h_i(x)+\frac{1}{2}\sum_{i=1}^{N}\gamma^2\|d_i\|^2-\frac{1}{2}\sum_{i=1}^{N}\|u_i\|^2\\ &+\frac{1}{2}\sum_{i=1}^{N}\left\|u_i+g_i^T(x_i)\frac{\partial^T V_i}{\partial x_i}(x_i)\right\|^2-\frac{1}{2}\gamma^2\sum_{i=1}^{N}\left\|d_i-\frac{1}{\gamma^2}k_i^T(x_i)\frac{\partial^T V_i}{\partial x_i}(x_i)\right\|^2\\ &+\frac{1}{2}\sum_{i=1}^{N}\frac{\partial V_i}{\partial x_i}(x_i)\left[\frac{1}{\gamma^2}k_i(x_i)k_i^T(x_i)-g_i(x_i)g_i^T(x_i)\right]\frac{\partial^T V_i}{\partial x_i}(x_i).\end{aligned} \quad \text{(A1.9)}$$

According to the condition in (A1.7) and considering the large-scale model in (A1.1), one has:

$$\begin{aligned}&\sum_{i=1}^{N}\frac{\partial V_i}{\partial x_i}(x_i)f_i(x_i)+\sum_{i=1}^{N}\frac{\partial V_i}{\partial x_i}(x_i)h_i(x)+\\ &\frac{1}{2}\sum_{i=1}^{N}\frac{\partial V_i}{\partial x_i}(x_i)\left[\frac{1}{\gamma^2}k_i(x_i)k_i^T(x_i)-g_i(x_i)g_i^T(x_i)\right]\frac{\partial^T V_i}{\partial x_i}(x_i)\\ &\le -\frac{1}{2}\sum_{i=1}^{N}z_i^T(x_i)z_i(x_i)\le -\frac{1}{2}\sum_{i=1}^{N}\|y_i\|^2.\end{aligned} \quad \text{(A1.10)}$$

So, it is concluded from (A1.9) and (A1.10) that:

$$\begin{aligned}\frac{dV}{dt} &\leq -\frac{1}{2}\sum_{i=1}^{N}\|y_i\|^2 + \frac{1}{2}\gamma^2\sum_{i=1}^{N}\|d_i\|^2 - \frac{1}{2}\sum_{i=1}^{N}\|u_i\|^2 \\ &+ \frac{1}{2}\sum_{i=1}^{N}\left\|u_i + g_i^T(x_i)\frac{\partial^T V_i}{\partial x_i}(x_i)\right\|^2 - \frac{1}{2}\gamma^2\sum_{i=1}^{N}\left\|d_i - \frac{1}{\gamma^2}k_i^T(x_i)\frac{\partial^T V_i}{\partial x_i}(x_i)\right\|^2.\end{aligned} \tag{A1.11}$$

If the controller defined in (A1.8) is applied to each subsystem, the fourth term in the right-hand side of (A1.11) will vanish. One can integrate both sides of the inequality in (A1.11) from zero to some $T > 0$ as follows:

$$\begin{aligned}&\int_0^T\left(\frac{dV}{dt}\right)dt + \frac{1}{2}\sum_{i=1}^{N}\left(\int_0^T\left(\|y_i\|^2 + \|u_i\|^2\right)dt\right) \\ &+ \frac{\gamma^2}{2}\sum_{i=1}^{N}\left(\int_0^T\left\|d_i - \frac{1}{\gamma^2}k_i^T(x_i)\frac{\partial^T V_i}{\partial x_i}(x_i)\right\|^2 dt\right) \leq \frac{1}{2}\gamma^2\sum_{i=1}^{N}\left(\int_0^T\|d_i\|^2\,dt\right).\end{aligned} \tag{A1.12}$$

According to the definition of the Lyapunov function, one has:

$$\int_0^T\frac{dV}{dt}\,dt = \sum_{i=1}^{N}V_i(x_i(T)) - \sum_{i=1}^{N}V_i(x_i(0)) \geq 0.$$

Moreover, the third term in the left-hand side of (A1.12) is positive definite, too. Consequently:

$$\frac{1}{2}\int_0^T\sum_{i=1}^{N}\left(\|y_i\|^2 + \|u_i\|^2\right)dt \leq \frac{1}{2}\gamma^2\int_0^T\sum_{i=1}^{N}\|d_i\|^2\,dt. \tag{A1.13}$$

Furthermore, if the disturbance terms are ignored, it is concluded that:

$$\frac{dV}{dt} \leq -\frac{1}{2}\sum_{i=1}^{N}\|y_i\|^2 - \frac{1}{2}\sum_{i=1}^{N}\|u_i\|^2. \tag{A1.14}$$

This means that the derivative of the Lyapunov function is negative definite, which assures asymptotic stability of the equilibrium state of the large scale system. ■

From (A1.13) and (A1.14), it is obvious that the decentralized controllers guarantee both the stability and the robustness of the system. According to Theorem A1.2, the Lyapunov function in each subsystem should be a function of the same subsystem's states and it should not depend on the states of other subsystems. Moreover, the summation of all subsystems' states must satisfy the condition in (A1.7). It is very difficult to find such a Lyapunov function. In order to overcome this problem, the following theorem can be used. Theorem A1.3 states that although the Lyapunov function in each subsystem depends on the states of the other subsystems too, its partial derivative function can be implemented locally. This guarantees the implementation of the designed controller to be decentralized.

---

**Theorem A1.3: (Dehghani–Nikravesh Theorem)**

Consider the system model in (A1.1). Assume that there exists a smooth positive definite Lyapunov function $V_i(x): R^{n_i} \to R^+$ for each subsystem such that $V_i(x_0) = 0$ and $\frac{\partial V_i}{\partial x_i}(x)$ can be implemented in each subsystem locally. Define: $V(x) \triangleq \sum_{i=1}^{N} V_i(x)$. $N$ is the total number of subsystems and $V(x)$ is the Lyapunov function of the large-scale system, which is the summation of the Lyapunov functions of all subsystems.

If the Lyapunov functions satisfy the Hamilton–Jacobi inequality as follows:

$$\sum_{i=1}^{N} \frac{\partial V_i}{\partial x_i}(x) f_i(x_i) + \sum_{i=1}^{N} \frac{\partial V_i}{\partial x_i}(x) h_i(x) + \frac{1}{2}\sum_{i=1}^{N} z_i^T(x_i) z_i(x_i)$$
$$+\frac{1}{2}\sum_{i=1}^{N} \frac{\partial V_i}{\partial x_i}(x)\left[\frac{1}{\gamma^2} k_i(x_i) k_i^T(x_i) - g_i(x_i) g_i^T(x_i)\right]\frac{\partial^T V_i}{\partial x_i}(x) \le 0, \tag{A1.15}$$

then, the equilibrium state of the closed-loop system with the following state feedback controller is stable and it has $L_2$ gain less than $\gamma$.

$$u_i(x) = -g_i^T(x_i)\frac{\partial^T V_i}{\partial x_i}(x). \tag{A1.16}$$

**Proof:**

The proof of this theorem is the same as the proof of Theorem A1.2. The only change would be $V_i(x_i)$ to $V_i(x)$. ■

## A1.1 APPLICATION TO MULTI-MACHINE POWER SYSTEMS

For a large-scale power system consisting of $n$ generators interconnected through a transmission network, the model, which is derived in Dehghani and Nikravesh [d1] is used.

The model for each subsystem *i*, *(i = 1…n)*, can be written as follows. (Some information about the parameters definition is given in this appendix. For further details about the model, refer to Dehghani and Nikravesh [d1] and Dehghani and Nikravesh [d2].)

*Mechanical dynamics:*

$$\dot{\delta}_i(t) = \omega_i(t), \tag{A1.17}$$

$$\dot{\omega}_i(t) = -\frac{D_i}{J_i}.\omega_i(t) + \frac{1}{J_i}\left(P_{mi} - P_{mi}^o + P_{ei}^o - P_{ei}(t)\right) + w_i. \tag{A1.18}$$

*Electrical dynamic:*

$$\dot{E}'_{qi}(t) = \frac{1}{T'_{doi}}\left(E_{fi}(t) - E_{fi}^o - E_{qi}(t)\right). \tag{A1.19}$$

*Electrical equations:*

$$E_{qi}(t) = E'_{qi}(t) - E'^o_{qi} + \left(x_{di} - x'_{di}\right).\left(I_{di} - I_{di}^o\right), \tag{A1.20}$$

$$P_{ei}(t) = \sum_{j=1}^{n} E'_{qi}(t)E'_{qj}(t)B_{ij}\sin(\delta_{ij}(t)), \tag{A1.21}$$

$$Q_{ei}(t) = -\sum_{j=1}^{n} E'_{qi}(t)E'_{qj}(t)B_{ij}\cos(\delta_{ij}(t)), \tag{A1.22}$$

$$I_{di}(t) = -\sum_{j=1}^{n} E'_{qj}(t)B_{ij}\cos(\delta_{ij}(t)) = \frac{Q_{ei}(t)}{E'_{qi}(t)}, \tag{A1.23}$$

$$I_{qi}(t) = \sum_{j=1}^{n} E'_{qj}(t)B_{ij}\sin(\delta_{ij}(t)) = \frac{P_{ei}(t)}{E'_{qi}(t)}. \tag{A1.24}$$

Some observations about the above model are given below:

1. The base rotor speed is assumed to be one per unit.
2. A disturbance signal "$w_i$," which could be a sudden change on system mechanical power or a sudden change in the load demand, is added to the second state of the mathematical model. The disturbance effect on the power system frequency is studied in this appendix too.
3. In this study, the mechanical torque is considered to be constant and only the field voltage, which can be perturbed more easily than the mechanical torque, is considered as the input to the system.

According to (A1.17) through (A1.24), the following model for each subsystem is achieved:

State equation:

$$\dot{x}_i = f_i(x_i) + g_i(x_i)u_i + h_i(x) + k_i(x_i)d_i, \\ y_i = E_i x_i. \tag{A1.25}$$

Subsystem states:

$$x_i = \begin{bmatrix} x_{1i} & x_{2i} & x_{3i} \end{bmatrix}' = \begin{bmatrix} \delta_i - \delta_i^o & \omega_i & E'_{qi} - E'^o_{qi} \end{bmatrix}'.$$

Input: $u_i = E_{fi} - E^o_{fi}$ .

Measured output: $y_i = \omega_i$ ,
where:

$$f_i(x_i) + h_i(x) = \begin{bmatrix} \omega_i \\ \frac{1}{J_i}\left(P_{mi} - P^o_{mi} + P^o_{ei} - P_{ei} - D_i.\omega_i\right) \\ \frac{-1}{T'_{doi}}\left(E'_{qi} - E'^o_{qi} + (x_{di} - x'_{di}).(I_{di} - I^o_{di})\right) \end{bmatrix}, \tag{A1.26}$$

$$g_i(x_i) = \begin{bmatrix} 0 & 0 & \frac{1}{T'_{doi}} \end{bmatrix}^T, \; k_i(x_i) = [0 \quad 1 \quad 0]^T, \; E_i = [0 \quad 1 \quad 0]. \tag{A1.27}$$

Consider the following operating state for the above system:

$$\begin{bmatrix} \delta^o_1 & 0 & E'^o_{q1} & \cdots & \delta^o_i & 0 & E'^o_{qi} & \cdots & \delta^o_N & 0 & E'^o_{qN} \end{bmatrix}. \tag{A1.28}$$

Define the following Lyapunov function for each subsystem:

$$V_i(x) \triangleq \alpha_{1i}\omega_i^2 + \alpha_{2i}\left[Q_{ei} - Q^o_{ei} - (\delta_{ij} - \delta^o_{ij})P^o_{ei} - (E'_{qi} - E'^o_{qi})I^o_{di}\right] \\ + \alpha_{3i}\left(E'_{qi} - E'^o_{qi}\right)^2. \tag{A1.29}$$

A similar function which is derived by a systematic method based on a generalized Popov criterion is reported in Dehghani and Nikravesh [d1]. In order to derive this function, the following assumptions are made:

(a) Excitation voltage is constant, that is, all controls by AVR and PSS are neglected.
(b) Each internal voltage $E_q$ lags behind the q-axis by a constant angle all the time.
(c) Transfer conductances $G_{ij}$ are all negligible.

If one wants to design the control term '$E_{fi}$', the first assumption cannot be made. This point should be taken into consideration.

Now, a situation in which one can use Theorem A1.3 is to be found. It is concluded from the definition of the Lyapunov function that it is smooth and $V_i(0) = 0$. Furthermore, if '$\alpha_i$' parameters are assumed to be positive, according to Dehghani and Nikravesh [d1], the second term in the above function is positive definite, so the whole function is positive definite.

From (A1.19) and (A1.20):

$$E'_{qi} - E'^{o}_{qi} = -T'_{doi}\dot{E}'_{qi}(t) + E_{fi} - E^{o}_{fi} - (x_{di} - x'_{di}).(I_{di} - I^{o}_{di}). \tag{A1.30}$$

The derivative of $V_i(x)$ in (A1.29) is as follows (inserting from (A1.30) for $\left(E'_{qi} - E'^{o}_{qi}\right)$):

$$\begin{aligned}
\dot{V}_i(x) &= 2\alpha_{1i}\omega_i\dot{\omega}_i + \alpha_{2i}\left(P_{ei} - P^{o}_{ei}\right)\dot{\delta}_i \\
&\quad + \left(\alpha_{2i}\left(I_{di} - I^{o}_{di}\right) + 2\alpha_{3i}\left(E'_{qi} - E'^{o}_{qi}\right)\right)\dot{E}'_{qi} \\
&= \alpha_{2i}\left(P_{ei} - P^{o}_{ei}\right)\omega_i + \alpha_{2i}\left(I_{di} - I^{o}_{di}\right)\dot{E}'_{qi} \\
&\quad + 2\alpha_{1i}\omega_i\left(-\frac{D_i}{J_i}.\omega_i + \frac{1}{J_i}P^{o}_{ei} - \frac{1}{J_i}P_{ei}\right) \\
&\quad + 2\alpha_{3i}\left(-T'_{doi}\left(\dot{E}'_{qi}\right)^2 + \left(E_{fi} - E^{o}_{fi}\right)\dot{E}'_{qi}\right) \\
&\quad + 2\alpha_{3i}\left(-(x_{di} - x'_{di}).\left(I_{di} - I^{o}_{di}\right)\dot{E}'_{qi}\right) \\
&= \left(\alpha_{2i} - 2\frac{\alpha_{1i}}{J_i}\right)\left(P_{ei} - P^{o}_{ei}\right)\omega_i - 2\frac{D_i}{J_i}\alpha_{1i}\omega_i^2 \\
&\quad + \left(\alpha_{2i} - 2\alpha_{3i}.(x_{di} - x'_{di})\right).\left(I_{di} - I^{o}_{di}\right)\dot{E}'_{qi} \\
&\quad - 2\alpha_{3i}T'_{doi}\left(\dot{E}'_{qi}\right)^2 + 2\alpha_{3i}\left(E_{fi} - E^{o}_{fi}\right)\dot{E}'_{qi}.
\end{aligned} \tag{A1.31}$$

$a_{1i} > 0$, $a_{2i} = 2_{a1i}/J_i$, and $\alpha_{3i} = \alpha_{2i}/2(x_{di} - x'_{di})$, it can be shown that all the terms in the Lyapunov function in (A1.29) can be guaranteed to be positive definite and all the terms in its derivative in (A1.31), except the last one, are negative definite. If $E_{fi}$ is designed such that $(E_{fi} - E^{o}_{fi})\dot{E}'_{qi} \le 0$, the derivative of the Lyapunov function will be negative definite. Moreover:

$$\frac{\partial V_i}{\partial x_i}(x) = \begin{bmatrix} \alpha_{2i}\left(P_{ei} - P^{o}_{ei}\right) \\ 2\alpha_{1i}\omega_i \\ -2\alpha_{3i}T'_{doi}\dot{E}'_{qi} + 2\alpha_{3i}\left(E_{fi} - E^{o}_{fi}\right) \end{bmatrix}^T. \tag{A1.32}$$

On the other hand, using (A1.23) and (A1.24) yields:

$$I_i = \sqrt{I_{di}^2 + I_{qi}^2} = \frac{\sqrt{P_{ei}^2 + Q_{ei}^2}}{E'_{qi}} \qquad I_{di} = \frac{Q_{ei} I_i}{\sqrt{P_{ei}^2 + Q_{ei}^2}}. \tag{A1.33}$$

Considering (A1.19), (A1.20), and (A1.33), the following equations will be procured:

$$\begin{aligned}
\dot{E}'_{qi}(t) &= -\frac{1}{T'_{doi}}\left(E'_{qi}(t) - E'^{o}_{qi} - E_{fi} + E^{o}_{fi}\right) \\
&\quad - \frac{1}{T'_{doi}}(x_{di} - x'_{di}).\left(I_{di} - I^{o}_{di}\right) \\
&= \frac{E_{fi}}{T'_{doi}} - \frac{E^{o}_{fi}}{T'_{doi}} - \frac{\sqrt{P_{ei}^2 + Q_{ei}^2}}{T'_{doi}.I_i} + \frac{\sqrt{\left(P^{o}_{ei}\right)^2 + \left(Q^{o}_{ei}\right)^2}}{T'_{doi}.I^{o}_i} \\
&\quad - \frac{(x_{di} - x'_{di})}{T'_{doi}}.\left(\frac{I_i.Q_{ei}}{\sqrt{P_{ei}^2 + Q_{ei}^2}} - \frac{I^{o}_i.Q^{o}_{ei}}{\sqrt{\left(P^{o}_{ei}\right)^2 + \left(Q^{o}_{ei}\right)^2}}\right).
\end{aligned} \tag{A1.34}$$

From (A1.32) and (A1.34), $\frac{\partial V_i}{\partial x_i}(x)$ can be implemented using local measurements.

It can be seen that all conditions of Theorem A1.3 are satisfied. If the Lyapunov function satisfies the inequality in (A1.15), the nonlinear controller can be found from (A1.16). Thus:

$$\frac{\partial V_i}{\partial x_i}(x)\left[\frac{1}{\gamma^2} k_i(x_i) k_i^T(x_i) - g_i(x_i) g_i^T(x_i)\right]\frac{\partial^T V_i}{\partial x_i}(x)$$

$$= \begin{bmatrix} \alpha_{2i}\left(P_{ei} - P^{o}_{ei}\right) \\ 2\alpha_{1i}\omega_i \\ -2\alpha_{3i} T'_{doi}\dot{E}'_{qi} + 2\alpha_{3i}\left(E_{fi} - E^{o}_{fi}\right) \end{bmatrix}^T \begin{bmatrix} 0 & 0 & 0 \\ 0 & \frac{1}{\gamma^2} & 0 \\ 0 & 0 & -\frac{1}{T'^2_{doi}} \end{bmatrix}$$

$$\times \begin{bmatrix} \alpha_{2i}\left(P_{ei} - P^{o}_{ei}\right) \\ 2\alpha_{1i}\omega_i \\ -2\alpha_{3i} T'_{doi}\dot{E}'_{qi} + 2\alpha_{3i}\left(E_{fi} - E^{o}_{fi}\right) \end{bmatrix}$$

$$= 4\frac{\alpha_{1i}^2}{\gamma^2}\omega_i^2 - \frac{1}{T'^2_{doi}}\left(-2\alpha_{3i} T'_{doi}\dot{E}'_{qi} + 2\alpha_{3i}\left(E_{fi} - E^{o}_{fi}\right)\right)^2.$$

By selecting $y_i = z_i(x) = \omega_i$, the left-hand side of the Hamilton–Jacobi inequality in (A1.15) will be as follows:

$$\sum_{i=1}^{N} -2\frac{D_i}{J_i}\alpha_{1i}\omega_i^2 - 2\alpha_{3i}T'_{doi}\left(\dot{E}'_{qi}\right)^2 + \frac{1}{2}\omega_i^2 +$$
$$\sum_{i=1}^{N} \frac{1}{2}\left(4\frac{\alpha_{1i}^2}{\gamma^2}\omega_i^2 - \frac{1}{T'^2_{doi}}\left(-2\alpha_{3i}T'_{doi}\dot{E}'_{qi} + 2\alpha_{3i}\left(E_{fi} - E^o_{fi}\right)\right)^2\right)$$
$$= \sum_{i=1}^{N}\left(\frac{1}{2} + \frac{2}{\gamma^2}\alpha_{1i}^2 - 2\alpha_{1i}\frac{D_i}{J_i}\right)\omega_i^2 - 2\alpha_{3i}T'_{doi}\left(\dot{E}'_{qi}\right)^2$$
$$+ \sum_{i=1}^{N} -\frac{1}{2T'^2_{doi}}\left(-2\alpha_{3i}T'_{doi}\dot{E}'_{qi} + 2\alpha_{3i}\left(E_{fi} - E^o_{fi}\right)\right)^2. \tag{A1.35}$$

If one assumes $\alpha_{1i}$ is selected such that the inequality $1/2 + 2/\gamma^2\alpha_{1i}^2 - 2\alpha_{1i}$ $D_i/J_i < 0$ is satisfied, the whole Hamilton–Jacobi inequality will be negative definite. The range of variation of $\alpha_{1i}$ is given below:

$$\frac{1}{2} + \frac{2}{\gamma^2}\alpha_{1i}^2 - 2\alpha_{1i}\frac{D_i}{J_i} < 0 \quad \left(\frac{\sqrt{2}}{\gamma}\alpha_{1i} - \frac{D_i}{J_i}.\frac{\gamma}{\sqrt{2}}\right)^2 - \frac{D_i^2}{J_i^2}.\frac{\gamma^2}{2} + \frac{1}{2} < 0$$
$$\alpha_{1i} < \frac{D_i}{J_i}.\frac{\gamma^2}{2} + \frac{\gamma}{\sqrt{2}}\sqrt{\frac{D_i^2}{J_i^2}.\frac{\gamma^2}{2} - \frac{1}{2}}. \tag{A1.36}$$

It is obvious that the inequality $\gamma \geq J_i/D_i$ should be satisfied.

If the inequality is satisfied for a special $\gamma$, according to Theorem A1.3, the following feedback controller in each subsystem guarantees the large-scale system to have $L_2$ gain less than $\gamma$:

$$u_i = -\begin{bmatrix} 0 & 0 & \frac{1}{T'_{doi}} \end{bmatrix}\begin{bmatrix} \alpha_{2i}\left(P_{ei} - P^o_{ei}\right) \\ 2\alpha_{1i}\omega_i \\ -2\alpha_{3i}T'_{doi}\dot{E}'_{qi} + 2\alpha_{3i}\left(E_{fi} - E^o_{fi}\right) \end{bmatrix}$$
$$= 2\alpha_{3i}\dot{E}'_{qi} - \frac{2\alpha_{3i}}{T'_{doi}}\left(E_{fi} - E^o_{fi}\right). \tag{A1.37}$$

Note that $u_i = E_{fi} - E_{fi}^o$. Moreover, insert from (A1.34) for $E'_{qi}$ to procure the following equation for the input:

$$u_i = \frac{2\alpha_{3i}}{1+\dfrac{2\alpha_{3i}}{T'_{doi}}}\dot{E}'_{qi} = \frac{2\alpha_{3i}}{T'_{doi}+2\alpha_{3i}}\left(E_{fi} - E_{fi}^o - \frac{\sqrt{P_{ei}^2+Q_{ei}^2}}{I_i} + \frac{\sqrt{\left(P_{ei}^o\right)^2+\left(Q_{ei}^o\right)^2}}{I_i^o}\right)$$

$$-\frac{2\alpha_{3i}}{T'_{doi}+2\alpha_{3i}}\left(x_{di} - x'_{di}\right).\left(\frac{I_i.Q_{ei}}{\sqrt{P_{ei}^2+Q_{ei}^2}} - \frac{I_i^o.Q_{ei}^o}{\sqrt{\left(P_{ei}^o\right)^2+\left(Q_{ei}^o\right)^2}}\right). \tag{A1.38}$$

If the above control input is inserted into the Lyapunov function derivative in (A1.31), it is seen that (A1.31) will be changed as follows:

$$\dot{V}_i(x) = -2\frac{D_i}{J_i}\alpha_{1i}\omega_i^2 - 2\alpha_{3i}T'_{doi}\left(\dot{E}'_{qi}\right)^2 + \frac{4\alpha_{3i}^2}{1+\frac{2\alpha_{3i}}{T'_{doi}}}\left(\dot{E}'_{qi}\right)^2. \tag{A1.39}$$

According to the above formula, $\alpha_{3i}$ is chosen as follows:

$$-2\alpha_{3i}T'_{doi} + \frac{4\alpha_{3i}^2}{1+\frac{2\alpha_{3i}}{T'_{doi}}} < 0, \quad 4\alpha_{3i}^2 < 2\alpha_{3i}\left(T'_{doi}+2\alpha_{3i}\right), \quad 0 < 2\alpha_{3i}T'_{doi}, \tag{A1.40}$$

which means $\alpha_{3i} > 0$. This guarantees that the function defined in (A1.29) is a proper Lyapunov function for the power system model.

According to (A1.38), although the controller in each subsystem depends on the states of other subsystems, it can be implemented locally using local subsystem measurements.

# Appendix A2: Proof of Theorem 3.8

First, a new lemma, which is required for the proof of Theorem 3.8 is presented.

---

**Lemma A2.1:**

Under the assumptions of Theorem 3.8, a lower triangular matrix $Q = [q_{ij}]_{m\times m}$ with the same property as A, that is, (3.44) exists, such that:

(a) If $\|x(t)\| \geq b \geq 0$, $\forall t \in [t_0, t_0 + T]$, for given $t_0$ and $T < +\infty$ then for $i = 1,\ldots,m$:

$$\sum_{j=1}^{i} q_{ij} V_j(t,x) \leq -\phi_2(b)\tfrac{(t-t_0)^{m+1-i}}{(m+1-i)!} + \sum_{r=i}^{m} \tfrac{(t-t_0)^{r-i}}{(r-i)!} \sum_{j=1}^{r} q_{rj} V_j(t_0, x_0). \tag{A2.1}$$

Moreover,

$$q_{11}\phi_1(\|x\|) \leq -\phi_2(b)\tfrac{(t-t_0)^m}{m!} + \sum_{r=1}^{m} \tfrac{(t-t_0)^{r-1}}{(r-1)!} \sum_{j=1}^{r} q_{rj} V_j(t_0, x_0) \triangleq A_b(t). \tag{A2.2}$$

(b) A special use of (A2.2) for $b = 0$ yields:

$$q_{11}\phi_1(\|x\|) \leq \sum_{r=1}^{m} \tfrac{(t-t_0)^{r-1}}{(r-1)!} \sum_{j=1}^{r} q_{rj} V_j(t_0, x_0) = A_0(t), \quad \forall t \geq t_0. \tag{A2.3}$$

(c) No trajectory of $\dot{x} = f(t,x)$ escapes to infinity in finite time.

**Proof:**

The $Q_{m\times m}$ matrix is a lower triangular matrix, that is, $q_{ij} = 0$ for $I < j$. Other elements of Q are defined recursively, beginning from the last row:

$$q_{mj} \triangleq a_{mj}, \qquad j = 1,\ldots,m, \tag{A2.4}$$

where $a_{mj}$ was introduced in (3.43). The other rows of Q are defined recursively in the order from the bottom to the top of the matrix using the following algorithm:

For $i = m-1,\ldots,2,1$,

For $j = i,\ldots,1$,

$$q_{ij} \triangleq \sum_{r=j}^{i} q_{i+1,r+1}a_{rj} = [q_{i+1,j+1}\cdots q_{i+1,r+1}\cdots q_{i+1,i+1}][a_{jj}\cdots a_{rj}\cdots a_{ij}]^T. \tag{A2.5}$$

Next $j$

Next $i$

The operation of this equation is shown graphically in Figure A2.1. It is clear that $q_{ij} \geq 0$ for $i \geq j$. Moreover, $q_{ii} = \Pi_{r=i}^{m} a_{rr} > 0$ for $i = 1,\ldots,m$.

**Part (a):** For the proof of this part, let $\|x(t)\| \geq b \geq 0$ for $t \in [t_0, t_0 + T]$. The inequality (A2.1) is proved by backward mathematical induction on $i = m,\ldots,2,1$, that is, first (A2.1) is proved for $i = m$, then for $i = m-1$, $i = m-2$, and so on.

(1) For $i = m$, substitute $\|x(t)\| \geq b$ and (A2.4) for the last row of (3.43)

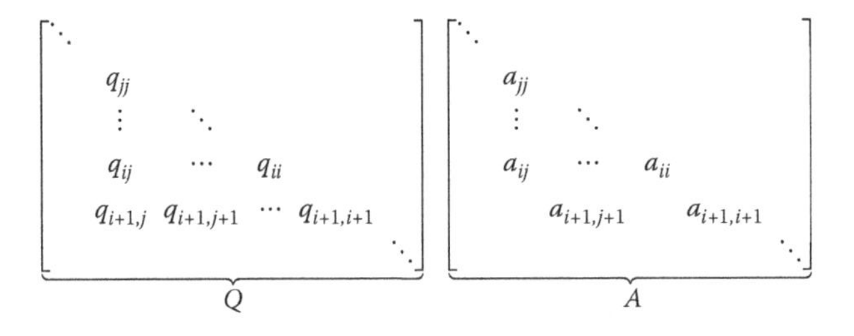


$$\sum_{j=1}^{m} q_{mj}\dot{V}_j(t,x) \leq -\phi_2(b), \quad \forall t \in [t_0, t_0 + T] \tag{A2.6}$$

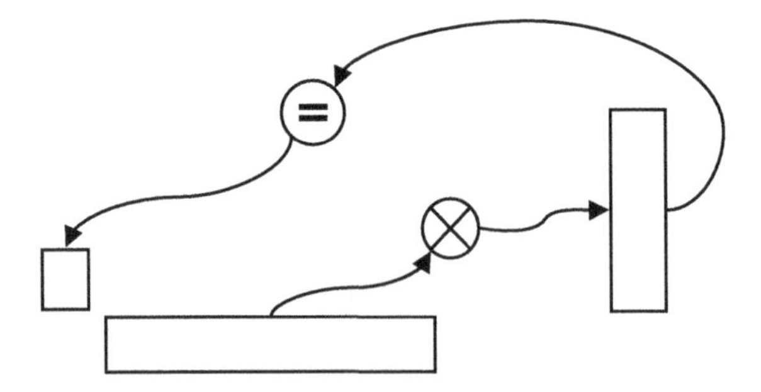


**FIGURE A2.1** Operation in (A2.1) to obtain the $Q$ matrix elements.

Then, integrate both sides of (A2.6) on the time interval $\tau \in [t_0, t]$ for some $t \in [t_0, t_0 + T]$ , thus:

$$\int_{t_0}^{t} \Big[\sum_{j=1}^{m} q_{mj} \dot{V}_j(\tau, x)\Big] d\tau \le \int_{t_0}^{t} -\phi_2(b) d\tau$$
$$\sum_{j=1}^{m} q_{mj} V_j(t,x) \le -\phi_2(b)(t - t_0) + \sum_{j=1}^{m} q_{mj} V_j(t_0, x_0) \tag{A2.7}$$

This relationship coincides with (A2.1) for $i = m$. The preceding lines of (3.43) are rewritten to continue:

$$V_{r+1}(t,x) \ge \sum_{j=1}^{r} a_{rj} \dot{V}_j(t,x), \qquad r = 1, \ldots, m-1 \tag{A2.8}$$

(2) Assume that (A2.1) holds for some $2 \le i \le m$, then the appropriate relationship for $(i-1)$ is achievable. Starting from the left-hand side of (A2.1) and using the fact that $q_{ij} \ge 0$ , and substituting $V_1(t,x) \ge 0$ and also substituting $V_2(t,x), V_3(t,x), \ldots,$ and $V_i(t,x)$ from (A2.8), exchanging the summations, and using (A2.5) yields:

$$\sum_{r=1}^{i} q_{ir} V_r(t,x) = q_{i1} V_1(t,x) + \sum_{r=1}^{i-1} q_{i,r+1} V_{r+1}(t,x)$$
$$\ge \sum_{r=1}^{i-1} q_{i,r+1} V_{r+1}(t,x) \ge \sum_{r=1}^{i-1} \sum_{j=1}^{r} q_{i,r+1} a_{rj} \dot{V}_j(t,x) \tag{A2.9}$$
$$= \sum_{j=1}^{i-1} \Big(\sum_{r=j}^{i-1} q_{i,r+1} a_{rj}\Big) \dot{V}_j(t,x) = \sum_{j=1}^{i-1} q_{i-1,j} \dot{V}_j(t,x).$$

Substituting (A2.9) for (A2.1), and using the dummy variable $\tau$ then, integrating on the time interval $\tau \in [t_0, t]$ for some $t \in [t_0, t_0 + T]$, evaluating the integrals and rearranging some terms from left to right, yields:

$$\sum_{j=1}^{i-1} q_{i-1,j} V_j(t,x) \le \sum_{j=1}^{i-1} q_{i-1,j} V_j(t_0, x_0) - \phi_2(b) \tfrac{(t-t_0)^{m+2-i}}{(m+2-i)!}$$
$$+ \sum_{r=i}^{m} \tfrac{(t-t_0)^{r+1-i}}{(r+1-i)!} \sum_{j=1}^{r} q_{rj} V_j(t_0, x_0). \tag{A2.10}$$

Combining the right-hand summations in (A2.10) yields (A2.1) for $(i-1)$, that is:

$$\sum_{j=1}^{i-1} q_{i-1,j} V_j(t,x) \le -\phi_2(b) \tfrac{(t-t_0)^{m+2-i}}{(m+2-i)!} + \sum_{r=i-1}^{m} \tfrac{(t-t_0)^{r+1-i}}{(r+1-i)!} \sum_{j=1}^{r} q_{rj} V_j(t_0, x_0). \tag{A2.11}$$

(3) Substituting $i = 1$ in (A2.1) then, the following relationship is obtained

$$q_{11}V_1(t,x) \le -\phi_2(b)\tfrac{(t-t_0)^m}{m!} + \sum_{r=1}^{m} \tfrac{(t-t_0)^{r-1}}{(r-1)!} \sum_{j=1}^{r} q_{rj}V_j(t_0,x_0). \tag{A2.12}$$

Then, substituting (3.41) into (A2.12) yields (A2.2).

**Part (b):** $\| x(t) \| \ge 0 \ \forall t \ge t_0$, thus repeating Part (a) with $b = 0$, implies (A2.3).

**Part (c):** The finiteness of $\|x(t)\|$ for all $t \in [0,+\infty)$ is implied from (A2.3), because $q_{11} > 0$, $\phi_1 \in K_\infty$, and the right-hand side of (A2.3) is a polynomial of time $t$, therefore, the solutions could not have any finite escape time. ■

---

**Proof of Theorem 3.8:**

**Part (a):** The trajectories of $\dot{x} = f(t,x)$ cannot stay far away from the zero equilibrium state for a long time. To show this, assume that some trajectories stay away from the equilibrium state even for a short interval, that is, $\|x(t)\| \ge b > 0$ for some interval $t \in [t_0, t_0 + T]$. In Lemma A2.1, this situation was considered. Replacing (3.42) for $t_0$ in (A2.2) yields:

$$\text{If } \begin{cases} \|x(t_0)\| \le c, \\ \|x(t)\| \ge b > 0, \quad \forall t \in [t_0, t_0 + T], \end{cases} \tag{A2.13}$$

then,

$$\begin{aligned} q_{11}\phi_1(b) \le q_{11}\phi_1(\| x(t) \|) &\le -\phi_2(b)\tfrac{(t-t_0)^m}{m!} \\ &+ \sum_{r=1}^{m} \tfrac{(t-t_0)^{r-1}}{(r-1)!} \sum_{j=1}^{r} q_{rj}\psi_j(c) \triangleq A_b^c(t-t_0). \end{aligned} \tag{A2.14}$$

Note that $A_b^c(t-t_0)$ in (A2.14) is a polynomial of $(t-t_0)$ with the following properties:

$$\forall c \ge b > 0, \quad \begin{cases} A_b^c(0) = q_{11}\psi_1(c) \ge q_{11}\phi_1(b) > 0, \\ \tfrac{d}{dt}[A_b^c(t-t_0)]\,|_{t=t_0} = q_{21}\psi_1(c) + q_{22}\psi_2(c) > 0, \\ \lim\limits_{(t-t_0)\to+\infty} A_b^c(t-t_0) = -\infty. \end{cases} \tag{A2.15}$$

The last line in (A2.15) implies:

$$T_1(b,c) > 0\,, \forall (t-t_0) > T_1(b,c), \quad A_b^c(t-t_0) < q_{11}\phi_1(b). \tag{A2.16}$$

It is known from (A2.14) that $q_{11}\phi_1(b) \leq A_b^c(t-t_0), \forall t \in [t_0, t_0+T]$. Comparing this with (A2.16) yields that $T \leq T_1(b,c)$, thus the relation (A2.13) and (A2.14) could not be true for $T > T_1(b,c)$. Therefore, $\|x(t)\|$ in (A2.13) should have a crossing point with the b level in some finite time and the following fact is proved:

$$\forall b,c > 0\,, \quad \|x(t_0)\| \leq c \qquad T \leq T_1(b,c), \quad \|x(t_0+T)\| \leq b. \tag{A2.17}$$

The uniform stability of the zero equilibrium state in the sense of Lyapunov could be proved by showing the existence of a function $\alpha(\cdot) \in K_\infty$ such that:

$$\|x(t)\| \leq \alpha(\|x_0\|), \quad \forall t \geq t_0. \tag{A2.18}$$

For the proof of (A2.18), see Lemma A2.2. Now it is sufficient to prove the globally uniform attraction of the zero equilibrium state, that is, to show that for each $\eta > 0$ and $c > 0$, there exists $T_2 = T_2(\eta, c) > 0$ such that:

$$\|x_0\| \leq c \qquad \|x(t)\| \leq \eta, \quad \forall t \geq t_0 + T_2(\eta, c). \tag{A2.19}$$

The time $t_0 + T$ might be considered as a new initial time, using (A2.18) to conclude that:

$$\|x(t_0+T)\| = b, \qquad \{\|x(t)\| \leq \alpha(b), \quad \forall t \geq t_0 + T\}. \tag{A2.20}$$

Combining the relations (A2.17) and (A2.20) yields:

$$\begin{aligned} &\forall b,c > 0\,, \|x(t_0)\| \leq c, \\ &\qquad T \leq T_1(b,c), \quad \{\|x(t)\| \leq \alpha(b), \quad \forall t \geq t_0 + T\}. \end{aligned} \tag{A2.21}$$

Then, using $t_0 + T \leq t_0 + T_1(b,c)$ in (A2.21), the following conclusion from (A2.21) is straightforward:

$$\forall b,c > 0\,, \|x(t_0)\| \leq c, \quad \|x(t)\| \leq \alpha(b), \quad \forall t \geq t_0 + T_1(b,c). \tag{A2.22}$$

Substituting $b = \alpha^{-1}(\eta) > 0$ for (A2.22) yields:

$$\forall \eta, c > 0\,, \|x(t_0)\| \leq c, \qquad \|x(t)\| \leq \eta, \quad \forall t \geq t_0 + T_1[\alpha^{-1}(\eta), c]. \tag{A2.23}$$

Therefore, (A2.19) is proved by $T_2(\eta,c) \triangleq T_1[\alpha^{-1}(\eta),c]$ and the zero equilibrium state of $\dot{x} = f(t,x)$ is uniformly globally asymptotically stable.

**Part (b):** The uniformly asymptotic stability of the zero equilibrium state is a local property of the nonlinear system $\dot{x} = f(t,x)$ and could be proved in a similar manner as above, because the relation (A2.18) guarantees that the trajectories starting very near to the equilibrium state do not escape the region of $\|x\|<r$. Note that all class $\mathtt{K}_\infty$ functions in Part (a) must be replaced by class $\mathtt{K}$ functions in Part (b). ■

---

**Lemma A2.2:**

Under the assumptions of Theorem 3.8, there exists a function $\alpha \in \mathtt{K}_\infty$ such that (A2.18) will be satisfied. ■

**Proof:**

Note that $\|x(t)\|$ could be an oscillating nonnegative function of time. Assume that the mean value of some rising interval is a given $b>0$. Choosing $t=t_0$ at the instant where $\|x(t)\| = b$:

$$\begin{cases} \|x(t_0)\| = b, \\ \|x(t)\| \geq b\,, \quad \forall t \in [t_0, t_0+T], \quad \text{for some finite time } T>0. \end{cases} \tag{A2.24}$$

The relationship (A2.24) is of the form of (A2.13) and implies (A2.14) for $c=b>0$, that is:

$$\begin{aligned} q_{11}\phi_1(b) \leq q_{11}\phi_1(\| x(t) \|) \leq -\phi_2(b)\tfrac{(t-t_0)^m}{m!} \\ +\sum_{r=1}^{m} \tfrac{(t-t_0)^{r-1}}{(r-1)!} \sum_{j=1}^{r} q_{rj}\psi_j(b) \triangleq A_b^b(t-t_0). \end{aligned} \tag{A2.25}$$

Then, using properties in (A2.15) implies that the function $A_b^b(t-t_0)$ has some maximum value $M(b)>0$ depending on the parameter $b>0$, that is:

$$M(b) \triangleq \sup_{t-t_0 \geq 0} A_b^b(t-t_0) \tag{A2.26}$$

Some properties of $M(b)$ are as follows:

(1) $M(b)$ is a continuous function of b, since the function $A_b^b(t-t_0)$ in (A2.25) and (A2.26) is continuous with respect to both b and $(t-t_0)$.
(2) $M(b) \geq q_{11}\phi_1(b) \geq 0$, while the equalities hold true only for $b = 0$.
(3) Using (A2.24) through (A2.26):

$$q_{11}\phi_1(\| x(t) \|) \leq M(b), \quad \forall t \in [t_0, t_0 + T] \tag{A2.27}$$

The inequality (A2.27) could be generalized as the following useful relation:

$$q_{11}\phi_1(\|x(t_1)\|) \leq M(\|x_0\|) \quad \forall t_1 \geq t_0, \forall x_0 \in R^n. \tag{A2.28}$$

Considering some finite time $t_1 \geq t_0$, we prove (A2.28) in either of the following two cases:

Case (1) $\|x(t_1)\| \leq \|x_0\|$: In this case: $q_{11}\phi_1(\|x(t_1)\|) \leq q_{11}\phi_1(\|x_0\|) \leq M(\|x_0\|)$ is trivial from the second property of M(b) above.
Case (2) $\|x(t_1)\| > \| x_0 \| \triangleq b$: Using the mean value theorem, there exists some time $t_1' \in [t_0, t_1)$ such that:

$$\begin{cases} \| x(t_1') \| = \| x_0 \| = b, \\ \| x(t) \| \geq b, \quad \forall t \in [t_1', t_1]. \end{cases} \tag{A2.29}$$

The relationship (A2.29) coincides with (A2.24) by choosing $T = t_1 - t_1'$ and new initial time $t_1'$. Then, use the resulting relation (A2.27) for $t = t_1$, that is, $q_{11}\phi_1(\|x(t_1)\|) \leq M(\|x_0\|)$.

After proving (A2.28), a new function $\bar{M}(b) \triangleq \sup_{0 \leq a \leq b} M(a)$ can be defined, which is continuous, nondecreasing, and $\bar{M}(0) = 0$. Then, define $\alpha'(b) \triangleq \bar{M}(b) + b$. It is clear that $\alpha' \in K_\infty$ and $M(\cdot) \leq \bar{M}(\cdot) \leq \alpha'(\cdot)$. Using (A2.28), it is concluded that $q_{11}\phi_1(\|x(t)\|) \leq \alpha'(\|x_0\|)$ for all $t \geq t_0$. Thus, (A2.18) is proved by defining $\alpha \triangleq \phi_1^{-1} \circ (\alpha'/q_{11}) \in K_\infty$. ■

# Appendix A3: Stability Analysis of Nonlinear Systems via Δ-Homogeneous Approximation

It is well known that any n-dimensional function *f(x)* can be expanded by Δ- homogeneous functions with different order with respect to an arbitrarily dilation $^{r}_{\alpha}$ as follows:

$$f(x)=\sum_{k=-r_n}^{+\infty} f^{(k)}(x), \quad f^{(k)} \in n_k, \quad \forall k \geq -r_n. \tag{A3.1}$$

A Δ-homogeneous approximation* of a nonlinear system is obtained using the first nonzero term of the expansion (A3.1). It is obvious that, if the zero equilibrium state of a Δ-homogeneous approximation of a nonlinear system is globally asymptotically stable, then the equilibrium state of the original nonlinear system is locally asymptotically stable using the same Δ-homogeneous Lyapunov function.

The importance of this method is appreciated when the linearized system is not asymptotically stable, that is, when the first Lyapunov method is not applicable, but a Δ-homogeneous approximation of this nonlinear system might be stable. This is why one could consider the Δ-homogeneous approximation as a generalized method of linearization technique for nonlinear systems as far as stability analysis is concerned.

The Δ-homogeneous approximation is mostly used in local nonlinear system stabilization. If a linearized system is uncontrollable and in the meantime unstable, then, it is not possible to stabilize the system using linear control design methods. Instead, if a Δ-homogeneous approximation of this nonlinear system is controllable† then, asymptotically stabilizing this Δ-homogeneous approximation could stabilize the original nonlinear system.

The selection of a suitable $^{r}_{\alpha}$ is not an easy task. Different researchers have done this selection based on local necessary and sufficient conditions for controllability using Lie brackets for following a nonlinear affine control system; for example, see Bianchini and Stefani [b4]:

$$\dot{x}=f_0(x)+\sum_{i=1}^{k} u_i f_i(x) \tag{A3.2}$$

* Some authors call it Nilpotent Approximation.

† For a nonlinear controllability definition and related topics see Bianchini and Stefani [b4].

The following example demonstrates the nonlinear system's controller design method using a Δ-homogeneous approximation.

### Example A3.1:

For the following nonlinear system, design a controller such that its zero equilibrium state of the system is locally asymptotically stable:

$$\dot{x}_1 = u, \qquad \dot{x}_2 = -x_1^3 + \varepsilon x_2 + h(x), \tag{A.3.3}$$

where $h(x)$ contains higher-order terms with respect to a dilation (which will be specified later), for example $h(x) = x_1^4 + x_1x_2 + x_2^2$. This control system could be represented as $\dot{x} = f(x) + ug(x)$ where $f = (-x_1^3 + \varepsilon x_2 + h(x))\frac{\partial}{\partial x_2}$, $g = \partial/\partial x_1$ and $f(0) = 0$.

If $\varepsilon < 0$, then the linearized system is uncontrollable and unstable as well, therefore, one has to use Δ-homogeneous approximation. Let us select $r = (2,7)$ and the following dilation:

$$\Delta_\alpha^r(x_1, x_2) = (\alpha^2 x_1, \alpha^7 x_2). \tag{A3.4}$$

The homogeneous approximation is as follows:

$$f_0 = -x_1^3 \frac{\partial}{\partial x_2}, \tag{A3.5}$$

because the term $-x_1^3$ is a homogeneous function of order 6 and the term $\varepsilon x_2$ is a homogeneous function of order 7, while $h(x)$ is assumed to be a homogeneous function of higher order. Therefore, the homogeneous approximation of this nonlinear system is given in the following form:

$$\begin{pmatrix} \dot{x}_1 \\ \dot{x}_2 \end{pmatrix} = \begin{pmatrix} 0 \\ -x_1^3 \end{pmatrix} + \begin{pmatrix} 1 \\ 0 \end{pmatrix} u. \tag{A3.6}$$

First, let us find the feedback control law, which globally stabilizes the zero equilibrium state of the approximated system $\dot{x} = f_0(x) + ug(x)$ given in (A3.6), which in turn locally stabilizes the zero equilibrium state of the original nonlinear system. To do this, let us choose the following feedback control law:

$$u(x) = -x_1^{1/2} + x_2^{1/7}, \tag{A3.7}$$

in which the following notation is used:

$$y^{\beta} \triangleq |y|^{\beta}\,\mathrm{sgn}(y). \tag{A3.8}$$

Using the feedback law of (A3.7), the approximation system of (A3.6) becomes as follows:

$$\begin{pmatrix} x_1 \\ x_2 \end{pmatrix} = \begin{pmatrix} -x_1^{1/2} + x_2^{1/7} \\ -x_1^3 \end{pmatrix} = F_0(x). \tag{A3.9}$$

This system is homogeneous of order −1 since:

$$F_0(\ {}^r_\alpha x) = \begin{bmatrix} -(\alpha^2 x_1)^{1/2} + (\alpha^7 x_2)^{1/7} \\ -(\alpha^2 x_1)^3 \end{bmatrix} = \alpha^{-1} \begin{bmatrix} \alpha^2 & 0 \\ 0 & \alpha^7 \end{bmatrix} \begin{bmatrix} -x_1^{1/2} + x_2^{1/7} \\ -x_1^3 \end{bmatrix}$$
$$= \alpha^{-1}\ {}^r_\alpha F_0(x). \tag{A3.10}$$

Using the LaSalle invariant theorem together with the following Lyapunov function:

$$V(x) = \frac{1}{4}x_1^4 + \frac{7}{8}|x_2|^{8/7} \in H_8,$$

it can be shown that the zero equilibrium state of the system (A3.10) is globally asymptotically stable.

If one uses the feedback control law (A3.7) for the original nonlinear system (A3.3), its closed loop has the following representation:

$$\begin{pmatrix} \dot{x}_1 \\ \dot{x}_2 \end{pmatrix} = \begin{pmatrix} -x_1^{1/2} + x_2^{1/7} \\ -x_1^3 + \varepsilon x_2 + h(x) \end{pmatrix} = \begin{pmatrix} -x_1^{1/2} + x_2^{1/7} \\ -x_1^3 \end{pmatrix} + \begin{pmatrix} 0 \\ \varepsilon x_2 + h(x) \end{pmatrix}$$
$$= F_0(x) + \ F(x), \tag{A3.11}$$

in which $\Delta F(x)$ contains higher-order terms with a repeat to the mentioned dilation. Thus, this feedback control law also locally asymptotically stabilizes the original nonlinear system as well. ■

# Appendix A4: Stabilization of Model Predictive Control of Nonlinear Time-Delayed Systems

In this appendix, stabilization of a model predictive control scheme for a class of nonlinear time-delayed systems is investigated. Model predictive control (MPC), also known as *receding horizon control*, has become a popular control strategy. However, a predictive controller without additional considerations does not naturally guarantee the closed-loop stability in the sense of Lyapunov. For delay-free systems, many approaches have been developed to avoid this problem. In most cases, closed-loop stability is guaranteed by using terminal state constraints and an appropriately chosen terminal cost.

In this study, based on the general framework for designing the stabilizing predictive controllers, an MPC scheme is proposed for input constrained and unconstrained nonlinear time-delayed systems [m8,m9]. Then, the details of the design method are presented and closed-loop stability proof is given.

## A4.1 STABILIZING PREDICTIVE CONTROL OF INPUT CONSTRAINED NONLINEAR TIME-DELAYED SYSTEMS

To guarantee the closed-loop stability of predictive control, an appropriate terminal cost functional as well as a suitably defined terminal inequality constraint are added into the finite horizon optimal control problem. As in the delay-free case, a local state feedback controller is used to determine the terminal cost and terminal inequality constraint offline. Simple conditions for the terminal cost functional and terminal region are derived in terms of linear matrix inequalities (LMIs).

Before introducing the idea, the existing methods in the literature are briefly reviewed. In [r6], an expanded zero equality constraint of terminal states was used to assure the closed-loop system stability. This method needs heavy online computations. In our method, we use terminal inequality constraint, which is more attractive computationally than an expanded zero constraint in [r6]. In addition, in [k7], a stabilizing MPC was proposed based on suitably defined terminal cost functional and terminal region. These conditions require the delayed state of the system to be bounded above by a linear function and also a global stabilizing controller has to exist in order to calculate this terminal cost. In our approach, without imposing any confining assumption on system dynamics, only a local stabilizer is used to calculate the design parameters. So, the presented result is more general compared to previous MPC schemes for nonlinear time-delayed systems [r6,k7].

### A4.1.1 Problem Statement

Consider the nonlinear time-delayed system described by the following relation:

$$\begin{aligned} \dot{x}(t) &= f(x(t), x(t-\tau), u(t)), \\ x(\theta) &= \quad (\theta)\,,\ \forall\theta \in [-\tau, 0]\ , \end{aligned} \tag{A4.1}$$

where $x(t) \in R^n$ is the state vector and $u(t) \in R^m$ is the control input vector subject to input constraints $u(t) \in U \in R^m$. The constraint set $U$ is compact, convex, and contains the origin in its interior. $\varphi \in C[-\tau, 0]$ is the initial vector function. The function $f$ is assumed to be continuously differentiable with $f(0,0,0) = 0$. The time delay $\tau$ is assumed to be known and constant. The issue here is to study the stability of the zero equilibrium state of the system (A4.1) via a model predictive controller.

### A4.1.2 Proposed Model Predictive Controller

The model predictive control is formulated by solving, online, a finite horizon optimal control problem. Based on measurements obtained at time $t$, the controller predicts the future behavior of the system over a prediction horizon $T$ and determines the control input such that a predetermined cost functional $J$ is minimized. To incorporate a feedback mechanism, the obtained control signal is implemented only until the next measurement becomes available.

It is well known that inappropriate determination of a finite horizon optimal control problem may cause instability, specially if the horizon is too short. Even if the resulting optimal cost behaves well, the closed-loop control system might be unstable. To guarantee the closed-loop stability, certain conditions have to be met in the finite horizon optimization problem to assure that the associated optimal cost can be used as a Lyapunov function for the closed-loop control system.

In the proposed scheme, a locally asymptotically stabilizing controller is first designed in some defined neighborhood $\Omega$ of the equilibrium state, and then an upper bound on the infinite horizon cost, using this local controller is computed and used as a terminal cost. In addition, a constraint is added to the open-loop optimal control problem that requires the final state to lie within $\Omega$. The finite horizon optimal control problem at time $t$ is formulated as follows:

$$\min_{u} \{J(x_t, u; t, T) = \int_t^{t+T} q(x(t'), u(t'))\,dt' + V(x_{t+T})\}. \tag{A4.2}$$

subject to:

$$\begin{aligned} &\dot{x}(t') = f(x(t'), x(t'-\tau_1), u(t')), \\ &u(t') \in U, \\ &x_{t+T} \in \Omega, \end{aligned} \tag{A4.3}$$

where $x(t')$ is the predicted trajectory starting from $x_t = x(t+\theta)$ for $-\tau \le \theta \le 0$ and driven by $u(t')$ for $t' \in [t, t+T]$. The region $\Omega$ which is called a *terminal region* is a closed set and contains $\mathbf{0} \in C\,[-\tau, 0]$ in its interior. $V$ is a suitably defined terminal cost functional. The stage cost $q: R^n \times R^m \to R^+$, which is continuous in its arguments penalizes states and control inputs according to:

$$q(x,u) \ge c_q\,(\|x\|^2 + \|u\|^2), \tag{A4.4}$$

for $c_q > 0$ and $q(0,0) = 0$. The notation $\|\cdot\|$ is used to denote the Euclidean norm of a vector. We assume that the optimal control which optimizes $J(x_t, u; t, T)$ is given by $u^*(t', x_t, t)$ and the associated optimal cost is denoted by $J^*(x_t, u^*; t, T)$. For the sake of simplicity in what follows, $J^*(x_t, u^*; t, T)$ is denoted by $J^*(x_t; t, T)$. The control input to the system is defined by the optimal solution of the problem at the sampling instants $t = k \quad t,\ k \in Z^+$ as follows:

$$u(t) = u^*(t'; x_t, t)\,,\ t \le t' \le t + \quad t. \tag{A4.5}$$

The implicit feedback controller resulting from the application of the (A4.5) is asymptotically stabilizing provided that the following crucial stabilizing conditions are satisfied:

(a) Prediction horizon $T$ is chosen such that the finite horizon problem (A4.2) together with (A4.3) has a feasible solution at $t = 0$,
(b) For the nonlinear time-delayed system (A4.1), a locally asymptotically stabilizing controller $u(t) = k(x_t) \in U$ is designed and a continuously differentiable positive definite functional $V(x_t)$ is found such that in the region of attraction $\Omega$, the following is held:

$$\frac{d}{dt}V(x_t) \le -q(x(t), k(x_t)) \qquad \forall x_t \in \Omega. \tag{A4.6}$$

In the following, it is shown that the proposed predictive controller stabilizes system (A4.1), providing an appropriate selection of terminal cost $V$ and terminal region $\Omega$.

### A4.1.3 Stability Analysis of a Closed-Loop System

Before addressing the asymptotic stability of the closed-loop system, regarding condition (a), the feasibility of repeated solution of the finite horizon optimal control problem is shown, that is, $\forall t > 0$ there is a control input $u(t'),\ t' \in [t, t+T]$ that results in a bounded value for the cost problem of (A4.2) and satisfies all the constraints of (A4.3).

**Lemma A4.1:**

The finite horizon problem (A4.2) together with (A4.3) is feasible for every $t > 0$, if it has a feasible solution at $t = 0$. ■

**Proof:**

Suppose at time $t$, a feasible solution of (A4.2) together with (A4.3) exists, that is, $u(t';x_t,t)$ , $t' \in [t, t+T]$. At the next instant $t+\delta$ , $\delta > 0$ a feasible control input can be constructed by appending control values at the present time step on application of the local controller $k(x_{t'})$:

$$\hat{u}(t';x_{t+\delta},t+\delta) = \begin{cases} u(t';x_t,t) & t' \in [t+\delta, t+T], \\ k(x_{t'}) & t' \in [t+T, t+T+\delta]. \end{cases}$$

It is obvious that $\hat{u}(t')$ consists of two parts with a part of feasible control that steers $x_{t+\delta}$ to $x_{t+T} \in \Omega$ where $k(x_t)$ keeps the system trajectory for $t+T \leq t' \leq t+T+\delta$ in $\Omega$ while satisfying all constraints. So, the feasibility of (A4.2) and (A4.3) at time $t$ results in feasibility at time $t + \delta$. Since the open-loop finite horizon problem (A4.2) together with (A4.3) admits a feasible solution at $t = 0$, by induction it is feasible for every $t > 0$. ■

To establish asymptotic stability, in Lemma A4.2 a characteristic of the optimal cost $J^*(x_t;t,T)$ is introduced. In Lemma A4.3, it is shown that the optimal cost of problem (A4.2) together with (A4.3) is indeed monotonically nonincreasing. Then in Lemma A4.4, it is proved that if the optimal cost is monotonically nonincreasing in the horizon $T$ then, the asymptotic stability of the proposed predictive controller is achieved. This enables us to state our final result about the asymptotic stability of the proposed scheme.

**Lemma A4.2:**

The optimal cost $J^*(x_t;t,T)$ of finite horizon in optimal control problem (A4.2) together with (A4.3) is continuous in $x_t$ at $x_t \equiv \mathbf{0}$ if stabilizing conditions (a) and (b) are met.* ■

**Proof:**

Let $\varphi$ belong to some neighborhood of the equilibrium state and $\neq \mathbf{0}$. $u(t') = 0$ , $t' \in [t, t+T]$ is chosen as a candidate solution to the finite horizon optimization problem. Now, we focus on the following system for $t' \in [t, t+T]$:

$$\dot{x}(t') = f(x(t'), x(t'-\tau)),$$

$$x_t = \quad .$$

* These conditions are given in Section A4.1.2.

Since $f$ is a continuously differentiable function of its arguments, a unique solution $x(t)$ exists on $[t,t+T]$ and is continuous in the initial state at $x_t \equiv \mathbf{0}$. We can choose the neighborhood such that the terminal region condition is satisfied. Now, let the associated cost functional be denoted by:

$$\bar{J}^*(\ ;t,t+T) = \int_T^{t+T} q(x(t'),u(t')=0)dt' + V(x_{t+T}).$$

Since $q$ and $V$ are continuous and $x(t)$ is continuous in $\varphi$ at $x_t = 0$, $\bar{J}^*$ is continuous at $x_t \equiv \mathbf{0}$, that is, $\forall\varepsilon>0$, there exits $\delta(\varepsilon)$ such that $\|\ \|<\delta$ implies $|\bar{J}^*|<\varepsilon$. Note that $|J^*| \le |\bar{J}^*| < \varepsilon$ since $j^*$ is obtained by applying $u^*(t)$, while $\bar{J}^*(t)$ is obtained by applying $u(t) \equiv 0$. So, $J^*$ is continuous at $x_t = \mathbf{0}$. ■

---

**Lemma A4.3:**

Assume that the stabilizing conditions (a) and (b) hold, then the optimal cost $J^*(x_t;t,T)$ of finite horizon problem (A4.2) together with (A4.3), satisfies the nonincreasing monotonicity property:

$$J^*(x_t;t,T+\delta) \le J^*(x_t;t,T)$$

■

**Proof: T**

The cost functional for the time interval $[t, t+T+\delta\,]$, beginning at initial state $X_t$ is $J(x_t, u;t,T+\delta)$. According to (A4.2), we have:

$$J^*(x_t;t,T+\delta) \le J(x_t,u;t,T+\delta) =$$

$$= \int_t^{t+T} q(x(t'),u^*(t'))dt' + + \int_{t+T}^{t+T+\delta} q(x(t'),k(x(t'))dt' + V(x_{t+T+\delta})$$

$$= J^*(x_t;t,T) - V(x_{t+T}) + V(x_{t+T+\delta}) + \int_{t+T}^{t+T+\delta} q(x(t'),k(x(t'))dt'.$$

On the other hand, stabilizing condition (b) yields:

$$-V(x_{t+T}) + V(x_{t+T+\delta}) + \int_{t+T}^{t+T+\delta} q(x(t'),k(x(t'))dt' \le 0.$$

Hence, $J^*(x_t;t,T+\delta) \le J^*(x_t;t,T)$ can be concluded, which completes the proof. ■

---

**Lemma A4.4:**

If the optimal cost of problem (A4.2) and (A4.3) satisfies $J^*(x_t,t,T+\delta) \le J^*(x_t,t,T)$, for every $\delta > 0$, then the equilibrium state of the system (4.4-1) with predictive control law (A4.5) is asymptotically stable. ■

**Proof:**

From (A4.2) we obtain:

$$J^*(x_t;t,T) = \int_t^{t+\delta} q(x(t'),u^*(t'))\,dt' + J^*(x_{t+\delta};t+\delta,T-\delta).$$

Since $J^*(x_t;t,T+\delta) \le J^*(x_t;t,T)$, utilizing Lemma A4.3, the following is concluded:

$$J^*(x_t;t,T) \ge \int_t^{t+\delta} q(x(t'),u^*(t'))\,dt' + J^*(x_{t+\delta};t+\delta,T).$$

Rearranging the above inequality and dividing both sides by $\delta$ yields:

$$\frac{J^*(x_{t+\delta};t+\delta,T) - J^*(x_t;t,T)}{\delta} \le -\frac{1}{\delta}\int_t^{t+\delta} q(x(t'),u^*(t'))\,dt'.$$

Now, if $\delta \to 0$, using inequality (A4.4) yields:

$$\frac{d}{dt} J^*(x_t;t,T) \le -q(x,u^*) \le -c_q(\|x\|^2 + \|u^*\|),$$

which means $J^*(x_t;t,T)$ is nonincreasing. Since $J^* \ge 0$, then $J^* = c$ as $t \to \infty$, where $c$ is a nonnegative constant and therefore $dJ^*/dt = 0$ as $t \to \infty$. Hence, from the above equation it is clear that $x(t)$ and $u^*(t)$ both approach zero as $t \to \infty$. Therefore, utilizing Lemma A4.2 proves that the closed-loop system is asymptotically stable. ■

Now, using Lemmas A4.3 and A4.4, the main result can be stated as Theorem A4.1.

---

**Theorem A4.1: (Mahboobi–Nikravesh Theorem 1)**

Assume that in the finite horizon optimal control problem (A4.2) and (A4.3), the design parameters $T$, $V$, and $\Omega$ are selected and/or determined such that the stabilizing conditions (a) and (b) hold. Then, the closed-loop system resulting from the application of the predictive control strategy to the system (A4.1) is asymptotically stabilizing. ■

**Example A4.1:**

As an example for demonstrating the proposed method, consider a nonlinear system with delayed state described by:

$$\dot{x}(t) = x^3(t-1) + u(t),$$

$$x(t) = 0.5, \qquad -1 \le t \le 0.$$

with input constraint $|u| \leq 1$. A Lyapunov–Krasovskii functional for the terminal cost can be defined as follows:

$$V(x_t) = 5x^2(t) + \int_{-1}^{0} x^2(t+\zeta)\, d\zeta.$$

Hereafter, the stage cost $q$ is chosen to be:

$$q(x(t), u(t)) = x^2(t) + u^2(t).$$

The local feedback controller in terminal region is selected as follows:

$$u(t) = -x(t).$$

So, the stability condition (A4.6) is written as follows:

$$\frac{dV}{dt} + q = [10x(t) + x^2(t) - x^2(t-1)][x^3(t-1) - x(t)]$$

$$+ x^2(t) + 0.2x^2(t) \leq 0.$$

which is negative definite for $|x_t| < \frac{1}{2}$, so the region $\Omega$ which is defined as $\Omega = \{x_t \in C_\tau : \sup|x_t| < \frac{1}{2}\}$ is chosen as the terminal region, which is also the region of attraction for $\dot{x} = x^3(t-1) + u(t)$ with $u(t) = -x(t)$, because $\dot{V}(x_t) \leq 0$ for every $x_t \in \Omega$.

### A4.1.4 Calculation of Terminal Region and Terminal Cost

Although, a suitable definition of the finite horizon problem in predictive control guarantees closed-loop system stability, the determination of the problem parameters, including terminal cost and terminal region is not a trivial task. In this subsection, structured approaches are proposed to calculate mentioned design parameters for input constrained nonlinear time-delayed systems.

As stated before, in the proposed scheme, a locally asymptotically stabilizing controller is first designed in some defined neighborhood $\Omega$ of the equilibrium state, and then an upper bound on the infinite horizon cost is computed using this local controller and the input is used as a terminal cost. In addition, a constraint is added to the optimal control problem that requires the final state to lie within $\Omega$.

A structured approach to obtain the terminal cost and terminal region for the nonlinear time-delayed system (A4.1) is given below. The basic idea is to use Razumikhin-type arguments to find the region of attraction as:

$$\Omega_\alpha = \left\{ x_t : \max_{\theta \in [-\tau, 0]} x(t+\theta)^T P\, x(t+\theta) \leq \alpha \right\}, \tag{A4.7}$$

for some $\alpha > 0$ and for some symmetric positive definite matrix $P$. In addition, a Lyapunov–Krasovskii functional satisfying (A4.6) is used as an upper bound of the infinite horizon cost.

The goal of this section is to derive simple conditions for a local linear control law $u(t) = K\,x(t)$ and find its region of attraction $\Omega_\alpha$ for system (A4.1) such that the input constraints are satisfied. Also, a simple continuously differentiable positive definite functional $V(x_t)$ is derived such that using the local control law, the following condition holds for all the states inside the terminal region:

$$\dot{V}(x_t) \leq -q(x(t), K\,x(t)).$$

For this, consider the Jacobean linearization:

$$\dot{x} = A\,x(t) + A_1\,x(t-\tau) + B u(t),$$

of the system given by (A4.1) about the origin in which:

$$A = \frac{\partial f}{\partial x}(0), \quad A_1 = \frac{\partial f}{\partial x(t-\tau)}(0) \quad \text{and} \quad B = \frac{\partial f}{\partial u}(0).$$

Define $\Phi$ as the difference between the nonlinear system (A4.1) and its Jacobean linearization:

$$\Phi(x(t), x(t-\tau), u(t)) = f(x(t), x(t-\tau), u(t)) - [A\,x(t) + A_1\,x(t-\tau) + B u(t)].$$

For the sake of brevity, $\Phi(x_t, u)$ is used as a shorthand notation for $\Phi(x(t), x(t-\tau), u(t))$. Since $f$ is assumed to be continuously differentiable and $\Omega_\alpha$ in (A4.7) is compact, there exist $L_1$ and $L_2$ such that:

$$\|\Phi(x_t, K\,x(t))\| \leq L_1 \|x(t)\| + L_2 \|x(t-\tau)\|, \tag{A4.8}$$

for all $x_t \in \Omega_\alpha$ and stabilizing controller $u(t) = Kx(t)$. Also, $L_1 = 0, L_2 = 0$ for $\alpha \to 0$.

First, the conditions are derived in terms of Linear Matrix Inequalities (LMIs) to obtain the linear controller and the terminal region of the form (A4.1).

---

**Lemma A4.5:**

Consider the system (A4.1). If there exist symmetric matrices $\Lambda > 0$, $\Lambda_i > 0$, $i=$ 1,2,3 and a matrix $\Gamma$ satisfying the following LMIs:

$$\begin{bmatrix} E_1 + 2\tau\Lambda & \tau A_1(A\Lambda + B\Gamma) & \tau A_1^2\,\Lambda & \tau A_1\,\Lambda \\ * & -\tau\Lambda_1 & 0 & 0 \\ * & * & -\tau\Lambda_2 & 0 \\ * & * & 0 & -\tau\Lambda_3 \end{bmatrix} < 0, \tag{A4.9}$$

$$\Lambda_i - \Lambda \leq 0, \quad i = 1,2,3,$$

in which $E_1 = \Lambda(A+A_1)^T + (A+A_1)\Lambda + \Gamma^T B^T + B\Gamma$, then the control law is $u(t) = K\,x(t)$ with $K = \Gamma\Lambda^{-1}$, and region $\Omega_\alpha$ in (A4.7) is the region of attraction for $P = \Lambda^{-1}$ and some $\alpha > 0$. ■

**Proof:**

Consider the following time-delayed system:

$$\dot{\xi}(t) = (A_K + A_1)\xi(t) + \Phi(\xi_t, K\,\xi(t)) - A_1\int_{-\tau}^{0}[A_K\xi(t+\theta) + A_1\,\xi(t-\tau+\theta) + \Phi(\xi_{t+\theta}, K\,\xi(t+\theta))]\,d\theta. \quad \text{(A4.10)}$$

$$\xi(\theta) = \psi(\theta)\ :\ \forall\theta \in [-2\tau, 0]\ ,$$

where $A_K = A + BK$. Because:

$$x(t-\tau) = x(t) - \int_{-\tau}^{0}(t+\theta)\,d\theta$$

$$= x(t) - \int_{-t}^{0}[A\,x(t+\theta) + A_1\,x(t+\theta-\tau) + B\,u(t+\theta)] + \Phi(x_{t+\theta}, K\,x(t+\theta))]\,d\theta\ ,$$

any solution of system (A4.1) is also a solution of (A4.10). Hence, if $\Omega_\alpha$ is the region of attraction for the latter system, then it is also region of attraction for the original system. Now, define:

$$V_1(\xi) \triangleq \xi^T(t)P\xi(t), \quad \text{(A4.11)}$$

in which the symmetric positive definite matrix $P = \Lambda^{-1}$. The time derivative of $V_1$ along the trajectories of (A4.10) is calculated as:

$$\dot{V}_1(\xi) = \xi^T(t)[(A_K + A_1)^T P + P(A_K + A_1)]\xi(t) + 2\xi^T(t)P\,\Phi(\xi_t, K\,\xi(t)) + \eta_1(\xi,t) + \eta_2(\xi,t) + \eta_3(\xi,t), \quad \text{(A4.12)}$$

in which:

$$\eta_1(\xi,t) \triangleq -2\int_{-\tau}^{0}\xi^T(t)PA_1A_K\xi(t+\theta)\,d\theta, \quad \text{(A4.13)}$$

$$\eta_2(\xi,t) \triangleq -2\int_{-\tau}^{0}\xi^T(t)PA_1^2\xi(t-\tau+\theta)\,d\theta, \quad \text{(A4.14)}$$

$$\eta_3(\xi,t) \triangleq -2\int_{-\tau}\xi^T(t)PA_1\,\Phi(\xi_{t+\theta}, K\,\xi(t+\theta))\,d\theta, \quad \text{(A4.15)}$$

For the symmetrical matrices $P_i = \Lambda^{-1}\Lambda_i\Lambda^{-1} > 0$ , $i = 1,2,3$ inequality $\Lambda_i - \Lambda \leq 0$ , $i = 1,2,3,$ in (A4.9) implies $P_i - P \leq 0$ .

It is clear that for any $v, w \in R^n$ and for any symmetric positive definite matrix $P_i \in R^{n\times n}$ :

$$-2v^T w \leq v^T P_i^{-1} v + w P_i w. \tag{A4.16}$$

Based on the Razumikhin theorem, assume that for some real number $\delta > 1$, one has:

$$V_1(\xi(t+\theta)) < \delta V_1(\xi(t)) \ : \ \theta \in [-2\tau, 0]. \tag{A4.17}$$

Thus, using inequalities (A4.16) and (A4.17) yields:

$$\eta_1(\xi,t) \leq \tau\xi^T(t) PA_1 A_K P_1^{-1} A_K^T A_1^T P\xi(t) + \tau\xi^T(t) P\xi(t),$$

$$\eta_2(\xi,t) \leq \tau\xi^T(t) PA_1^2 P_2^{-1} (A_1^2)^T P\xi(t) + \tau\xi^T(t) P\xi(t),$$

$$\eta_3(\xi,t) \leq \tau\xi^T(t) PA_1 P_3^{-1} A_1^T P\xi(t)$$

$$+ \int_{-\tau}^{0} \Phi(\xi_{t+\theta}, K\xi(t+\theta))^T P_3 \Phi(\xi_{t+\theta}, K\xi(t+\theta))\, d\theta.$$

Substituting the above inequalities into (A4.12) results in the following:

$$\begin{aligned} \dot{V}_1(\xi) = \xi^T(t)[&\tau PA_1(A_K P_1^{-1} A_K^T + A_1 P_2^{-1} A_1^T + P_3^{-1})^T A_1^T P \\ &+ 2\tau\delta P + (A_K + A_1)^T P + P(A_K + A_1)]\xi(t) \\ &+ \int_{-\tau}^{0} \Phi(\xi_{t+\theta}, K\xi(t+\theta))^T P_3^{-1} \Phi(\xi_{t+\theta}, K\xi(t+\theta))\, d\theta \\ &+ 2\xi^T(t) P\, \Phi(\xi_t, K\xi(t)). \end{aligned} \tag{A4.18}$$

Applying the Schur complement* to (A4.9) and using $\Lambda = P^{-1}$, $\Lambda_i = P^{-1} P_i P^{-1}$, $i = 1,2,3$, $K = \Gamma\Lambda^{-1}$ , and pre- and postmultiplication by $P$ yields:

$$-W_1 \triangleq (A_K + A_1)^T P + P(A_K + A_1) + 2\tau\delta P + \tau PA_1(A_K P_1^{-1} A_K^T + A_1 P_2^{-1} A_1^T + P_3^{-1})^T A_1^T P < 0.$$

By using (A4.16) and (A4.8), the following is obtained:

$$2\xi^T(t) P\Phi(\xi_t, K\xi(t)) \leq \underbrace{2\,(L_1 + L_2 + \delta L_2)\|P\|}_{\Sigma_1}\|\xi\|^2$$

---

* See Lemma 4.1.

and with $\gamma = \delta\, \lambda_{\max}(P)/\lambda_{\min}(P)$ ,

$$\int_{-\tau}^{0} \Phi(\xi_{t+\theta}, K\,\xi(t+\theta))^T P_3^{-1} \Phi(\xi_{t+\theta}, K\,\xi(t+\theta))\, d\theta \le \underbrace{\tau(L_1^2 + L_2^2 \gamma + L_1 L_2 \sqrt{\gamma})\|P_3\|}_{\Sigma_2} \|\xi\|^2 .$$

By choosing $\alpha$ small enough such that:

$$\Sigma_1 + \Sigma_2 < \lambda_{\min}(W_1)/2 ,$$

it is ensured that:

$$\dot{V}_1 < -\frac{\lambda_{\min}(W_1)}{2} \|\xi\|^2 . \tag{A4.19}$$

Note that this is possible since both $L_1$ and $L_2 \to 0$ for $\alpha \to 0$. Thus, by Razumikhin-type argument, it follows that $\Omega_\alpha$ is the region of attraction. ■

Now, a simple condition is derived for the choice of the terminal cost functional of the form:

$$V(x_t) = x^T(t) P\, x(t) + \int_{-\tau}^{0} x^T(t+\theta) S\, x(t+\theta)\, d\theta, \tag{A4.20}$$

in which $P$ and $S$ are $n \times n$ symmetric positive definite constant matrices.

---

**Lemma A4.6:**

Consider the system (A4.1). If there exist symmetric matrices $\Lambda > 0$, $\Upsilon > 0$, a matrix $\Gamma$, and a constant positive scalar $\varepsilon$ satisfying the following LMI:

$$\begin{bmatrix} E_2 + \quad + \varepsilon I & A_1 \Lambda & \Lambda Q^{1/2} & \Gamma^T R^{1/2} \\ * & - \quad + \varepsilon I_{\ 1} & 0 & 0 \\ * & * & -I & 0 \\ * & * & * & -I \end{bmatrix} < 0, \tag{A4.21}$$

in which $E_2 = \Lambda A^T + A\Lambda + \Gamma^T B^T + B\Gamma + \Lambda$, then the control law $u(t) = K\,x(t)$ with $K = \Gamma \Lambda^{-1}$ ensures that $\dot{V}(\mathbf{x}_t) \le -x(t) Q\, x(t) - x^T(t) K^T R K\, x(t)$ for all $x_t \in \Omega_\alpha$ for some $\alpha > 0$. The parameters in the functional $V$ are given by $P = \Lambda^{-1}$ and $S = \Lambda^{-1} \quad \Lambda^{-1}$. ■

**Proof:**

Applying the Schur complement to (A4.21), pre- and postmultiplying by $P = \Lambda^{-1}$, the following is obtained:

$$x^T(t)[A_K^T P + PA_K \Lambda + Q + K^T R K + P]x(t) + 2x^T(t)P A_1 x(t-\tau)$$
$$+ x^T(t) S x(t) - x^T(t-\tau) S x(t-\tau) < -\varepsilon(\|x(t)\|^2 + \|x(t-\tau)\|^2).$$

the derivative of $V$ along the solution of the system when using the local control law $u(t) = Kx(t)$ is as follows:

$$\dot{V}(x_t) = x^T(t)[A_K^T P + PA_K]x(t) + 2x^T(t)PA_1 x(t-\tau) + x^T(t) S x(t)$$
$$- x^T(t-\tau) S x(t-\tau) + 2x^T(t) P \Phi(x_t, K x(t)),$$

Comparing these recent relations, it is clear that the assertion is true if:

$$\varepsilon(\|x(t)\|^2 + \|x(t-\tau)\|^2) + x^T(t) P x(t) \geq 2x^T(t) P \Phi(x_t, K x(t)). \quad \text{(A4.22)}$$

Arguments similar to what is used in the proof of Lemma A4.5 yield:

$$\|x^T(t) P \Phi\| \leq x^T(t) P x(t) + \Phi^T P \Phi \leq x^T(t) P x(t) + (L_1 + L_1^2)\|x(t)\|^T$$
$$+ (L_2 + L_2^2)\|x(t-\tau)\|^2.$$

Clearly, if $\alpha$ is chosen such that $L_i + L_i^2 < \varepsilon$, $i = 1,2$, then (A4.22) holds and hence, the assertion is true. ■

---

**Theorem A4.2: (Mahboobi–Nikravesh Theorem 2)**

If there exist some symmetric positive definite matrices $\Lambda, \Lambda_i, Q, R,$ and a matrix $\Gamma$ such that LMIs (A4.9) and (A4.21) admit a feasible solution, then there exists a terminal region $\Omega_\alpha$ of the form (A4.7) and a functional of the form (A4.20) such that the stabilizing condition is satisfied for the stage cost: $q(x,u) = x^T Q x + u^T R u$. ■

**Proof:**

The proof follows directly from Lemma A4.5 and Lemma A4.6 and using the fact that it is always possible to choose sufficiently small $\alpha$ such that the input constraints are satisfied for all $x_t \in \Omega_\alpha$. ■

## A4.2 STABILIZING PREDICTIVE CONTROL OF UNCONSTRAINED NONLINEAR TIME-DELAYED SYSTEMS [M9]

In this section, a predictive control scheme which guarantees closed-loop stability is proposed for unconstrained nonlinear time-delayed systems utilizing Control Lyapunov–Krasovskii Functional (CLKF). In this approach, the stability can be achieved without imposing any terminal constraints. The absence of additional constraints results in significant improvements in computation speed.

### A4.2.1 Proposed Predictive Controller

Consider an input affine nonlinear time-delayed system with discrete and distributed delays described by:

$$\begin{aligned}\dot{x}(t) &= f(x(t), x(t-\tau_1), \ldots, x(t-\tau_l)) \\ &\quad + \int_{-r}^{0} G(\theta) F(x(t), x(t-\tau_1), \ldots, x(t-\tau_l), x(t+\theta))\, d\theta \\ &\quad + g(x(t), x(t-\tau_1), \ldots, x(t-\tau_l))\, u(t), \\ x(t) &= \ (t) : t \in [-\max(\tau_i, r), 0] \ ,\end{aligned} \tag{A4.23}$$

where $x(t) \in R^n$ and $u(t) \in R^m$ are system state and input vectors, respectively, and $(t)$, $t \in [-\max(\tau_i, r), 0]$ is a given initial vector function. $f$, $g$, $G$, and $F$ are bounded and smooth matrix functions of their arguments. Also, $f(0,0,\ldots,0) = 0$ and $F(0,0, \ldots 0,0) = 0$. The problem of interest is to stabilize the equilibrium state of system (A4.23) via model predictive controller.

The input signal which is applied to the real system between the sampling instants is given by the solution of the following finite horizon optimal control problem:

$$\min_{u} \ J(x_t, u, T) = \int_{t}^{t+T} q(x(t'), u(t'))\, dt' + E, \tag{A4.24}$$

subject to:

$$\begin{aligned}\dot{x}(t') &= f(x(t'), x(t'-\tau_1), \ldots, x(t'-\tau_l)) \\ &\quad + \int_{-r}^{0} G(\theta) F(x(t'), x(t'-\tau_1), \ldots, x(t'-\tau_l), x(t'+\theta))\, d\theta \\ &\quad + g(x(t'), x(t'-\tau_1), \ldots, x(t'-\tau_l))\, u(t'), \qquad x_t = \bar{x}_t,\end{aligned} \tag{A4.25}$$

where $x(t')$ and $u(t')$ are predicted variables (controller variables). The stage cost $q: R^n \times R^m \to R^+$ penalizes states and control inputs according to (A4.4), $E$ is an appropriately defined terminal cost. It is assumed that the optimal control that optimizes $J(x_t,u,T)$ is given by $u^*(t')$, $t \le t' \le t+T$ and for the sake of simplicity, the optimal cost is denoted by $J^*(x_t,T)$. The control input to the system is defined by (A4.5). The purpose of this section is to show that, using an appropriate CLKF in terminal cost $E$ is effective and strongly guarantees the stability of the closed-loop control system.

---

**Definition A4.1:**

A smooth functional $V(x_t)$ of the following form:

$$V(x_t) = V_1(x) + \sum_{i=1}^{l} \int_{-\tau_i}^{0} S_i(x(t+\ )) \, d\ \ + \int_{-r}^{0} \int_{t+\theta}^{t} W(\theta, x(\ )) \, d\ \ d\theta, \quad \text{(A4.26)}$$

where $V_1(x)$ is a smooth, positive definite, radially unbounded function of the current states. $S_i : R^n \to R$ and $W : R \times R^n \to R$ are smooth, nonnegative integrable functions. $V(x_t)$ is called a *Control Lyapunov–Krasovskii Functional* (CLKF) for the system (A4.23) if there exists a function $\alpha$ and $\alpha(s) > 0$ for $s > 0$ and two class $K_\infty$ functions $\beta_1$ and $\beta_2$ such that:

$$\beta_1(\|\chi_t(0)\|) \le V(\chi_t) \le \beta_2(\|\chi_t\|), \quad \text{(A4.27)}$$

$$\min_{\mathrm{u}} \dot{V} = \min_{\mathrm{u}} \{L_{f^*} V + L_g V_1 \, u\} < -\alpha(\|\chi_t(0)\|), \quad \text{(A4.28)}$$

for all piecewise continuous functions $\chi_t : [-\max(\tau_i, r), 0] \to R^n$, where the Lee derivatives are given below:

$$L_g V_1 = \frac{\partial V_1}{\partial x} g,$$

$$L_{f^*} V = \frac{\partial V_1}{\partial x}\left(f + \int_{-r}^{0} \Gamma(\theta) F \, d\theta\right) + \sum_{i=1}^{l} (S_i(x(t)) - S_i(x(t-\tau_i))) \quad \text{(A4.29)}$$

$$+ \int_{-r}^{0} (W(\theta, x(t)) - W(\theta, x(t+\theta))) \, d\theta.$$

■

To guarantee the closed-loop stability in predictive control of system (A4.23), the CLKF (A4.26) of the system is incorporated in the terminal cost of the optimal control problem (A4.24) together with (A4.25) as $E = \mu\, V(\mathbf{x}_{t+T})$, where $\mu \in R^+$.

### A4.2.2 Stability Analysis

Similar to Section A4.1.3, it can be shown that the proposed predictive controller stabilizes the equilibrium state of the system (A4.23). Similar to Lemma A4.1, it can easily be proved that repeated solution of the finite horizon optimal control problem (A4.24) together with (A4.25) is feasible. In addition, the optimal cost of problem (A4.24) together with (A4.25) have the characteristics proved in Lemmas A4.1 and A4.2. As in the case of Lemma A4.4, it can be shown that the nonincreasing monotonicity of the optimal cost implies the asymptotic stability of the proposed predictive controller. In Lemma 4.7, it is shown that the optimal cost of problem (A4.24) together with (A4.25) is monotonic nonincreasing functional. This property together with the continuity property of optimal cost enables stating the final result about the stability of the proposed scheme.

---

**Lemma A4.7:**

Assume that $x_t$ belongs to the feasible region of finite horizon problem (A4.24) together with (A4.25), then there exists a $\mu_0$ such that for all $\mu \geq \mu_0$ and $\delta > 0$, the optimal cost satisfies $J^*(x_t, T+\delta) \leq J^*(x_t, T)$. ■

**Proof:**

Let $\hat{x}(t')$ , $t' \in [t\ , t+T+\delta]$ be the trajectory obtained by using $\hat{u}(t')$ as a control input, where $\hat{u}(t')$ is given by:

$$\hat{u}(t') = \begin{cases} u^*(t') & t \leq t' < t+T, \\ \kappa(x_{t'}) & t+T \leq t' \leq t+T+\delta, \end{cases}$$

where $\kappa(x_{t'})$ is feedback law that can be obtained directly from the known CLKF, which stabilizes the equilibrium state of the system. The cost of using $\hat{u}$ for time interval $[t, t+T+\delta]$, beginning at initial state $x_t$ is $J(x_t, \hat{u}, T+\delta)$. Therefore:

$$J^*(x_t, T+\delta) \leq J(x_t, \hat{u}, T+\delta)$$

$$= \int_t^{t+T} q(x(t'), u^*(t'))dt' + \int_{t+T}^{t+T+\delta} q(\hat{x}(t'), \hat{u}(t'))dt' + \mu\, V(\hat{x}_{t+T+\delta})$$

$$= J^*(x_t, T) + \int_{t+T}^{t+T+\delta} q(\hat{x}(t'), \hat{u}(t'))dt' + \mu\, (V(\hat{x}_{t+T+\delta}) - V(x^*_{t+T})),$$

hence, $\mu = \mu_0 > 0$ can always be chosen such that:

$$\int_{t+T}^{t+T+\delta} q(\hat{x}(t'), \hat{u}(t'))\,dt' + \mu_0 \left(V(\hat{x}_{t+T+\delta}) - V(x^*_{t+T})\right) \le 0, \tag{A4.30}$$

leading to:

$$J^*(x_t, T+\delta) \le J^*(x_t, T),$$

which completes the proof. ■

Assurance about the existence of stabilizing controller $\kappa(x_{t'})$ is sufficient for the validity of the proof given for Lemma A4.7. This feedback control law is directly achievable from the known CLKF. Finally, the main result is stated as Theorem A4.3.

---

**Theorem A4.3: (Mahboobi–Nikravesh Theorem 3)**

Assume that there exists a CLKF for the system (A4.23) and finite horizon problem (A4.24) together with (A4.25) admits a feasible solution at $t = 0$, then the receding horizon predictive controller asymptotically stabilizes the equilibrium state of the system. ■

As an example for demonstrating the proposed method, consider the nonlinear system with delayed state described by:

$$\begin{aligned} \dot{x}(t) &= x^3(t-1) + [x(t) - \frac{1}{2}x(t-1)]u(t), \\ x(t) &= 0.5, \qquad -1 \le t \le 0. \end{aligned} \tag{A4.31}$$

The CLKF-based predictive control is applied to regulate the system (A4.31). Constructing a CLKF is the first step in design procedure. Select the following CLKF candidate:

$$V(x_t) = \frac{1}{2}x^2(t) + \int_{-1}^{0} x^4(t+\theta)\,d\theta. \tag{A4.32}$$

It is shown that the selected candidate CLKF satisfies the conditions of the Definition A4.1:

$$\underbrace{\frac{1}{2}x^2(t)}_{\beta_1(\|x_t(0)\|)} \le \frac{1}{2}x^2(t) + \int_{-1}^{0} x^4(t+\theta)\,d\theta \le \underbrace{\|x_t\|^2 + \|x_t\|^4}_{\beta_2(\|x_t\|)}$$

Also, the time derivative of $V$ is:

$$\dot{V}(x_t) = x(t)\left[x(t) - \frac{1}{2}x(t-1)\right]u(t) + x(t)x^3(t-1) + x^4(t) - x^4(t-1).$$

Minimizing $\dot{V}(x_t)$ with respect to $u$ yields:

$$\min_u \dot{V} = \begin{cases} -x^4(t-1) \le 0 = -\alpha(\|0\|), & \textit{if } x(t) = 0, \\ -7x^4(t) \le -\alpha(\|x(t)\|), & \textit{if } x(t) = \frac{1}{2}x(t-1). \end{cases}$$

Regarding (A4.28), one may choose $\alpha = 7s^4$, therefore, $V$ is a CLKF for system (A4.31) and is used as the terminal cost. The finite horizon cost function for the proposed predictive controller is represented by:

$$J = \int_t^{t+T} x^2(t') + 0.2u^2(t')dt' + 30\left\{\frac{1}{2}x^2(t+T) + \int_{-1}^{0} x^4(t+T+\theta)d\theta\right\}, \quad \text{(A4.33)}$$

The method proposed in Mahboobi et al. [m8] is utilized to implement the proposed controller to simulate the performance of a closed-loop system. The results obtained by the proposed method are compared to those obtained by the control law provided by Freeman's formula [j1]. Figure A4.1 shows the dynamic behavior of the

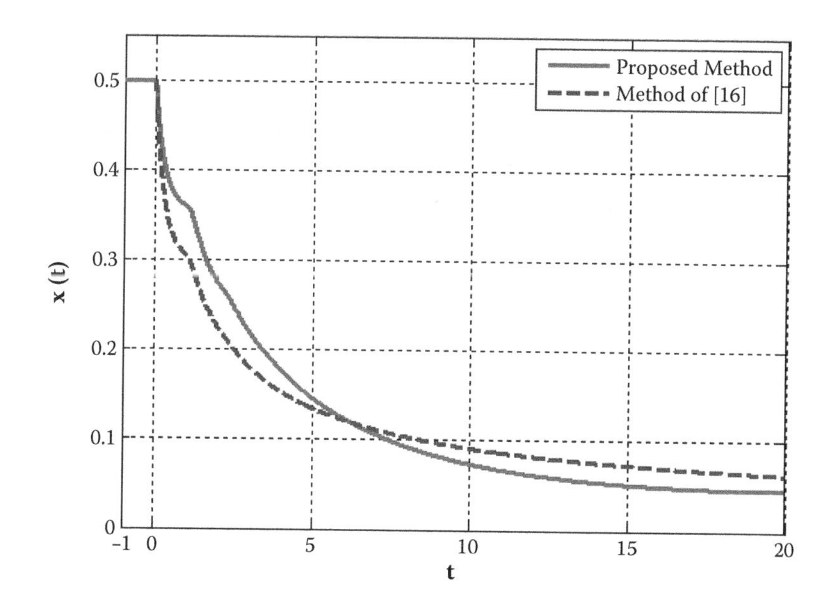


**FIGURE A4.1** State trajectory.

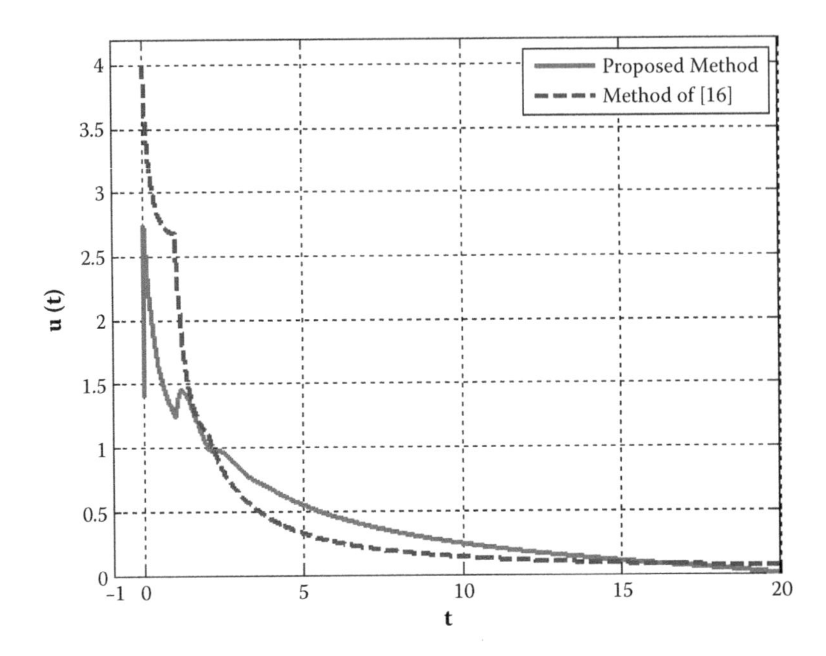


**FIGURE A4.2** Control input.

state and Figure A4.2 shows control input. The following cost function is used to measure performance for comparing the two mentioned procedures:

$$J_c = \int_0^{20} x^2(t') + 0.2u^2(t')dt'.$$

The value of Jc for the proposed method is 1.98 and for the control law provided with Freeman's formula is 2.97, which confirms that the proposed approach achieves closed-loop stability while improving the performance too.

# Appendix A5: Some New Notions for Symmetric Behavior of Matrices and Related Theorems

Throughout this appendix, it is assumed that $\mathbb{F}$ is an arbitrary field. Let $m,n \in N$, we denote the ring of all $n \times n$ matrices and the set of all $m \times n$ matrices over $\mathbb{F}$ by $M_n(\mathbb{F})$ and $M_{m\times n}(\mathbb{F})$, respectively, and for simplicity let $\mathbb{F}^n = M_{1\times n}(\mathbb{F})$. The zero matrix is denoted by 0. Also the element of an $m \times n$ matrix, $R$, on the $i$-th row and $j$-th column is denoted by $r(i,j)$. If $A = (a_{ij}) \in M_n(\mathbb{F})$, then $[A]_{ij} \in M_{(n-1)}(\mathbb{F})$ is a matrix obtained from omitting the $i$-th row and the $j$-th column of the matrix $A$. Also, $\text{cof}(A) \in M_n(\mathbb{F})$ is defined by $\text{cof}(A) = (a_{ij}(-1)^{i+j}\det([A]_{ij})$. For any $i,j$, $1 \le i, j \le n$, we denote by $E_{ij}$, that element in $M_n(\mathbb{F})$ whose $(i,j)$ entry is 1 and whose other entries are 0. Also, 0 and $I$ are the zero and identity matrix, also $0_r$ and $I_r$ denote the zero matrix and the identity matrix with dimension $r$, respectively.

The main purpose of this section is to define some kinds of symmetric matrices such as centrally symmetric, row-wise (column-wise) symmetric, row-wise (column-wise) negative symmetric, and plus symmetric, in section two.

## A5.1 LEMMAS AND THEOREMS

Symmetric matrices were discussed in Chapter 5. It can now easily be checked that if $F$ is a PS matrix, then it is also a CS, RWS, and CWS matrice as well. So every property of CS, RWS, or CWS matrices, is a property of a PS matrix, too.

---

**Lemma A5.1:**

Let $n \in \text{N}$ and $F$ be an $n$-th order matrix. Then,

1. If $F$ is both a RWS matrix and a CWS matrix, then, it is a PS matrix, and therefore, a CS matrix.
2. If $F$ is both RWNS matrix and CWNS matrix, then it is a CS matrix.
3. If $F$ is both CS matrix and RWS matrix, then it is a PS matrix.
4. If $F$ is both CS matrix and CWS matrix, then it is a PS matrix.

**Proof:**

Let $i,j$ $1 \leq i$ and $j \leq n$ be arbitrary scalars.

(a) Since $F$ is a RWS matrix, then, $r(i, j) = r(n + 1 - i, j)$, and since it is a CWS matrix, then, $r(i, j) = r(i,n + 1 - j)$ and $r(n + 1 - i,j) = r(n + 1 - i, n + 1 - j)$. Therefore:

$$r(i, j) = r(n + 1 - i, j) = r(i,j) = r(i,n + 1 - j) = r(n + 1 - i, n + 1 - j),$$

thus $F$ is a PS matrix.

(b) Since $F$ is a RWNS matrix, then, $r(i, j) = -r(n + 1 - i, j)$, and since it is a CWNS matrix, then

$$-r(n + 1 - i, j) = -(-r(n + 1 - i,n + 1 - j)) = r(n + 1 - i,n + 1 - j).$$

Therefore, $r(i, j) = r(n + 1 - i, n + 1 - j)$ thus $F$ is a CS matrix.

(c) Since $F$ is a CS matrix, then, $r(i, j) = r(n + 1 - i, n + 1 - j)$, and since it is a RWS matrix, then:

$$r(n + 1 - i, n + 1 - j) = r(n + 1 - (n + 1 - i), n + 1 - j) = r(i,n + 1 - j)$$

then, $r(i,j) = r(i,n + 1 - j)$ thus $F$ is a CWS matrix. Now, by (a), the proof is completed.

(d) The proof of this part is similar to the proof of part (c). ■

---

**Lemma A5.2:**

Let $m,n \in N$, $A,B \in M_n(\mathbb{F})$, and $C,D \in M_{m \times n}(\mathbb{F})$, therefore,

1. If $A$ and $B$ are CS matrices, then, $A + B$ is a CS matrix.
2 If $C$ and $D$ are RWS (CWS) matrices, then, $C + D$ is a RWS (CWS) matrix.
3. If $C$ and $D$ are RWNS (CWNS) matrices, then, $C + D$ is a RWNS (CWNS) matrix.
4. If $A$ and $B$ are PS matrices, then, $A + B$ is a PS matrix.

**Proof:**

(a) Let $i, j$, $1 \leq i$ and $j \leq n$ be arbitrary and $R = A + B$. Since $A$ and $B$ are CS matrices, then:

$$r(i,j) = a(i,j) + b(i,j) = a(n + 1 - i,n + 1 - j) + b(n + 1 - i, n + 1 - j) = r(n + 1 - i, n + 1 - j)$$

Thus, $A+B$ is a CS matrix and the proof is completed. Other statements can be proved similarly. ■

---

**Lemma A5.3:**

Let, $m,n,p,q \in N$, therefore:

1. If $A$ and $B \in \mathbb{F}$ are CS matrices, then, $AB$ is a CS matrix.
2. If $A \in M_{m\times n}(\mathbb{F})$ is a RWS (RWNS) matrix, then, $AB$ is a RWS (RWNS) matrix, for every $B \in M_{n\times p}(\mathbb{F})$.
3. If $B \in M_{n\times p}(\mathbb{F})$ is a CWS (CWNS) matrix, then, $AB$ is a CWS (CWNS) matrix, for every $A \in M_{m\times n}(\mathbb{F})$.
4. If $A \in M_{m\times n}(\mathbb{F})$ is RWS and $B \in M_{n\times m}(\mathbb{F})$ is CWS matrix, then, $AB$ is a PS matrix.
5. If $A \in M_{m\times n}(\mathbb{F})$ is RWNS and $B \in M_{n\times m}(\mathbb{F})$ is a CWNS matrix, then, $AB$ is a CS matrix.
6. If $A \in M_{n}(\mathbb{F})$ is a CS matrix and $B \in M_{n\times p}(\mathbb{F})$ is a RWS (RWNS) matrix, then $AB$ is a RWS (RWNS) matrix.
7. If $A \in M_{m\times n}(\mathbb{F})$ is a CWS (CWNS) matrix and $B \in M_{m}(\mathbb{F})$ is a CS matrix, then, $AB$ is a CWS (CWNS) matrix.

**Proof:**

(a) Let $i,j$, $1 \le i$ and $j \le n$ be arbitrary scalars and $R = AB$. Since $A$ and $B$ are CS matrices, then:

$$r(n+1-i,n+1-j) = \sum_{k=1}^{n} a(n+1-i,k)b(k,n+1-j)$$

$$= \sum_{k=1}^{n} a(i,n+1-k)b(n+1-k,j) = \sum_{k'=1}^{n} a(i,k')b(k',j) = r(i,j).$$

Therefore, $AB$ is a CS matrix.

(b) Let $i$, $1 \le i \le m$ and $j$, $1 \le j \le p$ be arbitrary scalars and $R = AB$. Since $A$ is a RWS matrix, then:

$$r(i,j) = \sum_{k=1}^{n} a(i,k)b(k,j) = \sum_{k=1}^{n} a(n+1-i,k)b(k,j) = r(n+1-i,j).$$

Thus, $AB$ is a RWS matrix. In the case that $A$ is a RWNS matrix the result follows in a similar way.

(c) The proof of this part is similar to the proof of part (b).

(d) Since $A$ is a RWS matrix, then, by (b) $AB$ is a RWS matrix, and since $B$ is a CWS matrix, then, by (c), $AB$ is a CWS matrix. Thus, by Lemma A5.1, $AB$ is a PS matrix.

(e) Since $A$ is a RWNS matrix, then, by part (b), $AB$ is a RWNS matrix, and since $B$ is a CWNS matrix, then, by part (c), $AB$ is a CWNS matrix. Thus, by Lemma A5.1, $AB$ is a CS matrix.

(f) Let $i$, $1 \le i \le n$, and $j$, $1 \le j \le p$, be arbitrary scalars and $R = AB$. Since $A$ is a CS matrix and $B$ is a RWS matrix, then:

$$r(i,j) = \sum_{k=1}^{n} a(i,k)b(k,j) = \sum_{k=1}^{n} a(n+1-i, n+1-k)\, b(n+1-k, j)$$

$$= \sum_{k'=1}^{n} a(n+1-i, k')b(k', j) = r(n+1-i, j).$$

Therefore, $AB$ is a RWS matrix. In the case that $A$ is a RWNS matrix, the result follows in a similar way.

(g) The proof of this part is similar to the proof of part (f). This completes the proof. ■

---

**Lemma A5.4:**

Let $m,n \in \mathrm{N}$, therefore:

1. If $A \in M_n(\mathbb{F})$ is a CS matrix, then, $A^T$ is a CS matrix, too.
2. If $A \in M_{m \times n}(\mathbb{F})$ is a RWS (RWNS) matrix, then, $A^T$ is a CWS (CWNS) matrix.
3. If $A \in M_{m \times n}(\mathbb{F})$ is a CWS (CWNS) matrix, then, $A^T$ is a RWS (RWNS) matrix.
4. If $A \in M_n(\mathbb{F})$ is a PS matrix, then, $A^T$ is a PS matrix, too.

**Proof:**

(a) Let $i$, $1 \le i \le n$, and $j$, $1 \le j \le n$ be arbitrary scalars, and $R = A^T$. Since $A$ is a CS matrix, then:

$$r(n+1-i, n+1-j) = a(n+1-j, n+1-i) = a(j,i) = r(i,j).$$

(b) Let $i$, $1 \leq i \leq m$ and $j$, $1 \leq j \leq n$ be arbitrary scalars and $R = A^T$. Since $A$ is a RWS matrix, then:

$$s(i,j) = a(j,i) = r(n+1-j,i) = s(i,n+1-j).$$

Thus, $A^T$ is a CWS matrix. In the case that $A$ is a RWNS matrix the result follows in a similar way.

(c) The proof of this part is similar to the proof of part (b).

(d) Since $A$ is a RWS matrix, then, by part (b), $A^T$ is a CWS matrix, and since $A$ is a CWS matrix, then, by part (c), $A^T$ is a RWS matrix. Therefore, by Lemma A5.1, $A^T$ is a PS matrix. This completes the proof. ■

---

**Theorem A5.1:**

Let $n \in N$, and $A \in M_n(\mathbb{F})$. Then:

1. If $A$ is a CS matrix, then, cof($A$) is a CS matrix, too.
2. If $A$ is a RWS (RWNS) matrix, then, cof($A$) is a RWNS (RWS) matrix. Furthermore, if $n \leq 4$, then, cof($A$) = 0.
3. If $A$ is a CWS (CWNS) matrix, then, cof($A$) is a CWNS (CWS) matrix. Furthermore, if $n \geq 4$, then, cof($A$) = 0.
4. If $A$ is a PS matrix, then, cof($A$) is a CS matrix. Furthermore, if $n \geq 4$, then, cof($A$) = 0.

**Proof:**

Let $i, j$, $1 \leq i, j \leq n$ be arbitrary scalars and $R$ = cof($A$).

(a) First, note that:

$$r(n+1-i, n+1-j) = (-1)^{n+1-i+n+1-j} \det([A]_{(n+1-i),(n+1-j)}).$$

On the other hand $[A]_{(n+1-i),(n+1-j)} = p[A]_{i,j} P$, where,

$$P = \begin{bmatrix} & 0 & 1 \\ & \cdots & \\ 1 & 0 & \end{bmatrix} \in M_{n-1}(\mathbb{F}).$$

Since $\det(P) = 1$ and $A$ is a CS matrix, then:

$$r(n+1-i, n+1-j) = (-1)^{n+1-i+n+1-j} \det([A]_{(n+1-i),(n+1-j)})$$

$$= (-1)^{2(n+1)-(i+j)} \det(P[A]_{i,j} P) = (-1)^{i+j} \det(P)\det(P[A]_{i,j})\det(P)$$

$$= (-1)^{i+j} \det(P[A]_{i,j} = r(i,j).$$

Therefore, cof($A$) is a CS matrix.

(b) Note that $r(n+1-i, j) = (-1)^{n+1-i+j} \det([A]_{(n+1-i,)j})$ .

If $n = 2$, then $[A]_{(3-i),j} = [A]_{i,j,}$ and therefore:

$$r(3-i,j) = (-1)^{3-i+j} \det([\mathrm{A}]_{(3-i),j}) = -(-1)^{i+j} [A]_{i,j} = -r(i,j).$$

It means that cof(A) is a RWNS matrix.
If $n = 3$, then it is easily seen that:

$$A_{(4-i),j} = \begin{bmatrix} 0 & 1 \\ 1 & 0 \end{bmatrix} A_{i,j}.$$

Thus:

$$r(4-i,j) = (-1)^{4-i+j} \det([A]_{(41-i),j}) = (-1)^{i+j} \det\left(\begin{bmatrix} 0 & 1 \\ 1 & 0 \end{bmatrix} A_{i,j}\right)$$

$$= -(-1)^{i+j} \det(A_{i,j}) = -r(i,j).$$

It means that cof($A$) is a RWNS matrix. Now if $n \geq 4$, then, it is easily seen that the matrix $[A]_{i,j}$ has at least two equivalent rows, therefore

$$\det(A_{i,j}) = 0, \text{ for every } i,j,\ 1 \leq i, j \leq n, \text{ thus, } \mathrm{cof}(A) = 0.$$

In the case that $A$ is a RWNS matrix, a similar way shows the proof.

(c) The proof of this part is similar to the proof of part (b).

(d) By Lemma A5.1, $A$ is a RWS matrix and a CWS matrix as well. Thus, by part (b), cof($A$) is a RWNS matrix and by part (c), cof($A$) is a CWNS matrix. Therefore, by Lemma A5.1, cof($A$) is a CS matrix. If $n \geq 4$, then, by part (b), cof($A$) = 0.

This completes the proof. ■

**Remark A5.1:**

Let $n \in \mathbb{N}$. If $n \geq 3$ then, every RWS, RWNS, CWS, CWNS, or PS matrix in $M_n$ ($\mathbb{F}$) is a noninvertible matrix, since it has at least two equivalent rows or two equivalent columns. ■

**Theorem A5.2:**

Let $n \in N$, and $A \in M_n$ ($\mathbb{F}$) be invertible. Then:

1. If $A$ is a CS matrix, then, $A^{-1}$ is a CS matrix, too.
2. If $A$ is a RWS (RWNS) matrix, then, $A^{-1}$ is a CWNS (CWS) matrix.
3. If $A$ is a CWS (CWNS) matrix, then, $A^{-1}$ is a RWNS (RWS) matrix.
4. If $A$ is a PS matrix, then, $A^{-1}$ is a CS matrix.

**Proof:**

Since $A^{-1} = \frac{1}{\det(A)}(cof(A))^T$, then, by Lemma A5.4 and Theorem A5.1, the result follows. ■

**Note A5.1:**

Consider a PS matrix as $A = E_{11} + E_{1n} + E_{nl} + E_{nn}$ and an invertible matrix $P = \sum_{i=1}^{n} \frac{1}{n} E_{ii}$, then, $P^{-1}AP = E_{11} + \frac{1}{n}E_{1n} + nE_{nl} + E_{nn}$ is neither PS nor any other introduced matrices. Therefore, by Remark A5.1, a similarity transformation does not preserve any of the defined symmetricities. ■

**Theorem A5.3:**

Let $n \in$ N, and $A \in M_n$ ($\mathbb{F}$) be a CS matrix which has $n$ distinct eigenvalues. Then, all eigenvectors are either RWS or RWNS. Furthermore, the number of RWS eigenvectors is $[\frac{n+1}{2}]$ and the number of RWNS eigenvectors is $[\frac{n}{2}]$. Note that $[\frac{n+1}{2}] + [\frac{n}{2}] = n$. ■

**Proof:**

Since $A$.adj $(A) =$ adj$(A).A = det(A)I$ for every $A \in M_n$ ($\mathbb{F}$), thus, if $\lambda$ is an eigenvalue of $R$, then.

$$(R - \lambda I)\text{adj}(R - \lambda I) = \det(R - \lambda I)\, I = 0\,.$$

Therefore, every column of adj$(R-\lambda I)$ is an eigenvector of $R$, which is associated with $\lambda$.

Write the columns of adj$(R-\lambda I)$ as $[c_1 \ldots c_n]$. Now, consider a pair of column vectors $c_j$ and $c_{n+1-j}$, for $j$, $1 \le j \le n$. Note that $R$ has $n$ distinct eigenvalue and hence, (adj$(R-\lambda I)$) = 1, so $c_j$ is proportional to $c_{n+1-j}$. Since adj$(R-\lambda I) = \text{cof}(R)^T$, then, by Theorem A5.1 and Lemma A5.4, adj$(R-\lambda I)$ is CS since $R$ is CS. Therefore, for any $j$, the two vectors $c_j$ and $c_{n+1-j}$ are a reverse form* of each other. Hence, the vector $c_j$ (and $c_{n+1-j}$ too) is proportional to its rotated form and this may not happen except in two ways:

1. When the vector $c_j$ (and also $c_{n+1-j}$) is RWS.
2. When the vector $c_j$ (and also $c_{n+1-j}$) is RWNS.

Assume that $R=[r_{ij}]$. First, let $n = 2m + 1$, for some $m \in \mathbb{N}$. Assume that:

$$x = [x_1 x_2 \cdots x_m y x_m x_{m-1} \cdots x_1]^T,$$

is a RWS eigenvector of $R$. Consider the equation $Rx = \lambda x$, which is equivalent to the following, since $R$ is a CS matrix:

$$\begin{bmatrix} r_{11} & \cdots & r_{1(m+1)} & r_{1(m+2)} & r_{1(m+3)} & \cdots & r_{1n} \\ \vdots & \ddots & \vdots & \vdots & \vdots & \ddots & \vdots \\ r_{(m+1)1} & \ddots & r_{(m+1)(m+1)} & r_{(m+1)m} & r_{(m+1)(m-1)} & \cdots & r_{(m+1)1} \\ \vdots & \ddots & \vdots & \vdots & \vdots & \ddots & \vdots \\ r_{1n} & \cdots & r_{1(m+1)} & r_{1n} & r_{1(m-1)} & \cdots & r_{11} \end{bmatrix}$$

$$\times \begin{bmatrix} x_1 \\ \vdots \\ x_m \\ y \\ x_m \\ \vdots \\ x_1 \end{bmatrix} = \lambda \begin{bmatrix} x_1 \\ \vdots \\ x_m \\ y \\ x_m \\ \vdots \\ x_1 \end{bmatrix}$$

Thus:

$$\begin{cases} r_{11}x_1 + r_{12}x_2 + \cdots + r_{1(m+1)}\, y + r_{1(m+2)}x_m + \cdots + r_{1n}x_1 = \lambda x_1 \\ \vdots \\ r_{m1}x_1 + r_{m2}x_2 + \cdots + r_{m(m+1)}y + r_{m(m+2)}x_m + \cdots + r_{mn}x_1 = \lambda x_m \\ r_{(m+1)1}x_1 + r_{(m+1)2}x_2 + \cdots + r_{(m+1)m}x_m + r_{(m+1)(m+1)}y + r_{(m+1)m}x_m + \cdots + r_{(m+1)1}x_1 = \lambda y \\ r_{mn}x_1 + r_{m(n-1)}x_2 + \cdots + r_{m(m+2)}x_m + r_{m(m+1)}y + r_{mm}x_m + \cdots + r_{m1}x_1 = \lambda x_m \\ \vdots \\ r_{1n}x_1 + r_{1(n-1)}x_2 + \cdots + r_{1(m+2)}x_m + r_{1(m+1)}y + r_{1m}x_m + \cdots + r_{11}x_1 = \lambda x_1 \end{cases}$$

* Here we define $c_i = (c_1,\ldots,c_n)^T$ to be the reverse form of $c_j = (c_n,\ldots c_1)^T$.

A straightforward manipulation shows that the first $m$ equations and the last $m$ equations are the same. Therefore, one has the following set of equations:

$$\begin{cases} (r_{11}+r_{1n}-\lambda)x_1+(r_{12}+r_{1(2m)})x_2+\cdots+(r_{1m}+r_{1(m+1)})x_m+r_{1(m+1)}y=0 \\ \vdots \\ (r_{m1}+r_{mn})x_1+(r_{m2}+r_{m(2m)})x_2+\cdots+(r_{mm}+r_{m(m+2)}-\lambda)x_m+r_{m(m+1)}y=0 \\ 2r_{(m+1)1}x_1+2r_{(m+1)2}x_2+\cdots+2r_{(m+1)m}x_m+(r_{(m+1)(m+1)}-\lambda)y=0 \end{cases}$$

Since a set of homogenous equations has a nonzero solution iff the determinant of the matrix of its coefficients is equal to zero and the determinant of the matrix of the coefficients of the above set of equations is a polynomial of degree $m + 1$ of $\lambda$ and it has at most $m + 1$ solutions, therefore, there are at most $m + 1$ eigenvalues of $R$ such that their associated eigenvectors are RWS.

Now, assume that:

$$x=[x_1,x_2,\ldots,x_m,0,-x_m,-x_{m-1},\ldots,-x_1]^T,$$

is a RWNS eigenvector of $R$. Consider the equation $Rx = \lambda x$, which is equivalent to the following, since $R$ is a CS matrix:

$$\begin{bmatrix} r_{11} & \cdots & r_{1(m+1)} & r_{1(m+2)} & r_{1(m+3)} & \cdots & r_{1n} \\ \vdots & \ddots & \vdots & \vdots & \vdots & \ddots & \vdots \\ r_{(m+1)1} & \ddots & r_{(m+1)(m+1)} & r_{(m+1)m} & r_{(m+1)(m-1)} & \cdots & r_{(m+1)1} \\ \vdots & \ddots & \vdots & \vdots & \vdots & \ddots & \vdots \\ r_{1n} & \cdots & r_{1(m+1)} & r_{1nm} & r_{1(m-1)} & \cdots & r_{11} \end{bmatrix}$$

$$\begin{bmatrix} x_1 \\ \vdots \\ x_m \\ y \\ -x_m \\ \vdots \\ -x_1 \end{bmatrix} = \lambda \begin{bmatrix} x_1 \\ \vdots \\ x_m \\ 0 \\ -x_m \\ \vdots \\ -x_1 \end{bmatrix}.$$

Thus:

$$\begin{cases} r_{11}x_1+r_{12}x_2+\cdots+r_{1m}x_m+0-r_{1(m+2)}x_m-\cdots-r_{1n}x_1=\lambda x_1 \\ \vdots \\ r_{m1}x_1+r_{m2}x_2+\cdots+r_{mm}x_m+0-r_{m(m+2)}x_m-\cdots-r_{mn}x_1=\lambda x_m \\ r_{(m+1)1}x_1+r_{(m+1)2}x_2+\cdots+r_{(m+1)m}x_m+0-r_{(m+1)m}x_m-\cdots-r_{(m+1)1}x_1=0 \\ r_{mn}x_1+r_{m(n-1)}x_2+\cdots+r_{m(m+2)}x_m+0-r_{mm}x_m-\cdots-r_{m1}x_1=-\lambda x_m \\ \vdots \\ r_{1n}x_1+r_{1(n-1)}x_2+\cdots+r_{1(m+2)}x_m+0-r_{1m}x_m-\cdots-r_{11}x_1=-\lambda x_1. \end{cases}$$

Again a straightforward manipulation shows that the first $m$ equations and the last $m$ equations are the same and the $m + 1st$ equation is trivial. Therefore, one has the following set of equations:

$$\begin{cases} (r_{11} - r_{1n} - \lambda)x_1 + (r_{12} - r_{1(2m)})x_2 + \cdots + (r_{1m} - r_{1(m+1)})x_m = 0 \\ \vdots \\ (r_{m1} - r_{mn})x_1 + (r_{m2} - r_{m(2m)})x_2 + \cdots + (r_{mm} - r_{m(m+2)} - \lambda)x_m = 0 \end{cases}$$

Since a set of homogenous equations has a nonzero solution iff the determinant of the matrix of its coefficients is equal to zero and the determinant of the matrix of coefficients of the above set of equations is a polynomial of degree $m$ of $\lambda$ and it has at most $m$ solutions; therefore, there are at most $m$ eigenvalues of $R$ such that their associated eigenvectors are RWNS.

Hence, the number of RWS eigenvectors are $m+1=[\frac{n+1}{2}]$ and the number of RWNS eigenvectors are $m=[\frac{n}{2}]$, when $n = 2m+1$, for some $m \in \mathrm{N}$.

If $n = 2m$, for some $m \in \mathrm{N}$, a similar method yields the result. The proof is completed. ■

---

**Note A5.2:**

Since $[\frac{n+1}{2}]+[\frac{n}{2}]=n$, then, by Theorem A5.3, every eigenvector of a CS matrix $R$ is either RWS or RWNS, if $R$ has $n$ distinct eigenvalues. ■

Let $R \in M_n(R)$ be a CS matrix. In the next theorem, $R^k$ is considered when $k \to \infty$. The $\lim_{k\to\infty} R^k$ does not necessarily exist, but it is interesting that $R^k$ tends toward a PS matrix as $k \to \infty$.

---

**Theorem A5.4:**

Let $n \in \mathrm{N}$, and $R \in M_n(R)$ be a CS matrix which has $n$ distinct eigenvalues, then, $R^k$ tends toward a PS matrix if all of its eigenvalues associated with RWNS eigenvectors are located in the unit circle. ■

**Proof:**

Let $M$ be the modal matrix of $R$ (i.e., the columns of $M$ are the eigenvectors of $R$). The form of $M$ is determined according to Theorem A5.3, that is, all eigenvectors of $R$ are either RWS or RWNS matrices. In addition, the relational matrix $R$ can be decomposed as $MJM^{-1}$, in which $J$ is a diagonal matrix containing all the eigenvalues of $R$. Here, just the form of these matrices are important. Let's denote $M^{-1}$ with $N$ for convenience.

First, assume that $n = 2m$, for some $m$, $m \in N$. Straightforward manipulation shows that:

$$M = \begin{bmatrix} A & B \\ PA & -PB \end{bmatrix}, \text{and } N = \begin{bmatrix} C & CP \\ D & -DP \end{bmatrix},$$

where A, B, C, D, P $\in M_m$ (R) and,

$$P = \begin{bmatrix} 0 & 1 \\ & \cdots \\ 1 & 0 \end{bmatrix} M_m(R).$$

By Theorem A5.3, it may be assumed that $\lambda_1,\ldots,\lambda_m$ are the eigenvalues of $R$ such that their associated eigenvectors are RWS and $\lambda_{m+1},\ldots,\lambda_n$ are the eigenvalues of $R$ such that their associated eigenvectors are RWNS. Let $k \in N$ and,

$$S = \begin{bmatrix} \lambda_1 & & 0 \\ & \ddots & \\ 0 & & \lambda_m \end{bmatrix} \text{and T} = \begin{bmatrix} \lambda_{m+1} & & 0 \\ & \ddots & \\ 0 & & \lambda_n \end{bmatrix}., \text{ thus } J^k = \begin{bmatrix} s^k & 0 \\ 0 & T^k \end{bmatrix}.$$

Therefore:

$$R^k = MJ^kN = \begin{bmatrix} AS^kC + BT^kD & AS^kCP - BT^kDP \\ PAS^kC - PBT^kD & PAS^kCP + PBT^kDP \end{bmatrix}$$

$$= \begin{bmatrix} AS^kC & AS^kCP \\ PAS^kC & PAS^kCP \end{bmatrix} + \begin{bmatrix} BT^kD & -BT^kDP \\ -PBT^kD & PBT^kDP \end{bmatrix}$$

Clearly, $\begin{bmatrix} AS^kC & AS^kCP \\ PAS^kC & PAS^kCP \end{bmatrix}$ is a matrix that is both a RWS as well as a CWS, and so by Lemma A5.1, it is a PS matrix. Also, an easy computation shows that every element of $\begin{bmatrix} BT^kD & -BT^kDP \\ -PBT^kD & PBT^kDP \end{bmatrix}$ is a polynomial of $\lambda_{m+1}^k,\ldots,\lambda_n^k$, which by hypothesis tends toward 0 as $k \to \infty$. Thus, $R^K$ tends toward a PS matrix.

Next, assume that $n = 2m + 1$, for some $m \in$N. In a similar way, it is shown that:

$$M = \begin{bmatrix} A & B \\ C & 0 \\ PA & -PB \end{bmatrix}, \text{and } N = \begin{bmatrix} D & F & DP \\ E & 0 & -EP \end{bmatrix},$$

where $A, D^T \in M_{m\times(m+1)}(R), B, E \in M_n(R)$, and:

$$C, F^T \in M_{1\times(m+1)}(R),$$

and:

$$P = \begin{bmatrix} 0 & & 1 \\ & \cdots & \\ 1 & & 0 \end{bmatrix} \in M_m(R).$$

By Theorem A5.3, one may assume that $\lambda_1, \ldots, \lambda_{m+1}$ are the eigenvalues of $R$ such that their associated eigenvectors are RWS and $\lambda_{m+2}, \ldots, \lambda_n$ are the eigenvalues of $R$ such that their associated eigenvectors are RWNS. Let $k \in N$ and:

$$S = \begin{bmatrix} \lambda_1 & & 0 \\ & \ddots & \\ 0 & & \lambda_{m+1} \end{bmatrix} \text{and } T = \begin{bmatrix} \lambda_{m+2} & & 0 \\ & \ddots & \\ 0 & & \lambda_n \end{bmatrix}. \text{ thus } J^k = \begin{bmatrix} s^k & 0 \\ 0 & T^k \end{bmatrix}.$$

Therefore:

$$R^k = MJ^kN = \begin{bmatrix} AS^kD + BT^kE & AS^kF & AS^kDP - BT^kEP \\ CS^kD & CD^kF & CS^kDP \\ PAS^kD - PBT^kE & PAS^kF & PAS^kDP + PBT^kEP \end{bmatrix}$$

$$= \begin{bmatrix} AS^kD & AS^kF & AS^kDP \\ CS^kD & CS^kF & CS^kDP \\ PAS^kD & PAS^kF & PAS^kDP \end{bmatrix} + \begin{bmatrix} BT^kE & 0 & -BT^kEP \\ 0 & 0 & 0 \\ -PBT^kE & 0 & +PAT^kEP \end{bmatrix}$$

Clearly, $\begin{bmatrix} AS^kD & AS^kF & AS^kDP \\ CS^kD & CS^kF & CS^kDP \\ PAS^kD & PAS^kF & PAS^kDP \end{bmatrix}$ is a matrix that is both a RWS and a CWS matrix, and so by Lemma A5.1, it is a PS matrix. Also, a straightforward manipulation shows that every element of $\begin{bmatrix} BT^kE & 0 & -BT^kEP \\ 0 & 0 & 0 \\ -PBT^kE & 0 & +PAT^kEP \end{bmatrix}$ is a polynomial of $\lambda_{m+2}^k, \ldots, \lambda_n^k$,, which by hypothesis approaches 0 as k $\to \infty$. Thus, $R^k$ tends toward a PS matrix. This completes the proof. ■

# Index

## B



## C



## D



## E



## F

## M



## N

## U



## V



## W



## Z

For Product Safety Concerns and Information please contact our
EU representative GPSR@taylorandfrancis.com Taylor & Francis
Verlag GmbH, Kaufingerstraße 24, 80331 München, Germany